ELASTICITY

SOLID MECHANICS AND ITS APPLICATIONS
Volume 12

---

*Series Editor:* G.M.L. GLADWELL
*Solid Mechanics Division, Faculty of Engineering*
*University of Waterloo*
*Waterloo, Ontario, Canada N2L 3G1*

*Aims and Scope of the Series*

The fundamental questions arising in mechanics are: *Why?*, *How?*, and *How much?* The aim of this series is to provide lucid accounts written by authoritative researchers giving vision and insight in answering these questions on the subject of mechanics as it relates to solids.

The scope of the series covers the entire spectrum of solid mechanics. Thus it includes the foundation of mechanics; variational formulations; computational mechanics; statics, kinematics and dynamics of rigid and elastic bodies; vibrations of solids and structures; dynamical systems and chaos; the theories of elasticity, plasticity and viscoelasticity; composite materials; rods, beams, shells and membranes; structural control and stability; soils, rocks and geomechanics; fracture; tribology; experimental mechanics; biomechanics and machine design.

The median level of presentation is the first year gràduate student. Some texts are monographs defining the current state of the field; others are accessible to final year undergraduates; but essentially the emphasis is on readability and clarity.

*For a list of related mechanics titles, see final pages.*

# Elasticity

by

J. R. BARBER

*Department of Mechanical Engineering and Applied Mechanics,*
*The University of Michigan, U.S.A.*

KLUWER ACADEMIC PUBLISHERS

DORDRECHT / BOSTON / LONDON

Library of Congress Cataloging-in-Publication Data

Barber, J. R.
Elasticity / by J.R. Barber.
p. cm. -- (Solid mechanics and its applications ; v. 12)
Includes index.
ISBN 0-7923-1609-6 (alk. paper)
1. Elasticity. I. Title. II. Series.
QA931.B23 1992
620.1'1232--dc20 91-46435
CIP

ISBN 0–7923–1610–X (Pb)
ISBN 0–7923–1609–6 (Hb)

Published by Kluwer Academic Publishers,
P.O. Box 17, 3300 AA Dordrecht, The Netherlands.

Kluwer Academic Publishers incorporates
the publishing programmes of
D. Reidel, Martinus Nijhoff, Dr W. Junk and MTP Press.

Sold and distributed in the U.S.A. and Canada
by Kluwer Academic Publishers,
101 Philip Drive, Norwell, MA 02061, U.S.A.

In all other countries, sold and distributed
by Kluwer Academic Publishers Group,
P.O. Box 322, 3300 AH Dordrecht, The Netherlands.



First printed in hardbound edition 1992
Second printing in hardbound en paperback edition 1993

*Printed on acid-free paper*



Printed in the Netherlands

# Contents


**Preface** **xiii**

**I GENERAL CONSIDERATIONS** **1**

**1 INTRODUCTION** **3**

1.1 Notation for stress and displacement . . . 4
1.1.1 Stress . . . 4
1.1.2 Index and vector notation . . . 6
1.1.3 Vectors, tensors and transformation rules . . . 7
1.1.4 Displacement . . . 10
1.2 Strains and their relation to displacements . . . 11
1.2.1 Tensile strain . . . 11
1.2.2 Rotation and shear strain . . . 13
1.2.3 Transformation of coördinates . . . 14
1.2.4 Definition of shear strain . . . 16
1.3 Stress-strain relations . . . 17
1.3.1 Lamé's constants . . . 18
1.3.2 Dilatation and bulk modulus . . . 19
PROBLEM . . . 19

**2 EQUILIBRIUM AND COMPATIBILITY** **21**

2.1 Equilibrium equations . . . 21
2.2 Compatibility equations . . . 22
2.2.1 The significance of the compatibility equations . . . 24
2.3 Equilibrium equations for displacements . . . 27
PROBLEMS . . . 28

**II TWO-DIMENSIONAL PROBLEMS** 29

**3 PLANE STRAIN AND PLANE STRESS** 31
3.1 Plane strain . . . . . 31
3.1.1 The corrective solution . . . . . 33
3.1.2 Saint-Venant's principle . . . . . 34
3.2 Plane stress . . . . . 34
3.2.1 Generalized plane stress . . . . . 36
3.2.2 Relationship between plane stress and plane strain . . . . . 36
PROBLEMS . . . . . 37

**4 STRESS FUNCTION FORMULATION** 39
4.1 The concept of a scalar stress function . . . . . 39
4.2 Choice of a suitable form . . . . . 40
4.3 The Airy stress function . . . . . 40
4.4 The governing equation . . . . . 41
4.4.1 The equilibrium equations . . . . . 41
4.4.2 Non-zero body forces . . . . . 41
4.4.3 The compatibility condition . . . . . 42
4.4.4 Method of solution . . . . . 42
4.4.5 Reduced dependence on elastic constants . . . . . 43
PROBLEM . . . . . 43

**5 PROBLEMS IN RECTANGULAR COÖRDINATES** 45
5.1 Biharmonic polynomial functions . . . . . 45
5.1.1 Second and third degree polynomials . . . . . 47
5.2 Rectangular beam problems . . . . . 49
5.2.1 Bending of a beam by an end load . . . . . 49
5.2.2 Higher order polynomials — symmetry considerations . . . . . 51
5.3 Series and transform solutions . . . . . 54
PROBLEMS . . . . . 56

**6 END EFFECTS** 59
6.1 Decaying solutions . . . . . 59
6.2 The corrective solution . . . . . 60
6.2.1 Separated variable solutions . . . . . 61
6.2.2 The eigenvalue problem . . . . . 62
6.3 Other Saint-Venant problems . . . . . 64

**7 BODY FORCES** 67
7.1 Stress function formulation . . . . . 67
7.1.1 Conservative vector fields . . . . . 68
7.1.2 The compatibility condition . . . . . 69

7.2 Particular cases . . . . . 69
7.2.1 Gravitational loading . . . . . 70
7.2.2 Inertia forces . . . . . 70
7.2.3 Quasi-static problems . . . . . 70
7.2.4 Rigid-body kinematics . . . . . 71
7.3 Solution for the stress function . . . . . 72
7.3.1 The rotating rectangular beam . . . . . 73
7.3.2 Solution of the governing equation . . . . . 75
7.4 Rotational acceleration . . . . . 76
7.4.1 The circular disk . . . . . 76
7.4.2 The rectangular bar . . . . . 78
PROBLEMS . . . . . 80

**8 PROBLEMS IN POLAR COÖRDINATES** **83**
8.1 Expressions for stress components . . . . . 83
8.2 Strain components . . . . . 84
8.3 Fourier series expansion . . . . . 85
8.3.1 Satisfaction of boundary conditions . . . . . 86
8.3.2 Circular hole in a shear field . . . . . 87
8.3.3 Degenerate cases . . . . . 90
8.4 The Michell solution . . . . . 92
8.4.1 Hole in a tensile field . . . . . 92
PROBLEMS . . . . . 95

**9 CALCULATION OF DISPLACEMENTS** **97**
9.1 The cantilever with an end load . . . . . 97
9.1.1 Rigid-body displacements and end conditions . . . . . 99
9.1.2 Deflection of the free end . . . . . 100
9.2 The circular hole . . . . . 101
9.3 Displacements for the Michell solution . . . . . 103
9.3.1 Equilibrium considerations . . . . . 103
PROBLEMS . . . . . 105

**10 CURVED BEAM PROBLEMS** **107**
10.1 Loading at the ends . . . . . 107
10.1.1 Pure bending . . . . . 107
10.1.2 Force transmission . . . . . 109
10.2 Eigenvalues and eigenfunctions . . . . . 112
10.3 The inhomogeneous problem . . . . . 112
10.3.1 Beam with sinusoidal loading . . . . . 112
10.3.2 The near-singular problem . . . . . 115
10.4 Some general considerations . . . . . 118
10.4.1 Conclusions . . . . . 119

PROBLEM . . . 119

**11 WEDGE PROBLEMS** **121**
11.1 Power law tractions . . . 121
11.1.1 Uniform tractions . . . 121
11.1.2 More general uniform loading . . . 124
11.1.3 Eigenvalues for the wedge angle . . . 125
11.2 Williams' asymptotic method . . . 125
11.2.1 Acceptable singularities . . . 126
11.2.2 Eigenfunction expansion . . . 128
11.2.3 Nature of the eigenvalues . . . 130
11.2.4 Other geometries . . . 132
11.3 General loading of the faces . . . 135
PROBLEMS . . . 136

**12 PLANE CONTACT PROBLEMS** **139**
12.1 Self-similarity . . . 139
12.2 The Flamant Solution . . . 140
12.3 The half-plane . . . 141
12.3.1 The normal force $F_y$ . . . 142
12.3.2 The tangential force $F_x$ . . . 143
12.3.3 Summary . . . 143
12.4 Distributed normal tractions . . . 144
12.5 Frictionless contact problems . . . 145
12.5.1 Method of solution . . . 146
12.5.2 The flat punch . . . 147
12.5.3 The cylindrical punch (Hertz problem) . . . 148
12.6 Problems with two deformable bodies . . . 151
12.7 Uncoupled problems . . . 154
12.7.1 Contact of cylinders . . . 155
12.8 Combined normal and tangential loading . . . 155
12.8.1 Mindlin's problem . . . 156
12.8.2 Steady rolling: Carter's solution . . . 159
PROBLEMS . . . 162

**13 FORCES, DISLOCATIONS AND CRACKS** **165**
13.1 The Kelvin solution . . . 165
13.1.1 Body force problems . . . 167
13.2 Dislocations . . . 168
13.2.1 Dislocations in Materials Science . . . 169
13.2.2 Similarities and differences . . . 170
13.2.3 Dislocations as Green's functions . . . 171
13.2.4 Stress concentrations . . . 172

13.3 Crack problems . . . . . 173
13.3.1 Linear Elastic Fracture Mechanics . . . . . 173
13.3.2 Plane crack in a tensile field . . . . . 174
PROBLEMS . . . . . 177

**14 THERMOELASTICITY 179**
14.1 The governing equation . . . . . 179
14.1.1 Example . . . . . 180
14.1.2 The method of strain suppression . . . . . 181
14.2 Heat conduction . . . . . 181
14.3 Steady-state problems . . . . . 182
14.3.1 Dundurs' Theorem . . . . . 183
PROBLEMS . . . . . 185

**III THREE DIMENSIONAL PROBLEMS 187**

**15 DISPLACEMENT FUNCTION SOLUTIONS 189**
15.1 The strain potential . . . . . 189
15.2 The Galerkin vector . . . . . 190
15.3 The Papkovich-Neuber solution . . . . . 192
15.4 Completeness and uniqueness . . . . . 193
15.4.1 Methods of partial integration . . . . . 194
PROBLEM . . . . . 197

**16 THE BOUSSINESQ POTENTIALS 199**
16.1 Solution A : The strain potential . . . . . 200
16.2 Solution B . . . . . 200
16.3 Solution E : Rotational deformation . . . . . 201
16.4 Solutions obtained by superposition . . . . . 202
16.4.1 Solution F : Frictionless isothermal contact problems . . . . . 204
16.4.2 Solution G: The surface free of normal traction . . . . . 204
16.5 The plane strain solution in complex variables . . . . . 206
16.5.1 Preliminary mathematical results . . . . . 206
16.5.2 Representation of vectors . . . . . 206
16.5.3 Representation of displacement . . . . . 207
16.5.4 Expressions for stresses . . . . . 208
PROBLEMS . . . . . 209

**17 THERMOELASTIC DISPLACEMENT POTENTIALS 211**
17.1 Plane problems . . . . . 212
17.1.1 Axisymmetric problems for the cylinder . . . . . 213
17.2 Solution T : Steady-state temperature . . . . . 214

17.2.1 Thermoelastic plane stress . . . . . 214
PROBLEM . . . . . 216

**18 SINGULAR SOLUTIONS** **217**
18.1 The source solution . . . . . 217
18.1.1 The centre of dilatation . . . . . 218
18.1.2 The Kelvin solution . . . . . 219
18.2 Dimensional considerations . . . . . 220
18.2.1 The Boussinesq solution . . . . . 221
18.3 Other singular solutions . . . . . 224
PROBLEMS . . . . . 225

**19 SPHERICAL HARMONICS** **227**
19.1 Fourier series solution . . . . . 227
19.2 Reduction to Legendre's equation . . . . . 228
19.3 Legendre functions . . . . . 229
19.3.1 Legendre polynomials . . . . . 229
19.3.2 Singular spherical harmonics . . . . . 230
19.4 Axisymmetric potentials . . . . . 230
19.5 Non-axisymmetric harmonics . . . . . 233

**20 AXISYMMETRIC PROBLEMS** **235**
20.1 The solid cylinder . . . . . 235
20.2 The spherical hole . . . . . 237
PROBLEMS . . . . . 239

**21 FRICTIONLESS CONTACT** **241**
21.1 Boundary conditions . . . . . 241
21.1.1 Mixed boundary-value problems . . . . . 242
21.2 Determining the contact area . . . . . 243
PROBLEM . . . . . 246

**22 THE BOUNDARY-VALUE PROBLEM** **249**
22.1 Hankel transform methods . . . . . 249
22.2 Collins' Method . . . . . 250
22.2.1 Indentation by a flat punch . . . . . 250
22.2.2 Integral representation . . . . . 252
22.2.3 Basic forms and surface values . . . . . 253
22.2.4 Reduction to an Abel equation . . . . . 254
22.2.5 Example – The Hertz problem . . . . . 258
22.3 Choice of form . . . . . 258
PROBLEMS . . . . . 259

**23 THE PENNY-SHAPED CRACK** **261**
23.1 The penny-shaped crack in tension . . . . . . . . . . . . . . . . 261
23.2 Thermoelastic problems . . . . . . . . . . . . . . . . . . . . . . 263
PROBLEMS . . . . . . . . . . . . . . . . . . . . . . . . . . . . 267

**24 THE INTERFACE CRACK** **269**
24.1 The uncracked interface . . . . . . . . . . . . . . . . . . . . . 269
24.2 The corrective solution . . . . . . . . . . . . . . . . . . . . . . 271
24.2.1 Global conditions . . . . . . . . . . . . . . . . . . . . . 272
24.2.2 Mixed conditions . . . . . . . . . . . . . . . . . . . . . 272
24.3 The penny-shaped crack in tension . . . . . . . . . . . . . . . . 274
24.3.1 Reduction to a single equation . . . . . . . . . . . . . . 276
24.3.2 Oscillatory singularities . . . . . . . . . . . . . . . . . . 277
24.4 The contact solution . . . . . . . . . . . . . . . . . . . . . . . 278
24.5 Implications for Fracture Mechanics . . . . . . . . . . . . . . . 279

**25 THE RECIPROCAL THEOREM** **281**
25.1 Maxwell's Theorem . . . . . . . . . . . . . . . . . . . . . . . . 281
25.2 Betti's Theorem . . . . . . . . . . . . . . . . . . . . . . . . . . 282
25.3 Use of the theorem . . . . . . . . . . . . . . . . . . . . . . . . 283
25.3.1 A tilted punch problem . . . . . . . . . . . . . . . . . . 284
25.3.2 Indentation of a half-space . . . . . . . . . . . . . . . . 287
PROBLEMS . . . . . . . . . . . . . . . . . . . . . . . . . . . . 288

**Index** **290**

# Preface

The subject of Elasticity can be approached from several points of view, depending on whether the practitioner is principally interested in the mathematical structure of the subject or in its use in engineering applications and in the latter case, whether essentially numerical or analytical methods are envisaged as the solution method. My first introduction to the subject was in response to a need for information about a specific problem in Tribology. As a practising engineer with a background only in elementary Strength of Materials, I approached that problem initially using the concepts of concentrated forces and superposition. Today, with a rather more extensive knowledge of analytical techniques in Elasticity, I still find it helpful to go back to these roots in the elementary theory and think through a problem physically as well as mathematically, whenever some new and unexpected feature presents difficulties in research. This way of thinking will be found to permeate this book. My engineering background will also reveal itself in a tendency to work examples through to final expressions for stresses and displacements, rather than leave the derivation at a point where the remaining manipulations would be routine.

With the practical engineering reader in mind, I have endeavoured to keep to a minimum any dependence on previous knowledge of Solid Mechanics, Continuum Mechanics or Mathematics. Most of the text should be readily intelligible to a reader with an undergraduate background of one or two courses in elementary Strength of Materials and a rudimentary knowledge of partial differentiation. Cartesian tensor notation and the summation convention are used in a few places to shorten the derivation of some general results, but these sections are carefully explained, so as to be self-contained

The book is based on a one semester graduate course on Linear Elasticity that I have taught at the University of Michigan since 1983. In such a restricted format, it is clearly necessary to make some difficult choices about which topics to include and, more significantly, which to exclude. To some extent, my choice is a personal one, being weighted towards topics which have arisen in connection with my own research in the fields of Contact Mechanics and Thermoelasticity. This also explains the preponderance of references to my own work, which in no way should be taken as a claim to a proportionate share in the development of this classical subject. The most significant *exclusion* is the complex variable solution of two-dimensional Elasticity, which, if it were to be adequately treated including essential mathematical preliminaries, would need most of a book of this length to itself. Instead, I have chosen to restrict the two-dimensional treatment to the more traditional real stress function approach, so as to leave room for a substantial amount of material on three-dimensional problems, which are arguably closer to the frontier of current research.

Modern practitioners of Elasticity are necessarily influenced by developments in numerical methods, which promise to solve all problems with no more information about the subject than is needed to formulate the description of a representative element of material in a relatively simple state of stress. As a researcher with a primary interest in the physical behaviour of systems, rather than in the mathematical structure of the solutions, I have frequently had recourse to numerical methods of all types and have tended to adopt the pragmatic criterion that the best method is that which gives the most convincing and accurate result in the shortest time. In this context, 'convincing' means that the solution should be capable of being checked against reliable closed-form solutions in suitable limiting cases and that it is demonstrably stable and in some sense convergent. Measured against these criteria, the 'best' solution to many practical problems is often not a direct numerical method, such as the finite element method, but rather one involving some significant analytical steps before the final numerical evaluation. This is particularly true in three-dimensional problems, where direct numerical methods are extremely computer-intensive if any reasonably accuracy is required, and in problems involving infinite or semi-infinite domains, discontinuities, bonded or contacting material interfaces or theoretically singular stress fields. The reader will therefore find considerable emphasis is placed on applications of this type.

I have provided a representative selection of problems suitable for class use at the end of most of the chapters. Many texts on Elasticity contain problems which offer a candidate stress function and invite the student to 'verify' that it defines the solution to a given problem. Students invariably raise the question 'How would we know to choose that form if we were not given it in advance?' I have tried wherever possible to avoid this by expressing the problems in the form they would arise in Engineering — i.e. as a body of a given geometry subjected to prescribed loading. This in turn has required me to write the text in such a way that the student can approach problems deductively. I have also generally opted for explaining difficulties that might arise in an 'obvious' approach to the problem, rather than steering the reader around them in the interests of brevity or elegance.

Even conceptually straightforward problems in three-dimensional Elasticity tend to be algebraically complicated, because of the dependence of the results on Poisson's ratio, which is absent in two-dimensions, except where body forces etc. are involved. With the greater general availablity of symbolic processors, this will soon cease to be a difficulty, but those without access to such methods might like to simplify some of the three-dimensional problems by restricting the solution to the case $\nu = 0$. At the University of Michigan, we are fortunate in having a very sophisticated computer environment, so I can ask my students to carry the problems as far as plotting graphs of stress distributions and performing parametric studies to determine optimum designs for components. From the engineering perspective, this leads to a significantly greater insight into the nature of the subject, as well as serving as an educational motivation for those students outside the classical Mechanics field.

Over the years, the students attending my course have been drawn from departments of Mechanical, Civil, Aerospace and Materials Engineering as well as Naval Architecture, Applied Mechanics and Mathematics and have therefore generally represented quite a diverse set of points of view. The questions raised by this very talented group of people have often resulted in vigorous classroom discussions which have had a major impact on my perspective on the subject. It is therefore to my graduate students, past and present, that this book is dedicated.

It is traditional at this point also to thank a multitude of colleagues, assistants and family members for their contributions and support. However, since I developed the text as part of my normal teaching duties and typset it myself using LaTeX, many of these credits would be out of place. My wife, Maria Comninou, may disagree, but as far as I can judge, my temperament was not significantly soured during the gestation period, but she does deserve my thanks for her technical input during times that we have collaborated on research problems.

I am particularly grateful to the Rector and Fellows of Lincoln College, Oxford, who provided me with a stimulating scholarly environment during an important stage in the writing, and to David Hills of that College who made my visit possible.

The line drawings were prepared by Rodney Hill and Figures 13.2, 13.3 are reproduced by kind permission of the Royal Society of London. I should also like to thank John Dundurs for permission to use Table 9.1, the usefulness of which has been proved by his students at Northwestern University and many others in the Elasticity community.

J.R.Barber<br>Ann Arbor<br>1991

[illegible] particular studies to [illegible] the engineering perspective, this [illegible] of the subject, as well as serving [illegible] the Classical Mechanics field.

[illegible] regarding my course have [illegible] Civil, Aerospace and Materials Engineering [illegible] Applied Mechanics and Mathematics [illegible] points of view. The questions [illegible] of people have often resulted in vigorous discussions [illegible] impact on my perspective on [illegible] graduate students [illegible] present that this book is indebted.

It is [illegible] to thank [illegible] contributions and support. However [illegible]

[illegible]

# Part I

# GENERAL CONSIDERATIONS

# Chapter 1

# INTRODUCTION

The subject of Elasticity is concerned with the determination of the stresses and displacements in a body as a result of applied mechanical or thermal loads, for those cases in which the body reverts to its original state on the removal of the loads. In this book, we shall further restrict attention to the case of linear infinitesimal elasticity, in which the stresses and displacements are linearly proportional to the applied loads and the displacements are small in comparison with the characteristic length dimensions of the body. These restrictions ensure that linear superposition can be used and enable us to employ a wide range of series and transform techniques which are not available for non-linear problems.

Most engineers first encounter problems of this kind in the context of the subject known as Strength of Materials, which is an important constituent of most undergraduate engineering curricula. Strength of Materials differs from Elasticity in that various plausible but unsubstantiated assumptions are made about the deformation process in the course of the analysis. A typical example is the assumption that plane sections remain plane in the bending of a slender beam. Elasticity makes no such assumptions, but attempts to develop the solution directly and rigorously from its first principles, which are Newton's laws of motion, Euclidian geometry and Hooke's law. Approximations are often introduced towards the end of the solution, but these are mathematical approximations used to obtain solutions of the governing equations rather than physical approximations that impose artificial and strictly unjustifiable constraints on the permissible deformation field.

However, it would be a mistake to draw too firm a distinction between the two approaches, since practitioners of each have much to learn from the other. Strength of Materials, with its emphasis on physical reasoning and a full exploration of the practical consequences of the results, is often able to provide insights into the problem that are less easily obtained from a purely mathematical perspective. Indeed, we shall make extensive use of physical parallels in this book and pursue many problems to conclusions relevant to practical applications, with the hope of deepening the reader's understanding of the underlying structure of the subject. Conversely, the

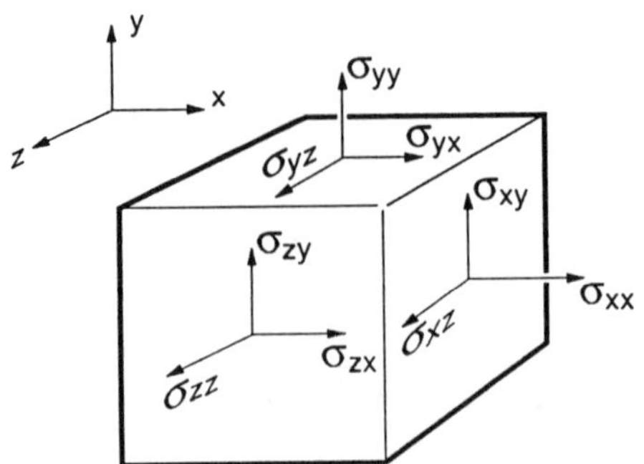


Figure 1.1: Notation for stress components

mathematical rigour of Elasticity gives us greater confidence in the results, since, even when we have to resort to an approximate solution, we can usually estimate its accuracy with some confidence — something that is very difficult to do with the physical approximations used in Strength of Materials[1]. Also, there is little to be said for using an *ad hoc* approach when, as is often the case, a more rigorous treatment presents no serious difficulty.

## 1.1 Notation for stress and displacement

It is assumed that the reader is more or less familiar with the concept of stress and strain from elementary courses on Strength of Materials. This section is intended to introduce the notation used, to refresh the reader's memory about some important ideas, and to record some elementary but useful results.

### 1.1.1 Stress

Components of stress will all be denoted by the symbol $\sigma$ with appropriate suffices. The second suffix denotes the direction of the stress component and the first the direction of the *outward* normal to the surface upon which it acts. This notation is illustrated in Figure 1.1 for the Cartesian coördinate system $x, y, z$.

Notice that one consequence of this notation is that normal (i.e. tensile and compressive) stresses have both suffices the same (e.g. $\sigma_{xx}, \sigma_{yy}, \sigma_{zz}$ in Figure 1.1) and

[1]In fact, the only practical way to examine the effect of these approximations is to relax them, by considering the same problem, or maybe a simpler problem with similar features, in the context of the theory of Elasticity.

are positive when tensile. The remaining six stress components in Figure 1.1 (i.e. $\sigma_{xy}, \sigma_{yx}, \sigma_{yz}, \sigma_{zy}, \sigma_{zx}, \sigma_{xz}$) have two *different* suffices and are shear stresses.

Books on Strength of Materials often use the symbol $\tau$ for shear stress, whilst retaining $\sigma$ for normal stress. However, there is no need for a different symbol, since the suffices enable us to distinguish normal from shear stress components. Also, we shall find that the use of a single symbol with appropriate suffices permits matrix methods to be used in many derivations and introduces considerable economies in the notation for general results.

The equilibrium of moments acting on the block in Figure 1.1 requires that

$$\sigma_{xy} = \sigma_{yx} \;\; ; \;\; \sigma_{yz} = \sigma_{zy} \;\text{ and }\; \sigma_{zx} = \sigma_{xz} \; . \tag{1.1}$$

This has the incidental advantage of rendering mistakes about the order of suffices harmless! (In fact, a few books use the opposite convention.) Readers who have not encountered three-dimensional problems before should note that there are *two* shear stress components on each surface and one normal stress component. There are some circumstances in which it is convenient to combine the two shear stresses on a given plane into a two-dimensional vector in the plane — i.e. to refer to the *resultant* shear stress on the plane. An elementary situation where this is helpful is in the Strength of Materials problem of determining the distribution of shear stress on the cross-section of a beam due to a transverse shear force. For example, we note that in this case, the resultant shear stress on the plane must be tangential to the edge at the boundary of the cross section, since the shear stress complementary to the component normal to the edge would have to act on the traction-free surface of the beam and must therefore be zero. This of course is why the shear stress in a thin-walled section tends to follow the direction of the wall.

We shall refer to a plane normal to the $x$-direction as an '$x$-plane' etc. The *only* stress components which act on an $x$-plane are those which have an $x$ as the first suffix (This is an immediate consequence of the definition).

Notice also that any $x$-plane can be defined by an equation of the form $x = C$, where $C$ is a constant. More precisely, we can define a 'positive $x$-plane' as a plane for which the positive $x$-direction is the outward normal and a 'negative $x$-plane' as one for which it is the inward normal. This distinction can be expressed mathematically in terms of inequalities. Thus, if part of the boundary of a solid is expressible as $x = C$, the solid must locally occupy one side or the other of this plane. If the domain of the solid is locally described by $x < C$, the bounding surface is a positive $x$-plane, whereas if it is described by $x > C$, the bounding surface is a negative $x$-plane.

The discussion in this section suggests a useful formalism for correctly defining the boundary conditions in problems where the boundaries are parallel to the coördinate axes. We first identify the equations which define the boundaries of the solid and then write down the three traction components which act on each boundary. For example, suppose we have a rectangular solid defined by the inequalities $0 < x < a$, $0 < y < b$, $0 < z < c$. It is clear that the surface $y = b$ is a positive $y$-plane

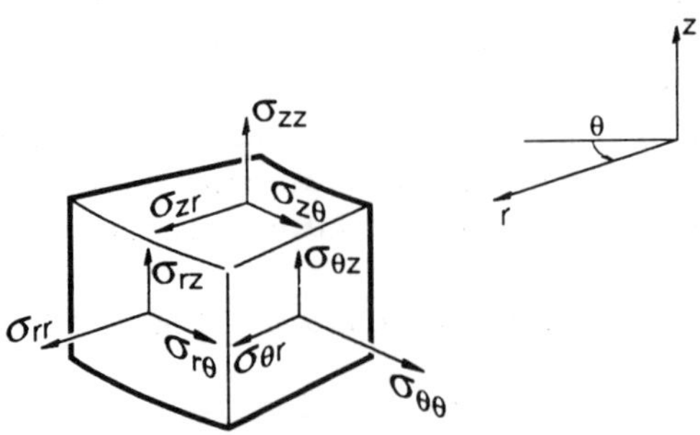


Figure 1.2: Stress components in polar coördinates

and we deduce immediately that the corresponding traction boundary conditions will involve the stress components $\sigma_{yx}, \sigma_{yy}, \sigma_{yz}$ — i.e. the three components that have $y$ as the first suffix. This procedure insures against the common student mistake of assuming (for example) that the component $\sigma_{xx}$ must be zero if the surface $y = b$ is to be traction-free. (*Note* : Don't assume that this mistake is too obvious for you to fall into it. When the problem is geometrically or algebraically very complicated, it is only too easy to get distracted.)

Stress components can be defined in the same way for other systems of orthogonal coördinates. For example, components for the system of cylindrical polar coördinates $(r, \theta, z)$ are shown in Figure 1.2. (This is a case where the definition of the '$\theta$-plane' through an equation, $\theta = C$, is easier to comprehend than 'the plane normal to the $\theta$-direction'. However, note that the $\theta$-direction is the direction in which a particle would move if $\theta$ were increased with $r, z$ constant.)

### 1.1.2 Index and vector notation

Many authors use the notation $x_1, x_2, x_3$, in place of $x, y, z$ for the Cartesian coördinate system, in which case the stress components are written $\sigma_{11}, \sigma_{12}$, etc. ($\sigma_{x_1x_1}$, $\sigma_{x_1x_2}$ etc. would obviously be too cumbersome). This notation has the particular advantage that in combination with the 'summation convention' it permits general results to be written and manipulated in a concise and elegant form. The summation convention lays down that any term in which the same latin suffix occurs twice stands for the sum of all the terms obtained by giving this suffix each of its possible values. For example $\sigma_{ii}$ is interpreted as

$$\sigma_{ii} \equiv \sum_{i=1}^{3} \sigma_{ii} = \sigma_{11} + \sigma_{22} + \sigma_{33} \tag{1.2}$$

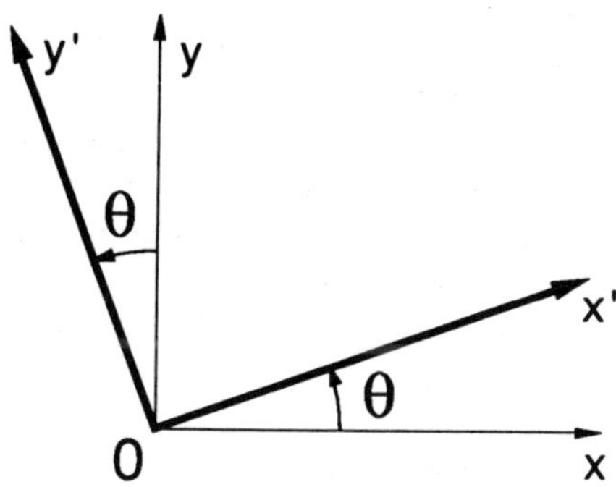


Figure 1.3: The coördinate systems $x, y$ and $x', y'$

and

$$\frac{\partial u_i}{\partial x_i} \equiv \frac{\partial u_1}{\partial x_1} + \frac{\partial u_2}{\partial x_2} + \frac{\partial u_3}{\partial x_3} . \tag{1.3}$$

On the other hand, some results are more conveniently manipulated or expressed in vector notation, for which we define the position vector

$$\boldsymbol{R} = \boldsymbol{i}x + \boldsymbol{j}y + \boldsymbol{k}z , \tag{1.4}$$

where $\boldsymbol{i}$, $\boldsymbol{j}$, $\boldsymbol{k}$ are unit vectors in directions $x, y, z$ respectively.

Important results will be given in each form wherever possible.

### 1.1.3 Vectors, tensors and transformation rules

*Vectors* can be conceived in a mathematical sense as ordered sets of numbers or in a physical sense as mathematical representations of quantities characterized by magnitude and direction. A link between these concepts is provided by the *transformation rules.* Suppose we know the components $(u_x, u_y)$ of the vector $\boldsymbol{u}$ in a given two-dimensional Cartesian coördinate system $(x, y)$ and we wish to determine the components $(u'_x, u'_y)$ in a new system $(x', y')$ which is inclined to $(x, y)$ at an angle $\theta$ in the counterclockwise direction as shown in Figure 1.3. The required components are

$$u'_x = u_x cos\theta + u_y sin\theta ; \tag{1.5}$$

$$u'_y = u_y cos\theta - u_x sin\theta . \tag{1.6}$$

We could *define* a vector as an entity, described by its components in a specified Cartesian coördinate system, which transforms into other coördinate systems according to rules like equations (1.5, 1.6) — i.e. as an ordered set of numbers which obey the transformation rules (1.5, 1.6). The idea of magnitude and direction could then be introduced by noting that we can always choose $\theta$ such that (i) $u'_y = 0$ and (ii) $u'_x > 0$. The corresponding direction $x'$ is then the direction of the resultant vector and the component $u'_x$ is its magnitude.

Now stresses have two suffices and components associated with all possible combinations of two coördinate directions, though we note that equation (1.1) shows that the order of the suffices is immaterial. (Another way of stating this is that the *matrix* of stress components $\sigma_{ij}$ is always symmetric). The stress components satisfy a more complicated set of transformation rules which in the two-dimensional case are those associated with Mohr's circle - i.e.

$$\sigma_{x'x'} = \sigma_{xx}cos^2\theta + \sigma_{yy}sin^2\theta + 2\sigma_{xy}sin\theta\cos\theta \; ; \quad (1.7)$$

$$\sigma_{x'y'} = \sigma_{xy}(cos^2\theta - sin^2\theta) + (\sigma_{yy} - \sigma_{xx})sin\theta\cos\theta \; ; \quad (1.8)$$

$$\sigma_{y'y'} = \sigma_{yy}cos^2\theta + \sigma_{xx}sin^2\theta - 2\sigma_{xy}sin\theta\cos\theta \; . \quad (1.9)$$

As in the case of vectors we can define a mathematical entity which has a matrix of components in any given Cartesian coördinate system and which transforms into other such coördinate systems according to rules like (1.7–1.9). Such quantities are called *second order Cartesian tensors.*

We know from Mohr's circle that we can always choose $\theta$ such that $\sigma_{x'y'} = 0$, in which case the directions $x', y'$ are referred to as principal directions and the components $\sigma_{x'x'}, \sigma_{y'y'}$ as principal stresses. Thus another way to characterize a second order Cartesian tensor is as a quantity defined by a set of orthogonal principal directions and a corresponding set of principal values.

As with vectors, a pragmatic motivation for abstracting the mathematical properties from the physical quantities which exhibit them is that many different physical quantities are naturally represented as second order Cartesian tensors. Apart from stress and strain, some commonly occurring examples are the second moments of area of a beam cross section $(I_{xx}, I_{xy}, I_{yy})$, the second partial derivatives of a scalar function $(\partial^2 f/\partial x^2; \partial^2 f/\partial x \partial y; \partial^2 f/\partial y^2)$ and the influence coefficient matrix $C_{ij}$ defining the displacement $\boldsymbol{u}$ due to a force $\boldsymbol{F}$ for a linear elastic system, i.e.

$$u_i = C_{ij}F_j \; , \quad (1.10)$$

where the summation convention is implied.

It is a fairly straightforward matter to prove that each of these quantities obeys transformation rules like (1.7–1.9). It follows immediately (for example) that every beam cross section has two orthogonal *principal axes* of bending about which the two principal second moments are respectively the maximum and minimum for the cross section.

A special tensor of some interest is that for which the Mohr's circle degenerates to a point. In the case of stresses, this corresponds to a state of *hydrostatic stress*, so-called because a fluid at rest cannot support shear stress (the constitutive law for

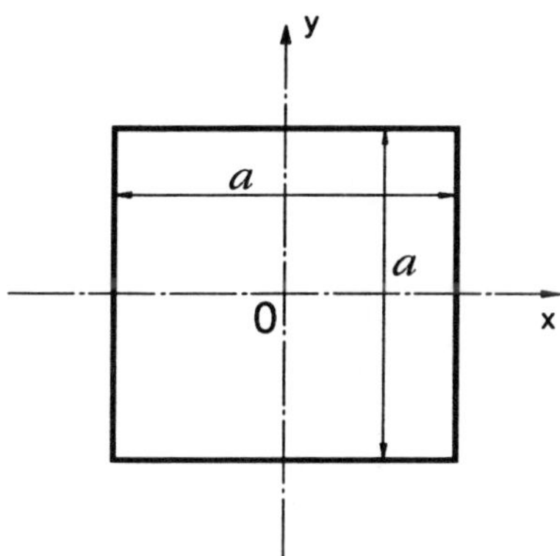


Figure 1.4: A beam of square cross-section

a fluid relates *velocity gradient* to shear stress) and hence $\sigma_{xy} = 0$ for all $x, y$. The only Mohr's circle which satisfies this condition is one of zero radius, from which we deduce immediately that all directions are principal directions and that the principal values are all equal. In the case of the fluid, we obtain the well-known result that the pressure in a fluid at rest is equal in all directions.

It is instructive to consider this result in the context of other systems involving tensors. For example, consider the second moments of area for the square cross section shown in Figure 1.4. By symmetry, we know that $Ox, Oy$ are principal directions and that the two principal second moments are both equal to $a^4/12$. It follows immediately that the Mohr's circle has zero radius and hence that the second moment about any other axis must also be $a^4/12$ — a result which is not obvious from an examination of the section.

As a second example, Figure 1.5 shows an elastic system consisting of three identical but arbitrary structures connecting a point, $P$, to a rigid support, the structures being inclined to each other at angles of $120^o$. The structures each have elastic properties expressible in the form of an influence function matrix as in equation (1.10) and are generally such that the displacement $\boldsymbol{u}$ is not colinear with the force $\boldsymbol{F}$. However, the *overall* influence function matrix for the system has the same properties in three different coördinate systems inclined to each other at $120^o$, since a rotation of the Figure through $120^o$ leaves the system unchanged. The only Mohr's circle which gives equal components after a rotation of $120^o$ is that of zero radius. We therefore conclude that the support system of Figure 1.5 is such that (i) the displacement of $P$ always has the same direction as the force $\boldsymbol{F}$ and (ii) the stiffness or compliance of

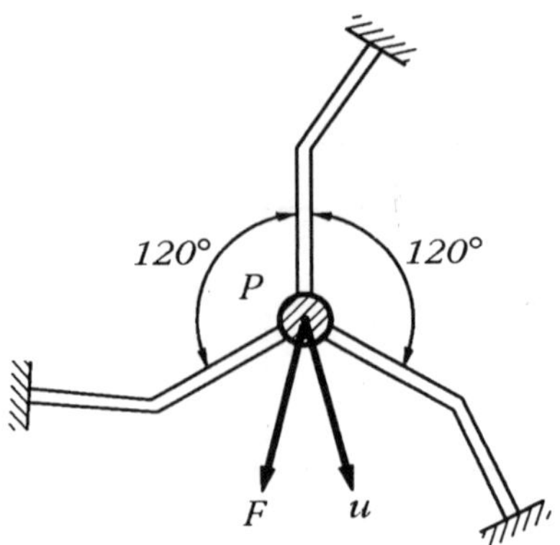


Figure 1.5: Support structure with 3 similar but unsymmetrical components

the system is the same in all directions[2].

These two examples illustrate that there is sometimes an advantage to be gained from considering a disparate physical problem that shares a common mathematical structure with that under investigation.

### 1.1.4 Displacement

The displacement of a particle $P$ is a vector $\boldsymbol{u}$ representing the difference between the final and the initial position of $P$ — i.e. it is the distance which $P$ moves during the deformation. The components of $\boldsymbol{u}$ are denoted by appropriate suffices — e.g. $u_x, u_y, u_z$, so that

$$\boldsymbol{u} = \boldsymbol{i}u_x + \boldsymbol{j}u_y + \boldsymbol{k}u_z \ . \tag{1.11}$$

The deformation of a body is completely defined if we know the displacement of its every particle. Notice however that there is a class of displacements which do not involve deformation — the so-called *rigid-body displacements.* A typical case is where all the particles of the body have the same displacement. The name arises, of course, because rigid-body displacement is the only class of displacement that can be experienced by a rigid body.

[2] A similar argument can be used to show that if a laminated fibre-reinforced composite is laid up with equal numbers of identical, but not necessarily symmetrical, laminas in each of 3 or more equispaced orientations, it must be elastically isotropic within the plane. This proof depends on the propoerties of the *fourth order* Cartesian tensor $c_{ijkl}$ describing the stress-strain relation $e_{ij} = c_{ijkl}\sigma_{kl}$ of the laminas.

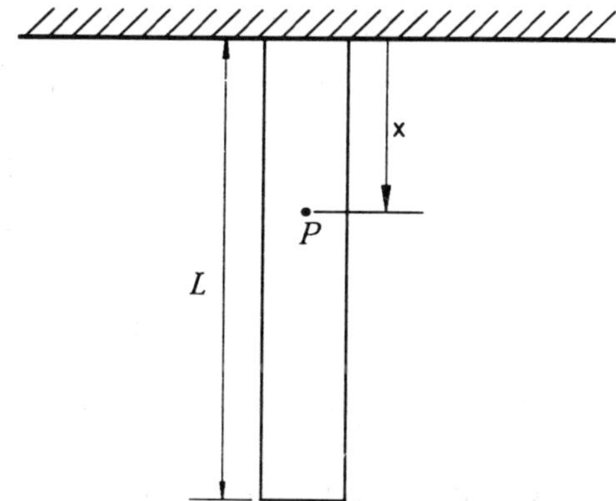


Figure 1.6: The bar suspended from the ceiling

## 1.2 Strains and their relation to displacements

Components of strain will be denoted by the symbol, $e$, with appropriate suffices (e.g. $e_{xx}, e_{xy}$). As in the case of stress, no special symbol is required for shear strain, though we shall see below that the quantity defined in most elementary texts (and usually denoted by $\gamma$) differs from that used in the mathematical theory of Elasticity by a factor of 2. A major advantage of this definition is that it makes the strain, $e$, a second order Cartesian Tensor (see §1.1.3 above). We shall demonstrate this by establishing transformation rules for strain similar to equations (1.7–1.9) in §1.2.4 below.

### 1.2.1 Tensile strain

Students usually first encounter the concept of strain in elementary Strength of Materials as the ratio of extension to original length and are sometimes confused by the apparently totally different definition used in more mathematical treatments of solid mechanics. We shall discuss here the connection between the two definitions — partly for completeness, and partly because the physical insight that can be developed in the simple problems of Strength of Materials is very useful if it can be carried over into more difficult problems.

Figure 1.6 shows a bar of original length $L$ and density $\rho$ hanging from the ceiling. Suppose we are asked to find how much it increases in length under the loading of its own weight.

It is easily shown that the tensile stress $\sigma_{xx}$ at the point $P$, distance $x$ from the ceiling, is

$$\sigma_{xx} = \rho g(L - x) \; , \tag{1.12}$$

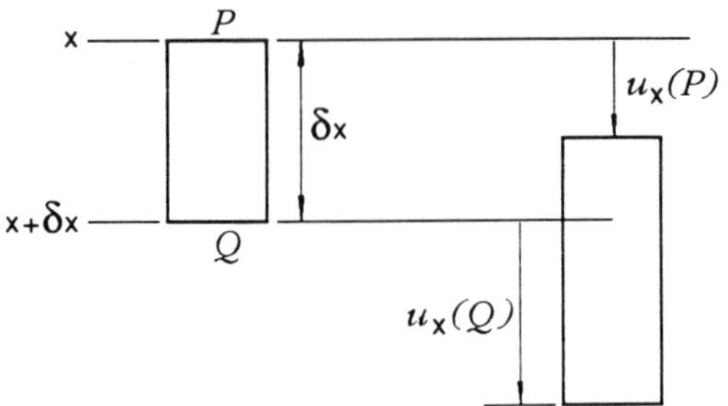


Figure 1.7: Infinitesimal section of the bar

where $g$ is the acceleration due to gravity, and hence from Hooke's law,

$$e_{xx} = \frac{\rho g(L-x)}{E} \; . \tag{1.13}$$

However, the strain varies continuously over the length of the bar and hence we can only apply the Strength of Materials definition if we examine an infinitesimal piece of the bar over which the strain can be regarded as sensibly constant.

We describe the deformation in terms of the downward displacement $u_x$ which depends upon $x$ and consider that part of the bar between $x$ and $x + \delta x$, denoted by $PQ$ in Figure 1.7.

After the deformation, $PQ$ must have extended by $u_x(Q) - u_x(P)$ and hence the local value of 'Strength of Materials' tensile strain is

$$e_{xx} = \frac{(u_x(Q) - u_x(P))}{\delta x} \; . \tag{1.14}$$

Taking the limit as $\delta x \to 0$, we obtain the definition

$$e_{xx} = \frac{\partial u_x}{\partial x} \; . \tag{1.15}$$

Corresponding definitions can be developed in three-dimensional problems for the other normal strain components. i.e.

$$e_{yy} = \frac{\partial u_y}{\partial y} \quad ; \quad e_{zz} = \frac{\partial u_z}{\partial z} \; . \tag{1.16}$$

Notice how easy the problem of Figure 1.6 becomes when we use these definitions. We get

$$\frac{\partial u_x}{\partial x} = \frac{\rho g(L-x)}{E} \; , \tag{1.17}$$

from (1.13, 1.15) and hence

$$u_x = \frac{\rho g(2Lx - x^2)}{2E} + A \; , \tag{1.18}$$

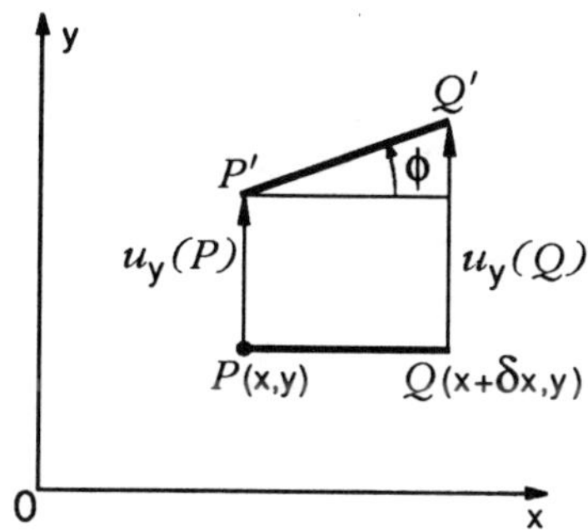


Figure 1.8: Rotation of a line segment

where $A$ is an arbitrary constant, which expresses the fact that our knowledge of the stresses and hence the strains in the body is not sufficient to determine its position in space. In fact, $A$ represents an arbitrary rigid-body displacement. In this case we need to use the fact that the top of the bar is joined to a supposedly rigid ceiling — i.e. $u_x(0) = 0$ and hence $A = 0$ from (1.18).

### 1.2.2 Rotation and shear strain

Noting that the two $x$'s in $e_{xx}$ correspond to those in its definition $\partial u_x/\partial x$, it is natural to seek a connection between the shear strain $e_{xy}$ and one or both of the derivatives $\partial u_x/\partial y, \partial u_y/\partial x$. As a first step, we shall discuss the geometrical interpretation of these derivatives.

Figure 1.8 shows a line segment $PQ$ of length $\delta x$, aligned with the $x$-axis, the two ends of which are displaced in the $y$-direction. Clearly if $u_y(Q) \neq u_y(P)$, the line $PQ$ will be rotated by these displacements and if the angle of rotation is small it can be written

$$\phi = \frac{(u_y(x+\delta x) - u_y(x))}{\delta x} \,, \tag{1.19}$$

(anticlockwise positive).

Proceeding to the limit as $\delta x \to 0$, we have

$$\phi = \frac{\partial u_y}{\partial x} \,. \tag{1.20}$$

Thus, $\partial u_y/\partial x$ is the angle through which a line originally in the $x$-direction rotates towards the $y$-direction during the deformation[3].

[3]Students of Strength of Materials will have already used a similar result when they express the slope of a beam as $du/dx$, where $x$ is the distance along the axis of the beam and $u$ is the transverse displacement.

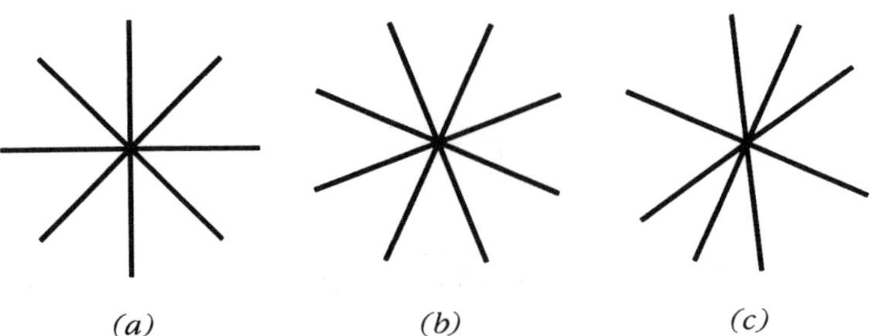


Figure 1.9: Rotation of lines at point P; ($a$) Original state, ($b$) Rigid-body rotation; ($c$) Rotation and deformation

Now, if $PQ$ is a line drawn on the surface of an elastic solid, the occurence of a rotation $\phi$ does not necessarily indicate that the solid is deformed - we could rotate the line simply by rotating the solid as a rigid body. To investigate this matter further, we imagine drawing a series of lines at different angles through the point $P$ as shown in Figure 1.9($a$).

If the vicinity of the point $P$ suffers merely a local rigid-body rotation, all the lines will rotate through the same angle and retain the same relative inclinations as shown in 1.9($b$). However, if different lines rotate through *different* angles, as in 1.9($c$), the body must have been deformed. We shall show in the next section that the rotations of the lines in Figure 1.9($c$) are not independent and a consideration of their interdependency leads naturally to a definition of shear strain.

### 1.2.3 Transformation of coördinates

Suppose we knew the displacement components $u_x, u_y$ throughout the body and wished to find the rotation, $\phi$, of the line $PQ$ in Figure 1.10, which is inclined at an angle $\theta$ to the $x$-axis.

We construct a new axis system $Ox'y'$ with $Ox'$ parallel to $PQ$ as shown, in which case we can argue as above that $PQ$ rotates anticlockwise through the angle

$$\phi(PQ) = \frac{\partial u'_y}{\partial x'} \ . \tag{1.21}$$

Furthermore, we have

$$\frac{\partial}{\partial x'} = \boldsymbol{\nabla}.\boldsymbol{i}' \tag{1.22}$$

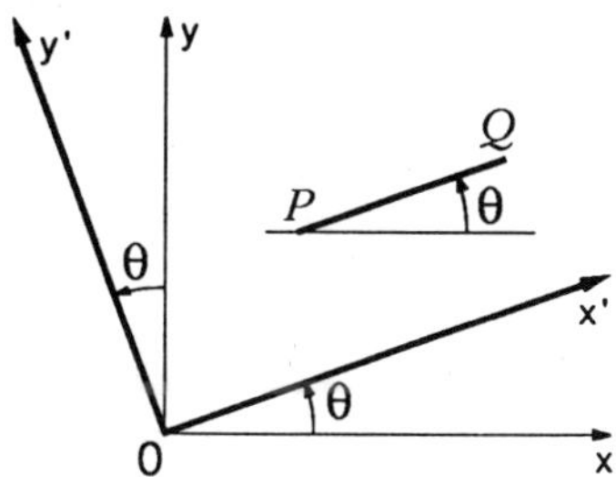


Figure 1.10: Rotation of a line inclined at angle $\theta$

$$\begin{aligned} &= \left(\boldsymbol{i}\frac{\partial}{\partial x}+\boldsymbol{j}\frac{\partial}{\partial y}\right).\boldsymbol{i}' = \boldsymbol{i}.\boldsymbol{i}'\frac{\partial}{\partial x}+\boldsymbol{j}.\boldsymbol{i}'\frac{\partial}{\partial y} \\ &= \cos\theta\frac{\partial}{\partial x}+\sin\theta\frac{\partial}{\partial y}\,, \qquad (1.23) \end{aligned}$$

and by similar arguments,

$$u_{x'} = u_x\cos\theta + u_y\sin\theta\,; \qquad (1.24)$$

$$u_{y'} = u_y\cos\theta - u_x\sin\theta\,. \qquad (1.25)$$

Substituting these results into equation (1.21), we find

$$\begin{aligned} \phi(PQ) &= \left(\cos\theta\frac{\partial}{\partial x}+\sin\theta\frac{\partial}{\partial y}\right)(u_y\cos\theta - u_x\sin\theta) \qquad (1.26) \\ &= \frac{\partial u_y}{\partial x}\cos^2\theta - \frac{\partial u_x}{\partial y}\sin^2\theta + \left(\frac{\partial u_y}{\partial y}-\frac{\partial u_x}{\partial x}\right)\sin\theta\cos\theta \\ &= \frac{1}{2}\left(\frac{\partial u_y}{\partial x}-\frac{\partial u_x}{\partial y}\right)+\frac{1}{2}\left(\frac{\partial u_y}{\partial x}+\frac{\partial u_x}{\partial y}\right)\cos 2\theta \\ &\quad +\frac{1}{2}\left(\frac{\partial u_y}{\partial y}-\frac{\partial u_x}{\partial x}\right)\sin 2\theta\,. \qquad (1.27) \end{aligned}$$

In the final expression (1.27), the first term is independent of the inclination $\theta$ of the line $PQ$ and hence represents a rigid-body rotation as in Figure 1.9(*b*). We denote this rotation by the symbol $\omega$, which with the convention illustrated in Figure 1.8 is anticlockwise positive.

Notice that as this term is independent of $\theta$ in equation (1.27), it is the same for any right-handed set of axes - i.e.

$$\omega = \frac{1}{2}\left(\frac{\partial u_y}{\partial x}-\frac{\partial u_x}{\partial y}\right) = \frac{1}{2}\left(\frac{\partial u'_y}{\partial x'}-\frac{\partial u'_x}{\partial y'}\right)\,, \qquad (1.28)$$

for any $x', y'$.

In three dimensions, $\omega$ represents a small positive rotation about the $z$-axis and is therefore more properly denoted by $\omega_z$ to distinguish it from the corresponding rotations about the $x$ and $y$ axes. i.e.

$$\omega_x = \frac{1}{2}\left(\frac{\partial u_z}{\partial y} - \frac{\partial u_y}{\partial z}\right) \; ; \; \omega_y = \frac{1}{2}\left(\frac{\partial u_x}{\partial z} - \frac{\partial u_z}{\partial x}\right) \; ; \; \omega_z = \frac{1}{2}\left(\frac{\partial u_y}{\partial x} - \frac{\partial u_x}{\partial y}\right) , \tag{1.29}$$

or in suffix notation

$$\omega_k = \frac{1}{2}\left(\epsilon_{ijk}\frac{\partial u_j}{\partial x_i}\right) , \tag{1.30}$$

where $\epsilon_{ijk}$ is the *alternating tensor* which is defined to be 1 if the suffices are in cyclic order (e.g. 2,3,1), -1 if they are in *reverse* cyclic order (e.g. 2,1,3) and zero if any two suffices are the same. (Notice, the alternating tensor is *not* a second order Cartesian tensor.)

The rotation $\omega$ is a vector in three-dimensional problems and can be defined by the equation

$$\boldsymbol{\omega} = \frac{1}{2}\text{curl } \boldsymbol{u} \equiv \frac{1}{2}\boldsymbol{\nabla} \times \boldsymbol{u} , \tag{1.31}$$

but in two dimensions it behaves as a scalar since two of its components degenerate to zero[4].

### 1.2.4 Definition of shear strain

We are now in a position to *define* the shear strain $e_{xy}$ as the difference between the rotation of a line drawn in the $x$-direction and the corresponding rigid body rotation, $\omega_z$, i.e.

$$e_{xy} = \frac{\partial u_y}{\partial x} - \omega_z = \frac{1}{2}\left(\frac{\partial u_y}{\partial x} + \frac{\partial u_x}{\partial y}\right) \tag{1.32}$$

and similarly

$$e_{yz} = \frac{1}{2}\left(\frac{\partial u_z}{\partial y} + \frac{\partial u_y}{\partial z}\right) \; ; \; e_{zx} = \frac{1}{2}\left(\frac{\partial u_x}{\partial z} + \frac{\partial u_z}{\partial x}\right) . \tag{1.33}$$

Note that $e_{xy}$ so defined is one half of the quantity $\gamma_{xy}$ used in Strength of Materials and in many older books on Elasticity.

The strain-displacement relations (1.15, 1.16, 1.32, 1.33) can be written in the concise form

$$e_{ij} = \frac{1}{2}\left(\frac{\partial u_i}{\partial x_j} + \frac{\partial u_j}{\partial x_i}\right) . \tag{1.34}$$

[4]In some books, $\omega_x, \omega_y, \omega_z$ are denoted by $\omega_{yz}, \omega_{zx}, \omega_{xy}$ respectively. This notation is not used here because it gives the erroneous impression that $\boldsymbol{\omega}$ is a second order tensor rather than a vector.

With the notation of equations (1.15, 1.16, 1.29, 1.32), we can now write (1.27) in the form

$$\phi(PQ) = \omega_z + e_{xy}(\cos^2\theta - \sin^2\theta) + (e_{yy} - e_{xx})\sin\theta\cos\theta \tag{1.35}$$

and hence

$$e_{x'y'} = \phi(PQ) - \omega_z$$

(by definition)

$$= e_{xy}(\cos^2\theta - \sin^2\theta) + (e_{yy} - e_{xx})\sin\theta\cos\theta\ . \tag{1.36}$$

This is of course one of the coördinate transformation relations for strain. The other one

$$e_{x'x'} = e_{xx}\cos^2\theta + e_{yy}\sin^2\theta + 2e_{xy}\sin\theta\cos\theta \tag{1.37}$$

being obtainable from equations (1.15, 1.23, 1.24) in the same way. A comparison of equations (1.36, 1.37) and (1.7, 1.8) confirms that, with these definitions, the strain $e_{ij}$ is a second order Cartesian tensor.

## 1.3 Stress-strain relations

We shall develop the various forms of the linear elastic stress-strain relations for the isotropic medium by regarding Young's modulus $E$ and Poisson's ratio $\nu$ as fundamental constants. Hence we regard as experimentally determined the equations

$$e_{xx} = \frac{\sigma_{xx}}{E} - \frac{\nu\sigma_{yy}}{E} - \frac{\nu\sigma_{zz}}{E}\ ; \tag{1.38}$$

$$e_{yy} = \frac{\sigma_{yy}}{E} - \frac{\nu\sigma_{zz}}{E} - \frac{\nu\sigma_{xx}}{E}\ ; \tag{1.39}$$

$$e_{zz} = \frac{\sigma_{zz}}{E} - \frac{\nu\sigma_{xx}}{E} - \frac{\nu\sigma_{yy}}{E}\ . \tag{1.40}$$

The relation between $e_{xy}$ and $\sigma_{xy}$ can then be obtained by using the transformation relations. We know that there are three principal directions such that if we align them with $x, y, z$, we have

$$\sigma_{xy} = \sigma_{yz} = \sigma_{zx} = 0 \tag{1.41}$$

and hence by symmetry

$$e_{xy} = e_{yz} = e_{zx} = 0\ . \tag{1.42}$$

Using a coördinate system aligned with the principal directions, we write

$$e_{x'y'} = (e_{yy} - e_{xx})\sin\theta\cos\theta \tag{1.43}$$

from equations (1.36, 1.42)

$$\begin{aligned} &= \frac{(\sigma_{yy}-\sigma_{xx})(1+\nu)\sin\theta\cos\theta}{E} \\ &= \frac{(1+\nu)\sigma_{x'y'}}{E} , \end{aligned} \tag{1.44}$$

from (1.38, 1.39, 1.8).

We write

$$\mu = \frac{E}{2(1+\nu)} , \tag{1.45}$$

so that equation (1.44) takes the form

$$e_{x'y'} = \frac{\sigma_{x'y'}}{2\mu} . \tag{1.46}$$

### 1.3.1 Lamé's constants

It is often desirable to solve equations (1.38–1.41) to express $\sigma_{xx}$ in terms of $e_{xx}$ etc. The solution is routine and leads to the equation

$$\sigma_{xx} = \frac{E\nu(e_{xx}+e_{yy}+e_{zz})}{(1+\nu)(1-2\nu)} + \frac{Ee_{xx}}{(1+\nu)} \tag{1.47}$$

and similar equations, which are more concisely written in the form

$$\sigma_{xx} = \lambda e + 2\mu e_{xx} \tag{1.48}$$

etc., where

$$\lambda = \frac{E\nu}{(1+\nu)(1-2\nu)} = \frac{2\mu\nu}{(1-2\nu)} \tag{1.49}$$

and

$$e \equiv e_{xx}+e_{yy}+e_{zz} \equiv e_{ii} \equiv \operatorname{div} \boldsymbol{u} \tag{1.50}$$

is known as the *dilatation*.

The stress-strain relations (1.46, 1.48) can be written more concisely in the index notation in the form

$$\sigma_{ij} = \lambda e_{mm}\delta_{ij} + 2\mu e_{ij} , \tag{1.51}$$

where $\delta_{ij}$ is the *Kronecker delta*, defined as 1 if $i=j$ and 0 if $i \neq j$. The constants $\lambda, \mu$ are known as Lamé's constants. Young's modulus and Poisson's ratio can be written in terms of Lamé's constants through the equations

$$E = \frac{\mu(3\lambda+2\mu)}{(\lambda+\mu)} ; \tag{1.52}$$

$$\nu = \frac{\lambda}{2(\lambda+\mu)} . \tag{1.53}$$

## 1.3.2 Dilatation and bulk modulus

The dilatation, $e$, is easily shown to be invariant as to coördinate transformation and is therefore a scalar quantity. In physical terms it is the local volumetric strain, since a unit cube increases under strain to a block of dimensions $(1+e_{xx}), (1+e_{yy}), (1+e_{zz})$ and hence the volume change is

$$\delta V = (1+e_{xx})(1+e_{yy})(1+e_{zz}) - 1 = e_{xx} + e_{yy} + e_{zz} + O(e_{xx}e_{yy}) \ . \qquad (1.54)$$

It can be shown that the dilatation $e$ and the rotation vector $\boldsymbol{\omega}$ are harmonic - i.e. $\nabla^2 e = \nabla^2 \boldsymbol{\omega} = 0$. For this reason, many early solutions of elasticity problems were formulated in terms of these variables, so as to make use of the wealth of mathematical knowledge about harmonic functions. We now have other more convenient ways of expressing elasticity problems in terms of harmonic functions, which will be discussed in Chapter 15 *et seq.*.

The dilatation is proportional to the mean stress $\bar{\sigma}$ through a constant known as the *bulk modulus* $K_b$. Thus, from equation (1.48),

$$\overline{\sigma} \equiv \frac{(\sigma_{xx} + \sigma_{yy} + \sigma_{zz})}{3} \equiv \frac{1}{3}\sigma_{ii} = K_b e \ , \qquad (1.55)$$

where

$$K_b = \lambda + \frac{2}{3}\mu = \frac{E}{3(1-2\nu)} \ . \qquad (1.56)$$

We note that $K_b \to \infty$ if $\nu \to 0.5$ — i.e. the material becomes incompressible.

### PROBLEM

1. Show that equations (1.38–1.40, 1.44) can be written in the concise form

$$e_{ij} = \frac{(1+\nu)\sigma_{ij}}{E} - \frac{\nu \sigma_{mm}\delta_{ij}}{E} \ . \qquad (1.57)$$

# Chapter 2

# EQUILIBRIUM AND COMPATIBILITY

We can think of an elastic solid as a highly redundant framework — each particle is built-in to its neighbours. For such a framework, we expect to get some equations from considerations of *equilibrium*, but not as many as there are unknowns. The deficit is made up by *compatibility* conditions — statements that the deformed components must fit together. These latter conditions will relate the dimensions and hence the strains of the deformed components and in order to express them in terms of the same unknowns as the stresses (forces) we need to make use of the stress-strain relations as applied to each component separately.

If we were to approximate the continuous elastic body by a system of interconnected elastic bars, this would be an exact description of the solution procedure. The only difference in treating the continuous medium is that the system of algebraic equations is replaced by partial differential equations describing the same physical or geometrical principles.

## 2.1 Equilibrium equations

We consider a small rectangular block of material — side $\delta x, \delta y, \delta z$ — as shown in Figure 2.1. We suppose that there is a body force[1], $\boldsymbol{p}$ per unit volume and that the stresses vary with position so that the stress components on opposite faces of the block differ by the differential quantities $\delta\sigma_{xx}, \delta\sigma_{xy}$ etc. Only those stress components which act in the $x$-direction are shown in Figure 2.1 for clarity.

Resolving forces in the $x$-direction, we find

$$(\sigma_{xx} + \delta\sigma_{xx} - \sigma_{xx})\delta y \delta z + (\sigma_{xy} + \delta\sigma_{xy} - \sigma_{xy})\delta z \delta x$$

[1]A body force is one that acts directly on every particle of the body, rather than being applied by tractions at its boundaries and transmitted to the various particles by means of internal stresses. This is an important distinction which will be discussed further in Chapter 7, below. The commonest example of a body force is that due to gravity.

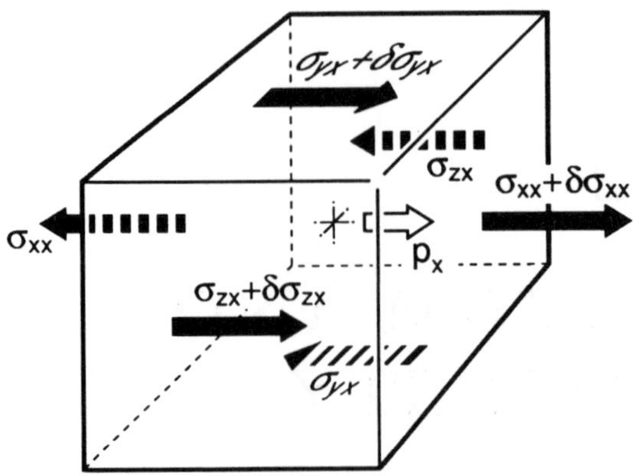


Figure 2.1: Forces in the $x$-direction on an elemental block

$$+(\sigma_{xz} + \delta\sigma_{xz} - \sigma_{xz})\delta x\delta y + p_x\delta x\delta y\delta z \quad = \quad 0 \ . \tag{2.1}$$

Hence, dividing through by $(\delta x\delta y\delta z)$ and proceeding to the limit as these infinitesimals tend to zero, we obtain

$$\frac{\partial\sigma_{xx}}{\partial x} + \frac{\partial\sigma_{xy}}{\partial y} + \frac{\partial\sigma_{xz}}{\partial z} + p_x = 0 \ . \tag{2.2}$$

Similarly, we have

$$\frac{\partial\sigma_{yx}}{\partial x} + \frac{\partial\sigma_{yy}}{\partial y} + \frac{\partial\sigma_{yz}}{\partial z} + p_y = 0 \ ; \tag{2.3}$$

$$\frac{\partial\sigma_{zx}}{\partial x} + \frac{\partial\sigma_{zy}}{\partial y} + \frac{\partial\sigma_{zz}}{\partial z} + p_z = 0 \ , \tag{2.4}$$

or in suffix notation

$$\frac{\partial\sigma_{ij}}{\partial x_j} + p_i = 0 \ . \tag{2.5}$$

These are the *differential equations of equilibrium.*

## 2.2 Compatibility equations

The easiest way to satisfy the equations of compatibility — as in framework problems — is to express all the strains in terms of the displacements. In a framework, this ensures that the components fit together by identifying the displacement of points in two links which are pinned together by the same symbol. If the framework is redundant, the number of pin displacements thereby introduced is less than the number of component lengths (and hence extensions) determined by them — and the number of unknowns is therefore reduced.

For a solid body, the process is essentially similar, but much more straightforward. We define the six components of strain in terms of displacements through equations (1.34). These six equations introduce only three unknowns $(u_x, u_y, u_z)$ and hence the latter can be eliminated to give equations constraining the strain components.

For example, from (1.15, 1.16) we find

$$\frac{\partial^2 e_{xx}}{\partial y^2} = \frac{\partial^3 u_x}{\partial x \partial y^2} \; ; \; \frac{\partial^2 e_{yy}}{\partial x^2} = \frac{\partial^3 u_y}{\partial y \partial x^2} \; ; \tag{2.6}$$

and hence

$$\frac{\partial^2 e_{xx}}{\partial y^2} + \frac{\partial^2 e_{yy}}{\partial x^2} = \frac{\partial^2}{\partial x \partial y}\left(\frac{\partial u_x}{\partial y} + \frac{\partial u_y}{\partial x}\right) = 2\frac{\partial^2 e_{xy}}{\partial x \partial y} \; , \tag{2.7}$$

i.e.

$$\frac{\partial^2 e_{xx}}{\partial y^2} - 2\frac{\partial^2 e_{xy}}{\partial x \partial y} + \frac{\partial^2 e_{yy}}{\partial x^2} = 0 \; . \tag{2.8}$$

Two more equations of the same form may be obtained by permuting suffices. It is tempting to pursue an analogy with algebraic equations and argue that, since the six strain components are defined in terms of three independent displacement components, we must be able to develop three (i.e. $6-3$) independent compatibility equations. However, it is easily verified that, in addition to the three equations similar to (2.8), the strains must satisfy three more equations of the form

$$\frac{\partial^2 e_{zz}}{\partial x \partial y} = \frac{\partial}{\partial z}\left(\frac{\partial e_{yz}}{\partial x} + \frac{\partial e_{zx}}{\partial y} - \frac{\partial e_{xy}}{\partial z}\right) \; . \tag{2.9}$$

The resulting six equations are independent in the sense that no one of them can be derived from the other five, which all goes to show that arguments for algebraic equations do not always carry over to partial differential equations.

A concise statement of the six compatibility equations can be written in the index notation in the form

$$\epsilon_{pks}\frac{\partial}{\partial x_k}\left(\frac{\partial e_{sj}}{\partial x_i} - \frac{\partial e_{si}}{\partial x_j}\right) = 0 \; . \tag{2.10}$$

The full set of six equations makes the problem very complicated. In practice, therefore, most three-dimensional problems are treated in terms of displacements instead of strains. This satisfies the requirement of compatibility automatically. However, in two dimensions, all except one of the compatibility equations degenerate to identities, so that a formulation in terms of stresses or strains is more practical.

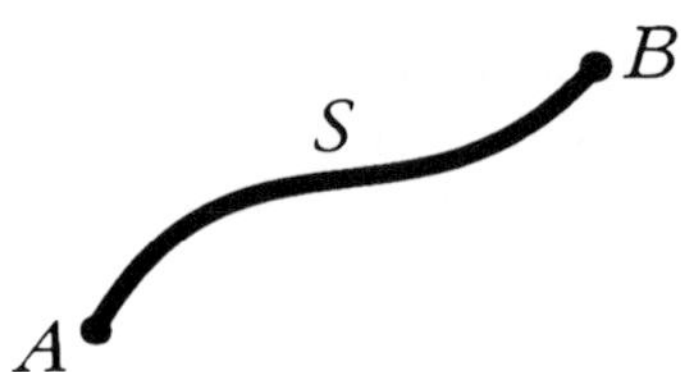


Figure 2.2: Path of the integral in equation (2.11)

### 2.2.1 The significance of the compatibility equations

The physical meaning of equilibrium is fairly straightforward, but people often get mixed up about just what is being guaranteed by the compatibility equations.

#### Single-valued displacements

Mathematically, we might say that the strains are compatible *when they are definable in terms of a single-valued, continuously differentiable displacement.*

We could imagine reversing this process — i.e. integrating the strains (displacement gradients) to find the relative displacement ($\boldsymbol{u}_A - \boldsymbol{u}_B$) of two points $A, B$ in the solid (see Figure 2.2).

Formally we can write[2]

$$\boldsymbol{u}_A - \boldsymbol{u}_B = \int_B^A \frac{\partial \boldsymbol{u}}{\partial S} dS \ . \tag{2.11}$$

The integral will be along the line, S, and if the displacements are to be single-valued, it mustn't make any difference if we change the line provided it remains within the solid. In other words, the integral should be *path-independent.*

*The compatibility equations are not quite sufficient to guarantee this.*

They *do* guarantee

$$\oint \frac{\partial \boldsymbol{u}}{\partial S} dS = 0 \tag{2.12}$$

around an *infinitesimal* closed loop.

[2]Explicit forms of the integral (2.11) in terms of the strain components were developed by E.Cesaro and are known as *Cesaro integrals.* See, for example, A.E.H.Love, *A Treatise on the Mathematical Theory of Elasticity*, 4th edn., Dover (1944), §156A.

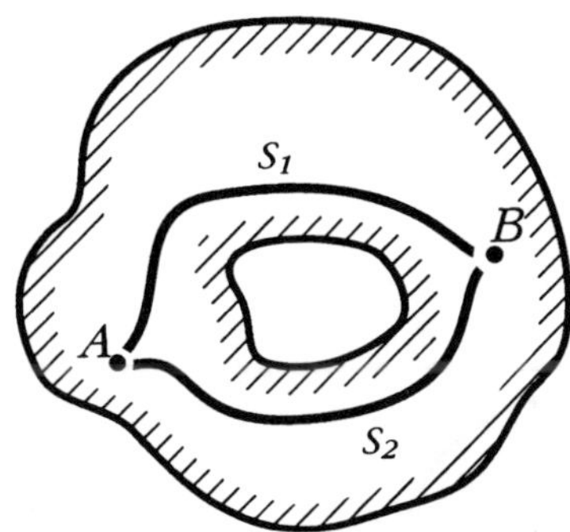


Figure 2.3: Qualitatively different integration paths in a multiply-connected body

Now, we could make infinitesimal changes in our line from $A$ to $B$ by taking in such small loops until the whole line was sensibly changed, thus satisfying the requirement that the integral (2.11) is path-independent, but the fact that the line is changed infinitesimally stops us from taking a qualitatively (topologically) different route through a multiply-connected body. For example, in Figure 2.3, it is impossible to move $S_1$ to $S_2$ by infinitesimal changes without passing outside the body[3].

In practice, if a solid is multiply-connected it is usually easier to work in terms of displacements and by-pass this problem. Otherwise the equivalence of topologically different paths in integrals like (2.11, 2.12) has to be *explicitly* enforced.

### Compatibility of deformed shapes

A more 'physical' way of thinking of compatibility is to state that *the separate particles of the body must deform under load in such a way that they fit together after deformation.* This interpretation is conveniently explored by way of a 'jig-saw' analogy.

We consider a two-dimensional body cut up as a jig-saw puzzle (Figure 2.4), of which the pieces are deformable. Figure 2.4($a$) shows the original puzzle and 2.4($b$) the puzzle after deformation by some external loads $\boldsymbol{F}_i$. (The pieces are shown as initially rectangular to aid visualization.)

The deformation of the puzzle must satisfy the following conditions:-

(i) The forces on any given piece (including external forces if any) must be in

[3]For a more rigorous discussion of this question, see A.E.H.Love, *loc. cit.* or B.A.Boley and J.H.Weiner, *Theory of Thermal Stresses*, John Wiley, New York, (1960), §§3.6–3.8.

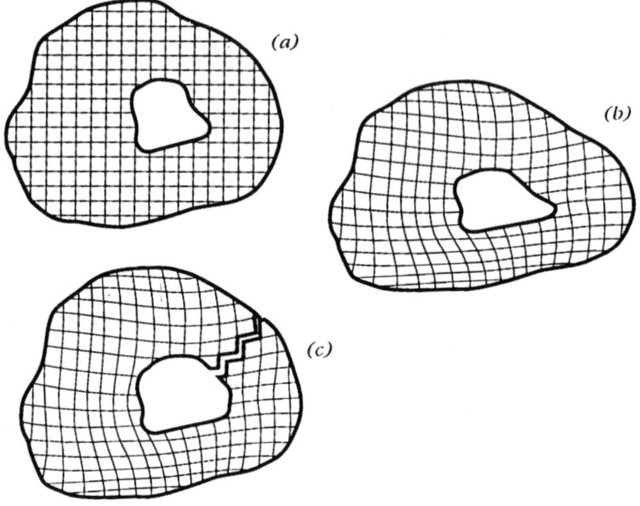


Figure 2.4: A multiply-connected, deformable jig-saw puzzle: ($a$) Before deformation; ($b$) After deformation; ($c$) Result of attempting to assemble the deformed puzzle if equation (2.12) is not satisfied for any closed path encircling the hole

equilibrium.

(ii) The deformed pieces must be the right *shape* to fit together to make the deformed puzzle (Figure 2.4(*b*)).

For the continuous solid, the compatibility condition guarantees that (ii) is satisfied, since shape is defined by displacement derivatives and hence by strains. However, if the puzzle is multiply-connected — e.g. if it has a central hole — the condition (ii) is not sufficient to ensure that the deformed pieces can be assembled into a coherent body. Suppose we imagine assembling the 'deformed' puzzle working from one piece outwards. The partially completed puzzle is simply-connected and the shape condition is sufficient to ensure the success of our assembly until we reach a piece which would convert the partial puzzle to a multiply-connected body. This piece may be the right *shape* for both sides, but the wrong *size* for the separation. If so, it will be possible to leave the puzzle in a state with a discontinuity as shown in Figure 2.4(*c*) at any arbitrarily chosen position — i.e. there is no part of the body at which continuity breaks down — but there will be no way in which Figure 2.4(*b*) can be constructed.

The lines defining the two sides of the discontinuity in Figure 2.4(*c*) are the same shape and hence, in the most general case, the discontinuity can be defined by six arbitrary constants corresponding to the three rigid-body translations and three rotations needed to move one side to coincide with the other[4]. We therefore get six additional algebraic conditions for each hole in a multiply-connected body.

## 2.3 Equilibrium equations for displacements

Remembering that it is often easier to work in terms of displacements to avoid complications with the compatibility conditions, it is convenient to express the eqilibrium equations in terms of displacements. This is done by substituting for the stresses from the stress-strain relations (1.51) into equation (2.2) giving

$$\lambda\frac{\partial e}{\partial x} + 2\mu\left(\frac{\partial e_{xx}}{\partial x} + \frac{\partial e_{xy}}{\partial y} + \frac{\partial e_{xz}}{\partial z}\right) + p_x = 0 \,. \qquad (2.13)$$

Also,

$$\begin{aligned} 2\left(\frac{\partial e_{xx}}{\partial x} + \frac{\partial e_{xy}}{\partial y} + \frac{\partial e_{xz}}{\partial z}\right) &= 2\frac{\partial^2 u_x}{\partial x^2} + \frac{\partial^2 u_x}{\partial y^2} + \frac{\partial^2 u_y}{\partial x \partial y} + \frac{\partial^2 u_x}{\partial z^2} + \frac{\partial^2 u_z}{\partial x \partial z} \\ &= \nabla^2 u_x + \frac{\partial e}{\partial x} \,, \qquad (2.14) \end{aligned}$$

[4]This can be proved by evaluating the relative displacements of corresponding points on opposite sides of the cut, using an integral of the form (2.11) whose path does not cross the cut.

and hence equation (2.13) can be written

$$(\lambda+\mu)\frac{\partial}{\partial x}\text{div } \boldsymbol{u}+\mu\nabla^2 u_x+p_x=0\ . \tag{2.15}$$

Two similar equations are obtained for the equilibrium conditions in the $y$- and $z$-directions.

We also note that

$$\lambda+\mu=\mu\left(1+\frac{2\nu}{1-2\nu}\right)=\frac{\mu}{(1-2\nu)}\ , \tag{2.16}$$

from (1.49) and hence the general equilibrium condition can be concisely written in the form

$$\boldsymbol{\nabla}\text{div } \boldsymbol{u}+(1-2\nu)\nabla^2\boldsymbol{u}+\frac{(1-2\nu)\boldsymbol{p}}{\mu}=0\ . \tag{2.17}$$

## PROBLEMS

1. Show that, if there are no body forces, the dilatation $e\equiv e_{xx}+e_{yy}+e_{zz}$ must satisfy the condition

$$\nabla^2 e=0\ .$$

2. Show that, if there are no body forces, the rotation $\boldsymbol{\omega}$ must satisfy the condition

$$\nabla^2\boldsymbol{\omega}=0\ .$$

3. One way of satisfying the compatibility equations in the absence of rotation is to define the components of displacement in terms of a potential function $\psi$ through the relations

$$u_x=\frac{\partial\psi}{\partial x}\ ;\ u_y=\frac{\partial\psi}{\partial y}\ ;\ u_z=\frac{\partial\psi}{\partial z}\ .$$

Use the stress-strain relations to derive expressions for the stress components in terms of $\psi$.

Hence show that the stresses will satisfy the equilibrium equations in the absence of body forces if and only if

$$\nabla^2\psi=\text{constant}\ .$$

# Part II

# TWO-DIMENSIONAL PROBLEMS

# Chapter 3

# PLANE STRAIN AND PLANE STRESS

A problem is two-dimensional if the field quantities such as stress and displacement depend on only two coördinates $(x, y)$ and the boundary conditions are imposed on a line $f(x, y) = 0$ in the $xy$-plane.

In this sense, there are strictly no two-dimensional problems in elasticity. There *are* circumstances in which the stresses are independent of the $z$-coördinate, but all real bodies must have some bounding surfaces which are not represented by a line in the $xy$-plane. The two-dimensionality of the resulting fields depends upon the boundary conditions on such surfaces being of an appropriate form.

## 3.1 Plane strain

It might be argued that a closed line in the $xy$-plane *does* define a solid body — namely an infinite cylinder of the appropriate cross-section whose axis is parallel to the $z$-direction. However, making a body infinite does not really dispose of the question of boundary conditions, since there are usually some implied boundary conditions 'at infinity'. For example, the infinite cylinder could be in a state of uniaxial tension, $\sigma_{zz} = C$, where $C$ is an arbitrary constant. However, a unique two-dimensional infinite cylinder problem can be defined by demanding that $u_x, u_y$ be independent of $z$ and that $u_z = 0$ for all $x, y, z$, in which case it follows that

$$e_{zx} = e_{zy} = e_{zz} = 0 \; . \tag{3.1}$$

This is the two-dimensional state known as *plane strain*.

In view of the stress-strain relations, an equivalent statement to equation (3.1) is

$$\sigma_{zx} = \sigma_{zy} = 0 \;\; ; \;\; u_z = 0 \; , \tag{3.2}$$

and hence a condition of plane strain will exist in a *finite* cylinder provided that (i) the boundary conditions on the sides of the cylinder are independent of $z$ and (ii) the

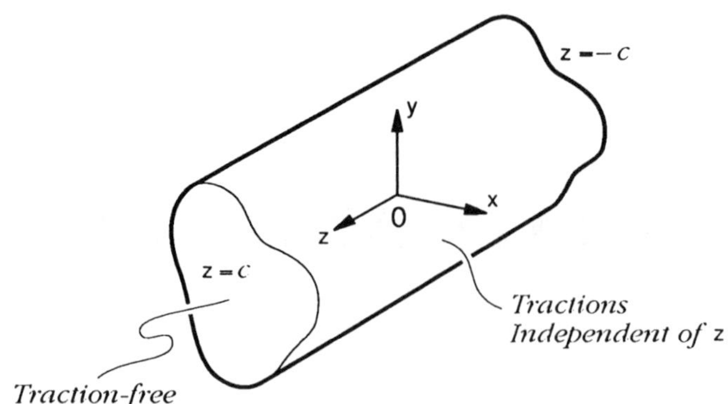


Figure 3.1: The long cylinder with traction-free ends

cylinder has plane ends (e.g. $z = \pm c$) which are in frictionless contact with two plane rigid walls.

From the condition $e_{zz} = 0$ (3.1) and the stress-strain relations, we can deduce

$$0 = \frac{\sigma_{zz}}{E} - \frac{\nu(\sigma_{xx} + \sigma_{yy})}{E}$$

i.e.

$$\sigma_{zz} = \nu(\sigma_{xx} + \sigma_{yy}) \tag{3.3}$$

and hence

$$\begin{aligned} e_{xx} &= \frac{\sigma_{xx}}{E} - \frac{\nu\sigma_{yy}}{E} - \frac{\nu^2(\sigma_{xx} + \sigma_{yy})}{E} \\ &= \frac{(1-\nu^2)\sigma_{xx}}{E} - \frac{\nu(1+\nu)\sigma_{yy}}{E} . \end{aligned} \tag{3.4}$$

Of course, there are comparatively few practical applications in which a cylinder with plane ends is constrained between frictionless rigid walls, but fortunately the plane strain solution can be used in an approximate sense for a cylinder with any end conditions, provided that the length of the cylinder is large compared with its cross-sectional dimensions.

We shall illustrate this with reference to the long cylinder of Figure 3.1, for which the ends, $z = \pm c$ are traction-free and the sides are loaded by tractions which are independent of $z$.

We first solve the problem under the plane strain assumption, obtaining an exact solution in which all the stresses are independent of $z$ and in which there exists a normal stress $\sigma_{zz}$ on all $z$-planes, which we can calculate from equation (3.3).

The plane strain solution satisfies all the boundary conditions of the problem *except* that $\sigma_{zz}$ also acts on the end faces, $z = \pm c$, where it appears as an unwanted normal traction. We therefore seek a *corrective* solution which, when superposed on the plane strain solution, removes the unwanted normal tractions on the surfaces $z = \pm c$, without changing the boundary conditions on the sides of the cylinder.

### 3.1.1 The corrective solution

The corrective solution must have zero tractions on the sides of the cylinder and a prescribed normal traction (equal and opposite to that obtained in the plane strain solution) on the end faces.

This is a fully three-dimensional problem which generally has no closed-form solution. However, if the prescribed tractions have the linear form

$$\sigma_{zz} = B + Cx + Dy \ , \tag{3.5}$$

the solution can be obtained in the context of Strength of Materials by treating the long cylinder as a beam subjected to an axial force

$$F = \int_A \sigma_{zz} dA \tag{3.6}$$

and bending moments

$$M_x = \int_A \sigma_{zz} y dA \ \ ; \ \ M_y = -\int_A \sigma_{zz} x da \tag{3.7}$$

about the axes $Ox, Oy$ respectively, where $O$ is chosen to coincide with the centroid of the cross-section[1].

If the original plane strain solution does not give a distribution of $\sigma_{zz}$ of this convenient linear form, we can still use equations (3.6, 3.7) to define the force and moments for an *approximate* Strength of Materials corrective solution. The error involved in using this approximate solution will be that associated with yet another corrective solution corresponding to the problem in which the end faces of the cylinder are loaded by tractions equal to the *difference* between those in the plane strain solution and the Strength of Materials linear form (3.5) associated with the force resultants (3.6, 3.7).

In this final corrective solution, the ends of the cylinder are loaded by *self-equilibrated tractions*, since the Strength of Materials approximation is carefully chosen to have the same force and moment resultants as the required exact solution and

[1]The Strength of Materials solution for axial force and pure bending is in fact *exact* in the sense of the Theory of Elasticity, but we shall not prove this here. See for example S.P.Timoshenko and J.N.Goodier, *Theory of Elasticity*, McGraw-Hill, New York, 3rd. edn. (1970) §102.

in such cases, we postulate that significant stresses will only be generated in the immediate vicinity of the ends — or more precisely, in regions whose distance from the ends is comparable with the cross-sectional dimensions of the cylinder. If the cylinder is many times longer than its cross-sectional dimensions, there will be a substantial portion near the centre where the final corrective solution gives negligible stresses and hence where the sum of the original plane strain solution and the Strength of Materials correction is a good approximation to the actual three-dimensional stress field.

### 3.1.2 Saint-Venant's principle

The thesis that a self-equilibrating system of loads produces only local effects is known as *Saint-Venant's principle.* It seems intuitively reasonable, but has not been proved rigorously except for certain special cases — some of which we shall encounter later in this course (see for example Chapter 6). It can be seen as a consequence of the rule that alternate load paths through a structure share the load in proportion with their stiffnesses. If a region of the boundary is loaded by a self-equilibrating system of tractions, the stiffest paths are the shortest — i.e. those which do not penetrate far from the loaded region. Hence, the longer paths — which are those which contribute to stresses distant from the loaded region — carry relatively little load.

Note that if the local tractions are *not* self-equilibrating, some of the load paths must go to other distant parts of the boundary and hence there will be significant stresses in intermediate regions. For this reason, it is important to superpose the Strength of Materials approximate corrective solution when solving plane strain problems for long cylinders with traction-free ends. By contrast, the final stage of solving the 'Saint-Venant problem' to calculate the correct stresses near the ends is seldom of much importance in practical problems, since the ends, being traction-free, are not generally points of such high stress as the interior.

As a point of terminology, we shall refer to problems in which the boundary conditions on the ends have been corrected only in the sense of force and moment resultants as being solved in the *weak* sense with respect to these boundaries. Boundary conditions are satisfied in the *strong* sense when the tractions are specified in a pointwise rather than a force resultant sense.

## 3.2 Plane stress

Plane stress is an approximate solution, in contrast to plane strain, which is exact. In other words, plane strain is a special solution of the complete three-dimensional equations of elasticity, whereas plane stress is only approached in the limit as the thickness of the loaded body tends to zero.

It is argued that if the two bounding $z$-planes of a thin plate are sufficiently close in comparison with the other dimensions, and if they are also free of tractions, the stresses on all parallel $z$-planes will be sufficiently small to be neglected in which case we write

$$\sigma_{zx} = \sigma_{zy} = \sigma_{zz} = 0 \ , \tag{3.8}$$

for all $x, y, z$.

It then follows that

$$e_{zx} = e_{zy} = 0 \ , \tag{3.9}$$

but $e_{zz} \neq 0$, being given in fact by

$$e_{zz} = -\frac{\nu}{E}(\sigma_{xx} + \sigma_{yy}) \ . \tag{3.10}$$

The two-dimensional stress-strain relations for normal strains are then

$$e_{xx} = \frac{\sigma_{xx}}{E} - \frac{\nu\sigma_{yy}}{E} \ ; \tag{3.11}$$

$$e_{yy} = \frac{\sigma_{yy}}{E} - \frac{\nu\sigma_{xx}}{E} \ . \tag{3.12}$$

The fact that plane stress is not an exact solution can best be explained by considering the compatibility equation

$$\frac{\partial^2 e_{yy}}{\partial z^2} - 2\frac{\partial^2 e_{yz}}{\partial y \partial z} + \frac{\partial^2 e_{zz}}{\partial y^2} = 0 \ . \tag{3.13}$$

Since *ex-hypothesi* none of the stresses vary with $z$, the first two terms in this equation are identically zero and hence the equation will be satisfied if and only if

$$\frac{\partial^2 e_{zz}}{\partial y^2} = 0 \ . \tag{3.14}$$

This in turn requires

$$\frac{\partial^2 (\sigma_{xx} + \sigma_{yy})}{\partial y^2} = 0 \ , \tag{3.15}$$

from equation (3.10).

Applying similar arguments to the other compatibility equations, we conclude that the plane stress assumption is exact if and only if $(\sigma_{xx} + \sigma_{yy})$ is a linear function of $x, y$. i.e. if

$$\sigma_{xx} + \sigma_{yy} = B + Cx + Dy \ . \tag{3.16}$$

The attentive reader will notice that this condition is exactly equivalent to (3.3, 3.5). In other words, the plane stress solution is exact if and only if the 'weak form' solution of the corresponding plane strain solution is exact. Of course this is not a coincidence. In this special case, the process of off-loading the ends of the long cylinder in the plane strain solution has the effect of making $\sigma_{zz}$ zero throughout the cylinder and hence the resulting solution also satisfies the plane stress assumption, whilst remaining exact.

### 3.2.1 Generalized plane stress

The approximate nature of the plane stress formulation is distasteful to elasticians of a more mathematical temperament, who prefer to preserve the rigour of an exact theory. This can be done by the contrivance of defining the *average* stresses across the thickness of the plate — e.g.

$$\bar{\sigma}_{xx} \equiv \frac{1}{2c}\int_{-c}^{c} \sigma_{xx} dz \ . \tag{3.17}$$

It can then be shown that the average stresses so defined satisfy the plane stress equations *exactly*. This is referred to as the *generalized plane stress* formulation.

In practice, of course, the gain in rigour is illusory unless we can also establish that the stress variation across the section is small, so that the local values are reasonably close to the average. A fully three-dimensional theory of thin plates under in-plane loading shows that the plane stress assumption is a good approximation except in regions whose distance from the boundary is comparable with the plate thickness[2].

### 3.2.2 Relationship between plane stress and plane strain

The solution of a problem under either the plane strain or plane stress assumptions involves finding a two-dimensional stress field, defined in terms of the components $\sigma_{xx}, \sigma_{xy}, \sigma_{yy}$, which satisfies the equilibrium equations (2.5), and for which the corresponding strains, $e_{xx}, e_{xy}, e_{yy}$, satisfy the only non-trivial compatibility equation (2.8). The equilibrium and compatibility equations are the same in both formulations, the only difference being in the relation between the stress and strain components, which for normal stresses are given by (3.4) for plane strain and (3.11, 3.12) for plane stress. The relation between the shear stress $\sigma_{xy}$ and the shear strain $e_{xy}$ is the same for both formulations and is given by equation (1.46). Thus, from a mathematical perspective, the plane strain solution simply looks like the plane stress solution for a material with different elastic constants. In fact, it is easily verified that equation (3.4) can be obtained from (3.11) by making the substitutions

$$E = \frac{E'}{(1-\nu'^2)} \ ; \ \nu = \frac{\nu'}{(1-\nu')} \ , \tag{3.18}$$

and then dropping the primes. This substitution also leaves the shear stress-shear strain relation unchanged as required.

In the following chapters, we shall treat problems generally with the plane stress assumptions, noting that results for plane strain can be recovered by the substitution (3.18) when required.

---

[2]See for example S.P.Timoshenko and J.N.Goodier, *loc. cit.*, §98.

## PROBLEMS

1. For a solid in a state of plane stress, show that if there are body forces $p_x, p_y$ per unit volume in the direction of the axes $x, y$ respectively, the compatibility equation can be expressed in the form

$$\nabla^2(\sigma_{xx}+\sigma_{yy}) = -(1+\nu)\left(\frac{\partial p_x}{\partial x}+\frac{\partial p_y}{\partial y}\right) .$$

Hence deduce that the stress distribution for any particular case is independent of the material constants and the body forces, provided the latter are constant.

2. (i). Derive from first principles the equation of compatibility which must be satisfied by the strains $e_{xx}$, $e_{xy}$, $e_{yy}$, in a state of plane stress.

(ii) Show that this equation is satisfied by unrestrained thermal expansion ($e_{xx} = e_{yy} = \alpha T, e_{xy} = 0$) provided that the temperature, $T$, is a two-dimensional harmonic function — i.e.

$$\frac{\partial^2 T}{\partial x^2}+\frac{\partial^2 T}{\partial y^2} = 0 .$$

(iii) Hence deduce that, subject to certain restrictions which you should explicitly list, no thermal stresses will be induced in a thin body with a steady-state, two-dimensional temperature distribution.

(iv) Show that an initially straight line on such a body will be distorted by the heat flow in such a way that its curvature is proportional to the local heat flux across it.

# Chapter 4

# STRESS FUNCTION FORMULATION

## 4.1 The concept of a scalar stress function

Newton's law of gravitation states that two heavy bodies attract each other with a force proportional to the inverse square of their distance — thus it is essentially a vector theory, being concerned with forces. However, the idea of a scalar gravitational potential can be introduced by defining the work done in moving a unit mass from infinity to a given point in the field. The principle of conservation of energy requires that this be a unique function of position and it is easy to show that the gravitational force at any point is then proportional to the gradient of this scalar potential. Thus, the original vector problem is reduced to a problem about a scalar potential and its derivatives.

In general, scalars are much easier to deal with than vectors. In particular, they lend themselves very easily to coördinate transformations, whereas vectors (and to an even greater extent tensors) require a set of special transformation rules (e.g. Mohr's circle).

In certain field theories, the scalar potential has an obvious physical significance. For example, in the conduction of heat, the temperature is a scalar potential in terms of which the vector heat flux can be defined. However, it is not necessary to the method that such a physical interpretation can be given. The gravitational potential can be given a physical interpretation as discussed above, but this interpretation may never feature in the solution of a particular problem, which is simply an excercise in the solution of a certain partial differential equation with appropriate boundary conditions. In the theory of elasticity, we make use of scalar potentials called *stress functions* or *displacement functions* which have no obvious physical meaning other than their use in defining stress or displacement components in terms of derivatives.

## 4.2 Choice of a suitable form

In the choice of a suitable form for a stress or displacement function, there is only one absolute rule — that the operators which define the relationship between the scalar and vector (or tensor) quantities should indeed define a vector (or tensor).

For example, it is appropriate to define the displacement in terms of the first derivatives (the gradient) of a scalar or to define the stress components in terms of the second derivatives of a scalar, since the second derivatives of a scalar form the components of a Cartesian tensor.

In effect, what we are doing in requiring this similarity of form between the definitions and the defined quantity is ensuring that the relationship is preserved in coördinate transformations. It would be quite possible to work out an elasticity problem in terms of the displacement components $u_x, u_y, u_z$, treating these as essentially scalar quantities which vary with position — indeed this was a technique which was used in early theories. However, we would then get into trouble as soon as we tried to make any statements about quantities in other coördinate directions. By contrast, if we define (for example) $u_x = \partial\psi/\partial x$; $u_y = \partial\psi/\partial y$; $u_z = \partial\psi/\partial z$, (i.e. $\boldsymbol{u} = \boldsymbol{\nabla}\psi$), it immediately follows that $u_{x'} = \partial\psi/\partial x'$ for any $x'$.

## 4.3 The Airy stress function

The two-dimensional problem of elasticity is most easily reduced to a tractable potential problem by representing the stress components in the form

$$\sigma_{xx} = \frac{\partial^2 \phi}{\partial y^2} \;\; ; \;\; \sigma_{yy} = \frac{\partial^2 \phi}{\partial x^2} \;\; ; \;\; \sigma_{xy} = -\frac{\partial^2 \phi}{\partial x \partial y} \; . \tag{4.1}$$

This representation was introduced by G.B.Airy[1] in 1862 and $\phi$ is therefore generally referred to as the *Airy stress function.* It is not the most obvious form. It would seem more natural to write $\sigma_{xx} = \partial^2\psi/\partial x^2$; $\sigma_{yy} = \partial^2\psi/\partial y^2$; $\sigma_{xy} = \partial^2\psi/\partial x\partial y$ and indeed this also leads to a representation which is widely used as part of the general three-dimensional solution (see Chapter 15 below).

It is easily verified that equation (4.1) transforms as a Cartesian tensor as required. For example, using (1.23) we can write

$$\begin{aligned}
\frac{\partial^2 \phi}{\partial x'^2} &= \left(\cos\theta\frac{\partial}{\partial x} + \sin\theta\frac{\partial}{\partial y}\right)^2 \phi \\
&= \frac{\partial^2 \phi}{\partial x^2}\cos^2\theta + \frac{\partial^2 \phi}{\partial y^2}\sin^2\theta + 2\frac{\partial^2 \phi}{\partial x \partial y}\sin\theta\cos\theta \; ,
\end{aligned} \tag{4.2}$$

from which using (1.9 and 4.1) we deduce that $\sigma_{y'y'} = \partial^2\phi/\partial x'^2$ as required.

[1]For a good historical survey of the development of potential function methods in Elasticity, see H.M.Westergaard, *Theory of Elasticity and Plasticity*, Dover, New York (1964), Chapter 2.

## 4.4 The governing equation

As discussed in Chapter 2, the stress and strain fields have to satisfy the equations of equilibrium and compatibility if they are to describe permissible states of an elastic body. In stress function representations, this generally imposes certain constraints on the choice of stress function, which can be expressed by requiring it to be a solution of a certain partial differential equation. We shall determine this governing equation by substituting the representation (4.1) into the equilibrium and compatibility equations in two dimensions.

### 4.4.1 The equilibrium equations

On substituting (4.1) into the first equilibrium equation, we obtain

$$\frac{\partial^3 \phi}{\partial y^2 \partial x} - \frac{\partial^3 \phi}{\partial x \partial y^2} + p_x = 0 \tag{4.3}$$

and hence $p_x = 0$. In the same way, the second equilibrium equation is satisfied as long as $p_y = 0$. Thus, *the Airy stress function automatically satisfies the equilibrium equations provided the body forces are zero.* In some books, it is stated that this is precisely the reason for preferring the Airy function to the more obvious form referred to in §4.3 above, but this is misleading, since the 'more obvious form' — although it requires some constraints on $\psi$ in order to satisfy equilibrium — can be shown to define displacements which automatically satisfy the compatibility condition[2]. The real reason for preferring the Airy function is that it is capable of describing all possible states of stress in an elastic body, whilst the alternative is more restrictive. This will not be proved here.

It is worth noting that the Airy stress function can be used in other applications, with inelastic constitutive laws — for example plasticity theory — where its satisfaction of the equilibrium equations remains an advantage.

### 4.4.2 Non-zero body forces

If the body force $\boldsymbol{p}$ is not zero, but is of a restricted form such that it can be written $\boldsymbol{p} = -\nabla V$, where $V$ is a scalar potential, equations (4.1) can be generalized by including an extra term $+V$ in each of the normal stress components whilst leaving the shear stress definition unchanged. It is easily verified that, with this modification, the equilibrium equations are again satisfied.

Problems involving body forces will be discussed in more detail in Chapter 7 below. For the moment, we restrict attention to the the case where $\boldsymbol{p} = 0$ and hence where the representation (4.1) is appropriate.

---

[2]See Problem 2.3.

### 4.4.3 The compatibility condition

In order to determine the constraints placed upon the choice of $\phi$ by the compatibility condition, we first express the latter in terms of stresses using the stress-strain relations, obtaining

$$\frac{\partial^2 \sigma_{xx}}{\partial y^2} - \nu \frac{\partial^2 \sigma_{yy}}{\partial y^2} - 2(1+\nu)\frac{\partial^2 \sigma_{xy}}{\partial x \partial y} + \frac{\partial^2 \sigma_{yy}}{\partial x^2} - \nu \frac{\partial^2 \sigma_{xx}}{\partial x^2} = 0 \ . \tag{4.4}$$

We then substitute for the stress components from (4.1) obtaining

$$\frac{\partial^4 \phi}{\partial y^4} - \nu \frac{\partial^4 \phi}{\partial x^2 \partial y^2} + 2(1+\nu)\frac{\partial^4 \phi}{\partial x^2 \partial y^2} + \frac{\partial^4 \phi}{\partial x^4} - \nu \frac{\partial^4 \phi}{\partial x^2 \partial y^2} = 0 \ , \tag{4.5}$$

i.e.

$$\frac{\partial^4 \phi}{\partial x^4} + 2\frac{\partial^4 \phi}{\partial x^2 \partial y^2} + \frac{\partial^4 \phi}{\partial y^4} = \left(\frac{\partial^2}{\partial x^2} + \frac{\partial^2}{\partial y^2}\right)^2 \phi = 0 \ . \tag{4.6}$$

This equation is known as the *biharmonic equation* and is usually written in the shortened form

$$\nabla^4 \phi = 0 \ . \tag{4.7}$$

The biharmonic equation is the governing equation for the Airy stress function in elasticity problems. Thus, by using the Airy stress function representation, the problem of determining the stresses in an elastic body is reduced to that of finding a solution of equation (4.7) (i.e. a *biharmonic function*) whose derivatives satisfy certain boundary conditions on the surfaces.

### 4.4.4 Method of solution

Historically, the boundary-value problem for the Airy stress function has been approached in a semi-inverse way — i.e. by using the variation of tractions along the boundaries to give a clue to the kind of function required, but then exploring the stress fields developed from a wide range of such functions and selecting a combination which can be made to satisfy the required conditions. The disadvantage with this method is that it requires a wide experience of particular solutions and even then is not guaranteed to be successful.

A more modern method which has the advantage of always developing an appropriate stress function if the boundary conditions are unmixed (i.e. all specified in terms of stresses or all in terms of displacements) is based on representing the stress function in terms of analytic functions of the complex variable. This method will be discussed later. However, although it is powerful and extremely elegant, it is an unfortunate fact that for most problems it leaves us with the task of evaluating a difficult contour integral, which may not be as convenient a numerical technique as a direct series or finite difference attack on the original problem using real stress functions. (There are some exceptions — notably for bodies which are susceptible of a simple conformal transformation into the unit circle.).

### 4.4.5 Reduced dependence on elastic constants

It is clear from dimensional considerations that, when the compatibility equation is expressed in terms of $\phi$, Young's modulus must appear in every term and can therefore be cancelled, but it is an unexpected bonus that the Poisson's ratio terms also cancel in (4.5), leaving an equation that is independent of elastic constants. It follows that the stress field in a simply-connected elastic body[3] in a state of plane strain or plane stress is independent of the material properties if the boundary conditions are expressed in terms of tractions and in particular, that the plane stress and plane strain fields are identical.

Dundurs[4] has shown that a similar reduced dependence on elastic constants occurs in plane problems involving interfaces between two dissimilar elastic materials. In such cases, three independent dimensionless parameters (e.g. $\mu_1/\mu_2, \nu_1, \nu_2$) can be formed from the elastic constants $\mu_1, \nu_1, \mu_2, \nu_2$ of the materials 1, 2 respectively, but Dundurs proved that the stress field can be written in terms of only two parameters which he defined as

$$\alpha = \left(\frac{\kappa_1+1}{\mu_1} - \frac{\kappa_2+1}{\mu_2}\right) \Big/ \left(\frac{\kappa_1+1}{\mu_1} + \frac{\kappa_2+1}{\mu_2}\right) \; ; \tag{4.8}$$

$$\beta = \left(\frac{\kappa_1-1}{\mu_1} - \frac{\kappa_2-1}{\mu_2}\right) \Big/ \left(\frac{\kappa_1+1}{\mu_1} + \frac{\kappa_2+1}{\mu_2}\right) \, , \tag{4.9}$$

where $\kappa = 3 - 4\nu$ for plane strain and $(3-\nu)/(1+\nu)$ for plane stress. It can be shown that Dundurs' parameters must lie in the range $-1 \leq \alpha \leq 1 \; ; \; -0.5 \leq \beta \leq 0.5$ if $0 \leq \nu_1, \nu_2 \leq 0.5$.

**PROBLEM**

1. (i) Show that the function

$$\phi = y\omega + \psi$$

satisfies the biharmonic equation provided that $\omega, \psi$ are both *harmonic* (i.e. $\nabla^2\omega = 0, \nabla^2\psi = 0$).

(ii) Develop expressions for the stress components in terms of $\omega, \psi$, based on the use of $\phi$ as an Airy stress function.

(iii) Show that a solution suitable for the half-plane $y > 0$ subject to normal surface tractions only (i.e. $\sigma_{xy} = 0$ on $y = 0$) can be obtained by writing

$$\omega = -\frac{\partial\psi}{\partial y}$$

[3]The restriction to simply-connected bodies is necessary, since in a multiply-connected body there is an implied displacement condition, as explained in §2.2 above.

[4]J.Dundurs, Discussion on 'Edge bonded dissimilar orthogonal elastic wedges under normal and shear loading, ASME J.Appl.Mech., Vol. 36 (1969), 650-652.

and hence that under these conditions the normal stress $\sigma_{xx}$ near the surface $y = 0$ is equal to the applied traction $\sigma_{yy}$.

(iv) Do you think this is a rigorous proof? Can you think of any exceptions?

# Chapter 5

# PROBLEMS IN RECTANGULAR COÖRDINATES

The Cartesian coördinate system $(x, y)$ is clearly particularly suited to the problem of determining the stresses in a rectangular body whose boundaries are defined by equations of the form $x = a, y = b$. A wide range of such problems can be treated using stress functions which are polynomials in $x, y$. In particular, polynomial solutions can be obtained for 'Strength of Materials' type beam problems in which a rectangular bar is bent by an end load or by a distributed load on one face.

## 5.1 Biharmonic polynomial functions

In rectangular coördinates, the biharmonic equation takes the form

$$\frac{\partial^4 \phi}{\partial x^4} + 2\frac{\partial^4 \phi}{\partial x^2 \partial y^2} + \frac{\partial^4 \phi}{\partial y^4} = 0 \tag{5.1}$$

and it follows that any polynomial in $x, y$ of degree less than four will be biharmonic and is therefore appropriate as a stress function. However, for higher order polynomial terms, equation (5.1) is not identically satisfied. Suppose, for example, that we consider just those terms in a general polynomial whose combined degree (the sum of the powers of $x$ and $y$) is $N$. We can write these terms in the form

$$P_N(x, y) = A_0 x^N + A_1 x^{N-1} y + A_2 x^{N-2} y^2 + \ldots + A_N y^N \tag{5.2}$$

$$= \sum_{i=0}^{N} A_i x^{N-i} y^i \, , \tag{5.3}$$

where we note that there are $N + 1$ independent coefficients, $A_i$ $(i = 0, N)$. If we now substitute $P_N(x, y)$ into equation (5.1), we shall obtain a new polynomial of degree $(N-4)$, since each term is differentiated four times. We can denote this new polynomial by $Q_{N-4}(x,y)$ where

$$Q_{N-4}(x,y) = \nabla^4 P_N(x,y) \tag{5.4}$$

$$= \sum_{i=0}^{N-4} B_i x^{(N-4-i)} y^i \,. \tag{5.5}$$

The $(N-3)$ coefficients $B_0, \ldots, B_{N-4}$ are easily obtained by expanding the right hand side of equation (5.4) and equating coefficients. For example,

$$B_0 = N(N-1)(N-2)(N-3)A_0 + 4(N-2)(N-3)A_2 + 24A_4 \,. \tag{5.6}$$

Now the original function $P_N(x,y)$ will be biharmonic if and only if $Q_{N-4}(x,y)$ is zero for all $x, y$ and this in turn is only possible if every term in the series (5.5) is identically zero, since the polynomial terms are all linearly independent of each other. In other words

$$B_i = 0 \;\; ; \;\; i = 0 \text{ to } N-4 \,. \tag{5.7}$$

These conditions can be converted into a corresponding set of $(N-3)$ equations for the coefficients $A_i$. For example, the equation $B_0 = 0$ gives

$$N(N-1)(N-2)(N-3)A_0 + 4(N-2)(N-3)A_2 + 24A_4 = 0 \,, \tag{5.8}$$

from equation (5.6). We shall refer to the $(N-3)$ equations of this form as *constraints* on the coefficients $A_i$, since the coefficients are constrained to satisfy them if the original polynomial is to be biharmonic.

One approach would be to use the constraint equations to eliminate $(N-3)$ of the unknown coefficients in the original polynomial — for example, we could treat the first four coefficients, $A_0, A_1, A_2, A_3$, as unknown constants and use the constraint equations to define all the remaining coefficients in terms of these unknowns. Equation (5.8) would then be treated as an equation for $A_4$ and the subsequent constraint equations would each define one new constant in the series. It may help to consider a particular example at this stage. Suppose we consider the fifth degree polynomial

$$P_5(x,y) = A_0x^5 + A_1x^4y + A_2x^3y^2 + A_3x^2y^3 + A_4xy^4 + A_5y^5 \,, \tag{5.9}$$

which has six independent coefficients. Substituting into equation (5.4), we obtain the first degree polynomial

$$Q_1(x,y) = (120A_0 + 24A_2 + 24A_4)x + (24A_1 + 24A_3 + 120A_5)y \,. \tag{5.10}$$

The coefficients of $x$ and $y$ in $Q_1$ must both be zero if $P_5$ is to be biharmonic and we can write the resulting two constraint equations in the form

$$A_4 = -5A_0 - A_2 \,; \tag{5.11}$$

$$A_5 = -A_1/5 - A_3/5 \,. \tag{5.12}$$

Finally, we use (5.11, 5.12) to eliminate $A_4, A_5$ in the original definition of $P_5$, obtaining the definition of the most general biharmonic fifth degree polynomial

$$\begin{aligned} P_5(x,y) &= A_0(x^5 - 5xy^4) + A_1(x^4y - y^5/5) \\ &\quad + A_2(x^3y^2 - xy^4) + A_3(x^2y^3 - y^5/5) \; . \end{aligned} \tag{5.13}$$

This function will be biharmonic for any values of the four independent constants $A_0, A_1, A_2, A_3$. We can express this by stating that the polynomial $P_5$ has four degrees of freedom.

In general, the polynomial $Q$ is of degree 4 less than $P$ because the biharmonic equation is of degree 4. It follows that there are always four fewer constraint equations than there are coefficients in the original polynomial $P$ and hence that they can be satisfied leaving a polynomial with 4 degrees of freedom. However, the process degenerates if $N < 3$.

In view of the above discussion, it might seem appropriate to write an expression for the general polynomial of degree $N$ in the form of equation (5.13) as a preliminary to the solution of polynomial problems in rectangular coördinates. However, as can be seen from equation (5.13), the resulting expressions are algebraically messy and this approach becomes unmanageable for problems of any complexity. Instead, it turns out to be more straightforward algebraically to define problems in terms of the simpler unconstrained polynomials like equation (5.2) and to impose the constraint equations at a later stage in the solution.

### 5.1.1 Second and third degree polynomials

We recall that the stress components are defined in terms of the stress function $\phi$ through the relations

$$\sigma_{xx} = \frac{\partial^2 \phi}{\partial y^2} \; ; \tag{5.14}$$

$$\sigma_{yy} = \frac{\partial^2 \phi}{\partial x^2} \; ; \tag{5.15}$$

$$\sigma_{xy} = -\frac{\partial^2 \phi}{\partial x \partial y} \; . \tag{5.16}$$

It follows that when the stress function is a polynomial of degree $N$ in $x, y$, the stress components will be polynomials of degree $(N-2)$. In particular, constant and linear terms in $\phi$ correspond to null stress fields (zero stress everywhere) and can be disregarded.

The second degree polynomial

$$\phi = A_0x^2 + A_1xy + A_2y^2 \tag{5.17}$$

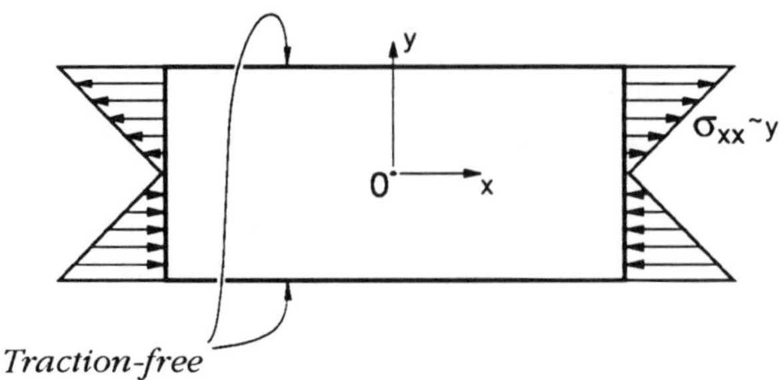


Figure 5.1: The rectangular beam in pure bending

yields the stress components

$$\sigma_{xx} = 2A_2 \ ; \quad \sigma_{xy} = -A_1 \ ; \quad \sigma_{yy} = 2A_2 \tag{5.18}$$

and hence corresponds to the most general state of biaxial uniform stress.

The third degree polynomial

$$\phi = A_0 x^3 + A_1 x^2 y + A_2 x y^2 + A_3 y^3 \tag{5.19}$$

yields the stress components

$$\sigma_{xx} = 2A_2 x + 6A_3 y \ ; \ \sigma_{xy} = -2A_1 x - 2A_2 y \ ; \ \sigma_{yy} = 6A_0 x + 2A_1 y \ . \tag{5.20}$$

If we arbitrarily set $A_0, A_1, A_2 = 0$, the only remaining non-zero stress will be

$$\sigma_{xx} = 6A_3 y \ , \tag{5.21}$$

which corresponds to a state of pure bending, when applied to the rectangular beam $-a < x < a, -b < y < b$, as shown in Figure 5.1.

The other terms in equation (5.19) correspond to a more general state of bending. For example, the constant $A_0$ describes bending of the beam by tractions $\sigma_{yy}$ applied to the boundaries $y = +b$, whilst the terms involving shear stresses $\sigma_{xy}$ could be obtained by describing a general state of biaxial bending with reference to a Cartesian coördinate system which is not aligned with the axes of the beam.

The above solutions are of course very elementary, but we should remember that, in contrast to the Strength of Materials solutions for simple bending, they are obtained without making any simplifying assumptions about the stress fields. For example, we have not assumed that plane sections remain plane, nor have we demanded that the beam be long in comparison with its depth. Thus, the present section could be taken as verifying the *exactness* of the Strength of Materials solutions for uniform stress and simple bending, as applied to a rectangular beam.

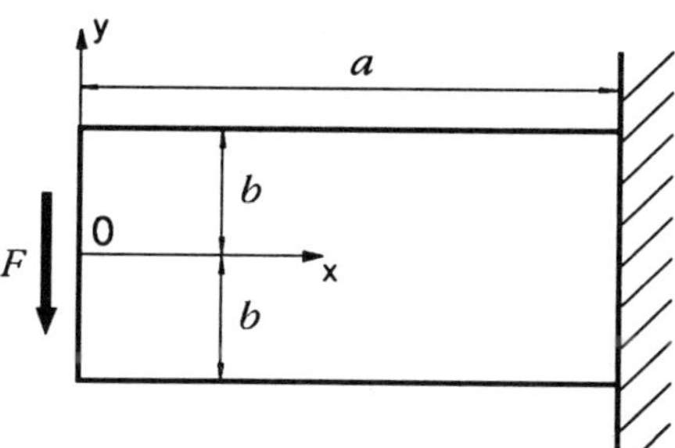


Figure 5.2: Cantilever with an end load

## 5.2 Rectangular beam problems

### 5.2.1 Bending of a beam by an end load

Figure 5.2 shows a rectangular beam, $0 < x < a, -b < y < b$, subjected to a transverse force, $F$ at the end $x = 0$, and built-in at the end $x = a$, the horizontal boundaries $y = +b$ being traction-free. The boundary conditions for this problem are most naturally written in the form

$$\sigma_{xy} = 0 \; ; \; y = \pm b \; ; \tag{5.22}$$

$$\sigma_{yy} = 0 \; ; \; y = \pm b \; ; \tag{5.23}$$

$$\sigma_{xx} = 0 \; ; \; x = 0 \; ; \tag{5.24}$$

$$\int_{-b}^{b} \sigma_{xy} dy = F \; ; \; x = 0 \; . \tag{5.25}$$

The boundary condition (5.25) is imposed in the *weak* form, which means that the value of the traction is not specified at each point on the boundary — only the force resultant is specified. In general, we shall find that problems for the rectangular beam have finite polynomial solutions when the boundary conditions on the ends are stated in the weak form, but that the *strong* (i.e. pointwise) boundary condition can only be satisfied on all the boundaries by an infinite series or transform solution. This problem is further discussed in Chapter 6.

Strength of Materials considerations suggest that the bending moment in this problem will vary with $x$ and hence that the stress component $\sigma_{xx}$ will have a leading term proportional to $xy$. This in turn suggests a fourth degree polynomial term $xy^3$ in the stress function $\phi$. Our procedure is therefore to start with the trial stress function

$$\phi = C_1 x y^3 \; , \tag{5.26}$$

examine the corresponding tractions on the boundaries and then seek a *corrective* solution which, when superposed on equation (5.26), yields the solution to the problem. Substituting (5.26) into (5.14–5.16), we obtain

$$\sigma_{xx} = 6C_1xy \; ; \tag{5.27}$$
$$\sigma_{xy} = -3C_1y^2 \; ; \tag{5.28}$$
$$\sigma_{yy} = 0 \; , \tag{5.29}$$

from which we note that the boundary conditions (5.23, 5.24) are satisfied identically, but that (5.22) is not satisfied, since (5.28) implies the existence of an unwanted uniform shear traction $-3C_1b^2$ on both of the edges $y = +b$. This unwanted traction can be removed by superposing an appropriate uniform shear stress, through the additional stress function term $C_2xy$. Thus, if we define

$$\phi = C_1xy^3 + C_2xy \; , \tag{5.30}$$

equations (5.27, 5.29) remain unchanged, whilst (5.28) is modified to

$$\sigma_{xy} = -3C_1y^2 - C_2 \; . \tag{5.31}$$

The boundary condition (5.22) can now be satisfied if we choose $C_2$ to satisfy the equation

$$C_2 = -3C_1b^2 \; , \tag{5.32}$$

so that

$$\sigma_{xy} = 3C_1(b^2 - y^2) \; . \tag{5.33}$$

The constant $C_1$ can be determined by substituting (5.33) into the remaining boundary condition (5.25), with the result

$$C_1 = \frac{F}{4b^3} \; . \tag{5.34}$$

The final stress field is therefore defined through the stress function

$$\phi = \frac{F(xy^3 - 3b^2xy)}{4b^3} \; , \tag{5.35}$$

the corresponding stress components being

$$\sigma_{xx} = \frac{3Fxy}{2b^3} \; ; \tag{5.36}$$
$$\sigma_{xy} = \frac{3F(b^2 - y^2)}{4b^3} \; ; \tag{5.37}$$
$$\sigma_{yy} = 0 \; . \tag{5.38}$$

We note that no boundary conditions have been specified on the built-in end, $x = a$. In the weak form, these would be

$$\int_{-b}^{b} \sigma_{xx} dy = 0 \; ; \; x = a \; ; \tag{5.39}$$

$$\int_{-b}^{b} \sigma_{xy} dy = F \; ; \; x = a \; ; \tag{5.40}$$

$$\int_{-b}^{b} \sigma_{xx} y dy = Fa \; ; \; x = a \; . \tag{5.41}$$

However, if conditions (5.22–5.25) are satisfied, (5.39–5.41) are merely equivalent to the condition that the whole beam be in equilibrium. Now the Airy stress function is so defined that whatever stress function is used, the corresponding stress field will satisfy equilibrium in the local sense of equations (2.5). Furthermore, if every particle of a body is separately in equilibrium, it follows that the whole body will also be in equilibrium. It is therefore not necessary to enforce equations (5.39–5.41), since if we were to check them, we should necessarily find that they are satisfied identically.

### 5.2.2 Higher order polynomials — symmetry considerations

In the previous section, we developed the solution by trial and error, starting from the leading term whose form was dictated by equilibrium considerations. A more general technique is to identify the highest order polynomial term from equilibrium considerations and then write down the most general polynomial of that degree and below. The constant multipliers on the various terms are then obtained by imposing boundary conditions and biharmonic constraint equations (see §5.1 above).

The only objection to this procedure is that it involves a lot of algebra. For example, in the problem of §5.2.1, we would have to write down the most general polynomial of degree 4 and below, which involves 12 separate terms even when we exclude the linear and constant terms as being null. However, the complexity of the problem can be greatly reduced by utilizing the natural symmetry of the rectangular beam. In most problems, the loading has some symmetry which can be exploited in limiting the number of independent polynomial terms and even when this is not the case, some saving of complexity can be achieved by representing the loading as the sum of symmetric and antisymmetric parts. We shall illustrate these ideas in the next example.

We consider the problem illustrated in Figure 5.3, in which the beam $-a < x < a, -b < y < b$ is simply supported at its ends, $x = \pm a$, and loaded by a uniform pressure, $p$, on the top surface, $y = b$.

Equilibrium considerations in this case dictate that the bending moment must vary with $x^2$ and hence that $\sigma_{xx}$ must contain terms like $x^2 y$ and $\phi$ terms like $x^2 y^3$ which is of degree 5. The most general polynomial of degree 5 and below involves 18 independent terms, so there is clearly some motivation here for making use of symmetry.

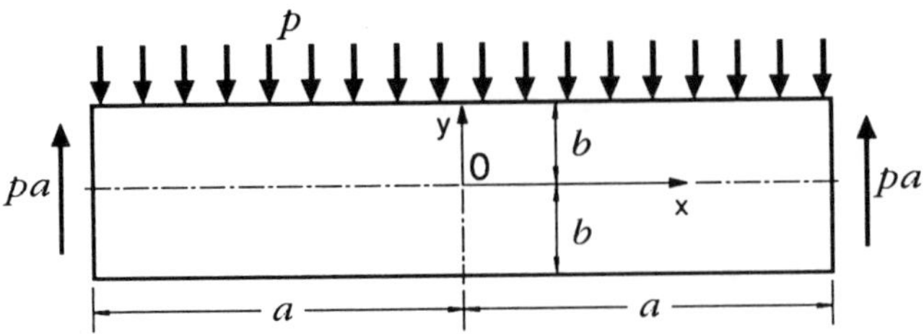


Figure 5.3: Beam with a uniform load

The problem is symmetrical about the mid-point of the beam and hence, by taking the origin there, we are able to deduce that the resulting stress function will contain only *even* powers of $x$. This immediately reduces the number of terms in the general stress function to 9.

The beam is also symmetrical about the axis $y = 0$, but the loading is not. We therefore decompose the problem into the two sub-problems illustrated in Figure 5.4($a, b$). The problem in Figure 5.4($a$) is *antisymmetric* in $y$ and hence requires a stress function with only *odd* powers of $y$, whereas that of Figure 5.4($b$) is *symmetric* and requires only *even* powers. (In fact, the problem of Figure 5.4($b$) clearly has the trivial solution corresponding to uniform uniaxial compression, $\sigma_{yy} = -p/2$, the appropriate stress function being $\phi = -px^2/4$.)

For the problem of Figure 5.4($a$), the most general fifth degree polynomial which is even in $x$ and odd in $y$ can be written

$$\phi = C_1x^4y + C_2x^2y^3 + C_3y^5 + C_4x^2y + C_5y^3 \;, \tag{5.42}$$

which has just five degrees of freedom. The appropriate boundary conditions for this sub-problem are

$$\sigma_{xy} = 0 \;;\; y = \pm b \;; \tag{5.43}$$

$$\sigma_{yy} = \mp\frac{p}{2} \;;\; y = \pm b \;; \tag{5.44}$$

$$\int_{-b}^{b} \sigma_{xx}dy = 0 \;;\; x = \pm a \;; \tag{5.45}$$

$$\int_{-b}^{b} \sigma_{xx}y\,dy = 0 \;;\; x = \pm a \;. \tag{5.46}$$

Notice that, in view of the symmetry, it is only necessary to satisfy these conditions on one of each pair of edges (e.g. on $y = b, x = a$). For the same reason, we do not

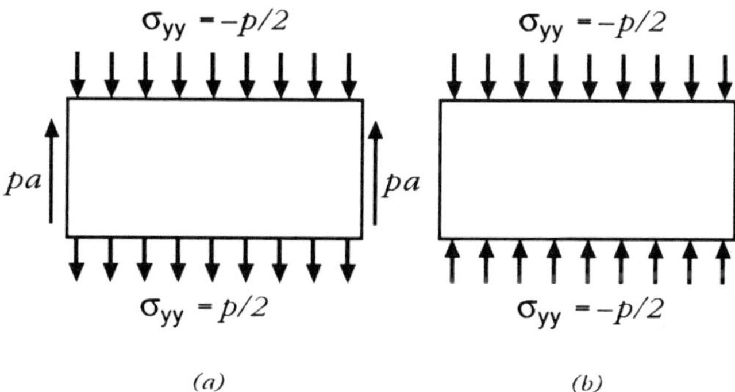


Figure 5.4: Decomposition of the problem into (*a*) antisymmetric and (*b*) symmetric parts

have to impose a condition on the vertical force at $x = \pm a$, since the symmetry demands that the forces be equal at the two ends and the *total* force must be $2pa$ to preserve global equilibrium, this being guaranteed by the use of the Airy stress function, as in the problem of §5.2.1.

As a general point of strategy, it is usually better to start the solution with the strong boundary conditions (equations (5.43, 5.44)), and in particular with those conditions that are homogeneous (in this case equation (5.43)). Substituting (5.42) into (5.15, 5.16), we find

$$\sigma_{xy} = -4C_1x^3 - 6C_2xy^2 - 2C_4x \; ; \tag{5.47}$$

$$\sigma_{yy} = 12C_1x^2y + 2C_2y^3 + 2C_4y \; . \tag{5.48}$$

Thus, condition (5.43) requires that

$$4C_1x^3 + (6C_2b^2 + 2C_4)x = 0 \;\; ; \;\; \text{for all } x \tag{5.49}$$

and this condition can only be satisfied if

$$C_1 = 0 \quad \text{and} \quad 6C_2b^2 + 2C_4 = 0 \; . \tag{5.50}$$

A similar procedure with equation (5.48) and boundary condition (5.44) gives the additional equation

$$2C_2b^3 + 2C_4b = -\frac{p}{2} \; . \tag{5.51}$$

Equations (5.50, 5.51) have the solution

$$C_2 = \frac{p}{8b^3} \;\; ; \;\; C_4 = -\frac{3p}{8b} \; . \tag{5.52}$$

We next determine $C_3$ from the condition that the function $\phi$ is biharmonic, obtaining

$$(24C_1 + 24C_2 + 120C_3)y = 0 \tag{5.53}$$

and hence

$$C_3 = -\frac{p}{40b^3} , \tag{5.54}$$

from (5.52, 5.53).

It remains to satisfy the two weak boundary conditions (5.45, 5.46) on the ends $x = \pm a$. The first of these is satsified identically in view of the antisymmetry of the stress field and the second gives the equation

$$4C_2a^2b^3 + 8C_3b^5 + 4C_5b^3 = 0 , \tag{5.55}$$

which, with equations (5.52, 5.54), serves to determine the remaining constant,

$$C_5 = \frac{p(2b^2 - 5a^2)}{40b^3} . \tag{5.56}$$

The final solution of the complete problem (the sum of that for Figures 5.4($a$) and ($b$)) is therefore obtained from the stress function

$$\phi = \frac{p}{40b^3}(5x^2y^3 - y^5 - 15b^2x^2y - 5a^2y^3 + 2b^2y^3 - 10b^3x^2) , \tag{5.57}$$

the corresponding stress field being

$$\sigma_{xx} = \frac{p}{20b^3}(15x^2y - 10y^3 - 15a^2y + 6b^2y) ; \tag{5.58}$$

$$\sigma_{xy} = \frac{3px}{4b^3}(b^2 - y^2) ; \tag{5.59}$$

$$\sigma_{yy} = \frac{p}{4b^3}(y^3 - 3b^2y - 2b^3) . \tag{5.60}$$

## 5.3 Series and transform solutions

Polynomial solutions can, in principle, be extended to more general loading of the beam edges, as long as the tractions are capable of a power series expansion. However, the practical use of this method is limited by the algebraic complexity encountered for higher order polynomials and by the fact that many important traction distributions do not have convergent power series representations.

A more useful method in such cases is to build up a general solution by components of Fourier form. For example, if we write

$$\phi = f(y)\cos(\lambda x) , \tag{5.61}$$

substitution in the biharmonic equation (5.1) shows that $f(y)$ must have the general form

$$f(y) = (A + By)e^{\lambda y} + (C + Dy)e^{-\lambda y} , \tag{5.62}$$

where $A, B, C, D$ are arbitrary constants.

A more general stress function, symmetric about $x = 0$, can then be written in the series form

$$\phi = \sum_{i=1}^{\infty}[(A_i + B_i y)e^{\lambda_i y} + (C_i + D_i y)e^{-\lambda_i y}]\cos(\lambda_i x) , \tag{5.63}$$

where $\lambda_i = i\pi x/a$. A corresponding antisymmetric solution can also be obtained by replacing $\cos(\lambda_i x)$ by $\sin((2i - 1)\pi x/2a)$.

Clearly the stresses obtained by substitution of (5.63) in (5.14–5.16) will be of similar Fourier form and the four constants $A_i, B_i, C_i, D_i$ for each Fourier component will permit us to satisfy arbitrary boundary conditions on the long edges $y = \pm b$, provided the tractions can be expanded as Fourier series[1]

The Fourier series method is also readily adapted for numerical solution, making use of the fact that the coefficients of a truncated Fourier series can be evaluated from the algebraic equations obtained by demanding equality between the tractions and their approximation at each of the zeros of the first neglected term of the series.

If the beam is infinite or semi-infinite ($a \to \infty$), the series must be replaced by the corresponding integral representation

$$\phi(x, y) = \int_0^{\infty} f(\lambda, y)\cos(\lambda x)d\lambda , \tag{5.64}$$

where

$$f(\lambda, y) = [A(\lambda) + yB(\lambda)]e^{\lambda y} + [C(\lambda) + yD(\lambda)]e^{-\lambda y} . \tag{5.65}$$

Equation (5.64) is introduced here as a generalization of (5.61) by superposition, but it is in fact the Fourier cosine transform of $\phi(x, y)$, the corresponding inversion being

$$f(\lambda, y) = \frac{2}{\pi}\int_0^{\infty} \phi(x, y)\cos(\lambda x)dx . \tag{5.66}$$

The boundary conditions on $y = \pm b$ will also lead to Fourier integrals, which can be inverted in the same way to determine the functions $A, B, C, D$. For a definitive treatment of the Fourier transform method, the reader is referred to the treatise by Sneddon[2]. Extensive tables of Fourier transforms and their inversions are given by Erdelyi[3]. The cosine transform (5.64) will lead to a symmetric solution. For more general loading, the complex exponential transform can be used.

---

[1]Notice however that the terms in (5.61) degenerate to trivial polynomial forms for $i = 0$. It is therefore generally necessary to include a low degree polynomial in the stress function to describe the uniform component in the tractions.

[2]I.N.Sneddon, *Fourier Transforms*, McGraw-Hill, New York, 1951.

[3]A.Erdelyi, ed., *Tables of Integral Transforms*, Bateman Manuscript Project, California Institute of Technology, Vol.1, McGraw-Hill, New York, 1954.

The Fourier transform method is also an efficient alternative in problems for the finite beam with localized self-equilibrated loading. In effect, we first use the transform method to solve the problem of an infinite beam with the same loading and then correct the boundary conditions on the ends $x = \pm a$ in the weak sense, using the results of §5.2. This method is of course particularly efficient when the loading is such that the transforms of the tractions in the infinite interval can be evaluated in closed form.

It is worth remarking on the way in which the series and transform solutions are natural generalizations of the elementary solution (5.61). One of the most powerful techniques in Elasticity — and indeed in any physical theory characterized by linear partial differential equations — is to seek a simple form of solution (often in separated variable form) that contains a parameter which can take a range of values. A more general solution can then be developed by superposing arbitrary multiples of the solution with different values of the parameter.

For example, if a particular solution can be written symbolically as $\phi = f(x, y, \lambda)$, where $\lambda$ is a parameter, we can develop a general series form

$$\phi(x, y) = \sum_{i=0}^{\infty} A_i f(x, y, \lambda_i) \tag{5.67}$$

or an integral form

$$\phi(x, y) = \int_a^b A(\lambda) f(x, y, \lambda) d\lambda \; . \tag{5.68}$$

The series form will naturally arise if there is a discrete set of *eigenvalues*, $\lambda_i$ for which $f(x, y, \lambda_i)$ satisfies some of the boundary conditions of the problem. Examples of this kind will be found in §§6.2, 11.2. In this case, the series (5.67) is most properly seen as an eigenfunction expansion. Integral forms arise most commonly (but not exclusively) in problems involving infinite or semi-infinite domains (see, for example, §§11.3, 22.2.2.).

Any particular solution containing a parameter can be used in this way and, since transforms are commonly named after their originators, the reader desirous of instant immortality might like to explore some of those which have not so far been used. Of course, the usefulness of the resulting solution depends upon its *completeness* — i.e. its capacity to represent all stress fields of a given class — and upon the ease with which the transform can be inverted.

## PROBLEMS

1. The beam $-c < y < c$, $0 < x < L$, is built-in at the end $x = 0$ and loaded by a uniform shear traction $\sigma_{xy} = S$ on the upper edge, $y = c$, the remaining edges, $x = L$, $y = -c$ being traction-free. Find a suitable stress function and the corresponding stress components for this problem, using the weak boundary conditions on $x = L$. Identify the nature of the approximation involved.

2. A large plate defined by $y > 0$ is subjected to a sinusoidally varying load

$$\sigma_{yy} = S \sin bx \quad ; \quad \sigma_{xy} = 0$$

at its plane edge $y = 0$.

Find the complete stress field in the plate and hence estimate the depth $y$ at which the amplitude of the variation in $\sigma_{yy}$ has fallen to 10% of $S$.

**Hint**: You might find it easier initially to consider the case of the layer $0<y<h$, with $y = h$ traction-free, and then let $h \to \infty$.

# Chapter 6

# END EFFECTS

The solution of §5.2.2 must be deemed approximate insofar as the boundary conditions on the ends $x = \pm a$ of the rectangular beam are satisfied only in the weak sense of force resultants, through equation (5.46). In general, if a rectangular beam is loaded by tractions of finite polynomial form, a finite polynomial solution can be obtained which satisfies the boundary conditions in the strong (i.e. pointwise) sense on two edges and in the weak sense on the other two edges.

The error involved in such an approximation corresponds to the solution of a corrective problem in which the beam is loaded by the difference between the actual tractions applied and those implied by the approximation. These tractions will of course be confined to the edges on which the weak boundary conditions were applied and will be self-equilibrated, since the weak conditions imply that the tractions in the approximate solution have the same force resultants as the actual tractions.

For the particular problem of §5.2.2, we note that the stress field of equations (5.58–5.60) satisfies the boundary conditions on the edges $y = \pm b$, but that there is a self-equilibrated normal traction

$$\sigma_{xx} = \frac{p}{10b^3}(3b^2y - 5y^3) \tag{6.1}$$

on the ends $x = \pm a$, which must be removed by superposing a corrective solution if we wish to satisfy the boundary conditions of Figure 5.3 in the strong sense.

## 6.1 Decaying solutions

In view of Saint-Venant's theorem, we anticipate that the stresses in the corrective solution will decay as we move away from the edges where the self-equilibrated tractions are applied. The decay rate is likely to be related to the width of the loaded region and hence we anticipate that the stresses in the corrective solution will be significant only in two regions near the ends, of linear dimensions comparable to the width of the beam. These regions are, shown shaded in Figure 6.1. It follows that

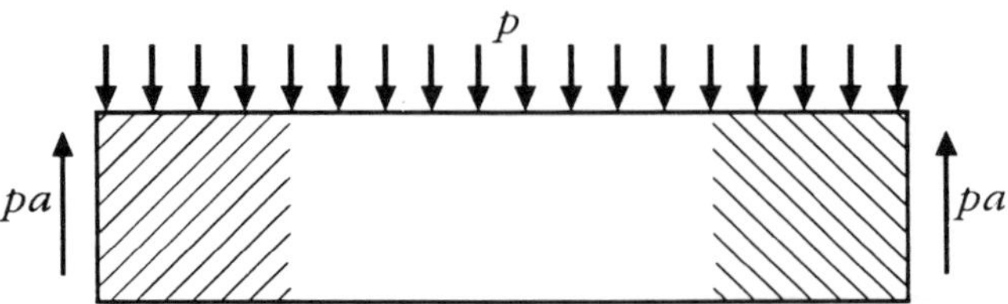


Figure 6.1: Regions of the beam influenced by end effects

the solution of §5.2.2 will be a good approximation in the *unshaded* region in Figure 6.1.

It also follows that the corrective solutions for the two ends are *uncoupled*, since the corrective field for the end $x = -a$ has decayed to negligible proportions before we reach the end $x = +a$ and vice versa. This implies that, so far as the left end is concerned, the corrective solution is essentially identical to that which would be required in the *semi-infinite* beam, $x > -a$. For the rest of this discussion, we shall therefore simplify the statement of the problem by considering the corrective solution for the left end only and shifting the origin to $-a$, so that the semi-infinite beam under consideration is defined by $x>0,\ -b<y<b$.

It is also now clear why we chose to satisfy the strong boundary conditions on the *long* edges, $y = \pm b$. If instead we had imposed strong conditions on $x = \pm a$ and weak conditions on $y = \pm b$, the shaded regions would have overlapped and there would be no region in which the finite polynomial solution was a good approximation to the stresses in the beam. It also follows that the approximation will only be useful when the beam has an aspect ratio significantly different from unity — i.e. $b/a >> 1$.

## 6.2 The corrective solution

We recall that the stress function, $\phi$, for the corrective solution must (i) satisfy the biharmonic equation (5.1), (ii) have zero tractions on the boundaries $y = \pm b$ and (iii) have prescribed non-zero tractions (such as those defined by equation (6.1)) on the end(s). We cannot generally expect to find a solution to satisfy all these conditions in closed form and hence we seek a series or transform (integral) solution as suggested in §5.3. However, the solution will be simpler if we can find a *class* of solutions, all of which satisfy some of the conditions. We can then write down a more general solution as a superposition of such solutions and choose the coefficients so as to satisfy the remaining condition(s).

Since the traction boundary condition on the end will vary from problem to problem, it is convenient to seek solutions which satisfy (i) and (ii) - i.e. biharmonic functions which define zero tractions on the surfaces $y = \pm b$. The traction-free condition requires

$$\sigma_{yy} = \sigma_{yx} = 0 \;\; ; \;\; y = \pm b \tag{6.2}$$

and hence

$$\frac{\partial^2 \phi}{\partial x^2} = \frac{\partial^2 \phi}{\partial x \partial y} = 0 \;\; ; \;\; y = \pm b \,. \tag{6.3}$$

### 6.2.1 Separated variable solutions

One way to obtain functions satisfying conditions (6.3) is to write them in the separated variable form

$$\phi = f(x)g(y) \,, \tag{6.4}$$

in which case, (6.3) will be satisfied for all $x$, provided that

$$g(y) = g'(y) = 0 \;\; ; \;\; y = \pm b \,. \tag{6.5}$$

Notice that the final corrective solution cannot be expected to be of separated variable form, but it is possible that it can be represented as the sum of such terms.

If the functions (6.4) are to be biharmonic, we must have

$$g\frac{d^4 f}{dx^4} + 2\frac{d^2 f}{dx^2}\frac{d^2 g}{dy^2} + f\frac{d^4 g}{dy^4} = 0 \,, \tag{6.6}$$

and this equation must be satisfied for all values of $x, y$. Now, if we consider the subset of points $x, c$, where $c$ is a constant, it is clear that $f(x)$ must satisfy an equation of the form

$$A\frac{d^4 f}{dx^4} + B\frac{d^2 f}{dx^2} + Cf = 0 \,, \tag{6.7}$$

where $A, B, C$, are constants, and hence $f(x)$ must consist of exponential terms such as $f(x) = exp(\lambda x)$. Similar considerations apply to the function $g(y)$. Notice incidentally that $\lambda$ might be complex or imaginary, giving sinusoidal functions, and there are also degenerate cases where $C$ and/or $B = 0$ in which case $f(x)$ could also be a polynomial of degree 3 or below.

Since we are seeking to represent a field which decays with $x$, we select terms of the form

$$\phi = g(y)e^{-\lambda x} \,, \tag{6.8}$$

in which case, (6.6) reduces to

$$\frac{d^4 g}{dy^4} + 2\lambda^2\frac{d^2 g}{dy^2} + \lambda^4 g = 0 \,, \tag{6.9}$$

which is a fourth order ordinary differential equation for $g(y)$ with general solution

$$g(y) = (A_1 + A_2 y)\cos\lambda y + (A_3 + A_4 y)\sin\lambda y \,. \tag{6.10}$$

### 6.2.2 The eigenvalue problem

The arbitrary constants $A_1, A_2, A_3, A_4$ are determined from the boundary conditions (6.2), which in view of (6.5) lead to the four simultaneous equations

$$(A_1 + A_2 b)\cos\lambda b + (A_3 + A_4 y)\sin\lambda b = 0\,; \tag{6.11}$$
$$(A_1 - A_2 b)\cos\lambda b - (A_3 - A_4 y)\sin\lambda b = 0\,; \tag{6.12}$$
$$(A_2 + A_3\lambda + A_4\lambda b)\cos\lambda b - (A_1\lambda + A_2\lambda b - A_4)\sin\lambda b = 0\,; \tag{6.13}$$
$$(A_2 + A_3\lambda - A_4\lambda b)\cos\lambda b + (A_1\lambda - A_2\lambda b - A_4)\sin\lambda b = 0\,. \tag{6.14}$$

This set of equations is homogeneous and will generally have only the trivial solution $A_1 = A_2 = A_3 = A_4 = 0$. However, there are some eigenvalues for the exponential decay rate, $\lambda$, for which the determinant of coefficients is singular and the solution is non-trivial.

A more convenient form of the equations can be obtained by taking sums and differences in pairs - i.e. by constructing the equations $(6.11+6.12), (6.11-6.12), (6.13+6.14), (6.13-6.14)$, which after rearrangement and cancellation of non-zero factors yields the set

$$A_1\cos\lambda b + A_4 b\sin\lambda b = 0\,; \tag{6.15}$$
$$A_1\lambda\sin\lambda b - A_4(\sin\lambda b + \lambda b\cos\lambda b) = 0\,; \tag{6.16}$$
$$A_2 b\cos\lambda b + A_3\sin\lambda b = 0\,; \tag{6.17}$$
$$A_2(\cos\lambda b - \lambda b\sin\lambda b) + A_3\lambda\cos\lambda b = 0\,. \tag{6.18}$$

What we have done here is to use the symmetry of the system to partition the matrix of coefficients. The terms $A_1\cos\lambda y$, $A_4 y\sin\lambda y$ are symmetric, whereas $A_2 y\cos\lambda y$, $A_3\sin\lambda y$ are antisymmetric. The boundary conditions are also symmetric and hence the symmetric and antisymmetric terms must *separately* satisfy them.

We conclude that the set of equations (6.15–6.18) has two sets of eigenvalues, for one of which the resulting eigenfunction is symmetric and the other antisymmetric. The symmetric eigenvalues are obtained by eliminating $A_1, A_4$ from (6.15, 6.16) with the result

$$\sin 2\lambda b + 2\lambda b = 0\,, \tag{6.19}$$

whilst the antisymmetric eigenvalues are obtained in the same way from (6.17, 6.18) with the result

$$\sin 2\lambda b - 2\lambda b = 0\,. \tag{6.20}$$

Figure 6.2 demonstrates graphically that the only *real* solution of equations (6.19, 6.20) is the trivial case $\lambda = 0$ (which in fact corresponds to the non-decaying solutions in which a force or moment resultant is applied at the end and transmitted along the beam). However, there is a denumerably infinite set of non-trivial *complex* solutions, corresponding to stress fields which oscillate whilst decaying along the beam. These solutions are fairly easy to find by writing $\lambda b = c + id$, separating real and imaginary

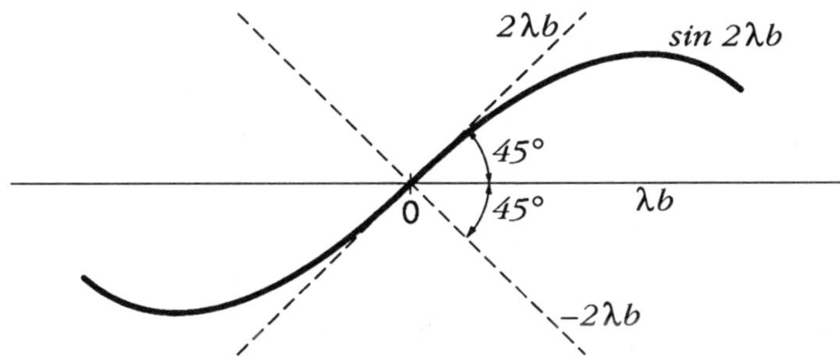


Figure 6.2: Graphical solution of equations (6.19, 6.20)

parts in the complex equation, and solving the resulting two simultaneous equations for the real numbers, $c, d$, using a suitable numerical algorithm.

Once the eigenvalues and the related eigenfunctions have been determined, the next stage is to establish a more general solution of the form

$$\phi = \sum_{i=1}^{\infty} C_i g_i(y) e^{-\lambda_i x} , \tag{6.21}$$

where $g_i(y)$ is the eigenfunction corresponding to the eigenvalue $\lambda_i$ and $C_i$ defines a set of as yet undetermined coefficients.

The final step is to choose the constants $C_i$ so as to satisfy the prescribed boundary condition on the end $x = 0$. This raises a question of completeness. It is fairly clear that (i) if we truncate the series (6.21), we can develop an approximate solution to the problem and (ii) that the accuracy of the solution (however defined) can always be improved by taking more terms. In particular cases, this is easily established numerically. However, it is more challenging to prove that the eigenfunction expansion is complete in the sense that an arbitrary prescribed self-equilibrated traction on $x = 0$ can be described to within any arbitrarily small accuracy by taking a sufficient number of terms in the series, though experience with other eigenfunction expansions (e.g. with expansion of elastodynamic states of a structure in terms of normal modes) suggests that this will always be true. In fact, although the analysis described in this section has been known since the investigations by Papkovich and Fadle in the early 1940s, the formal proof of completeness was only completed by Gregory[1] in 1980.

[1] R.D.Gregory, Green's functions, bi-linear forms and completeness of the eigenfunctions for the elastostatic strip and wedge, J.Elasticity, Vol. 9 (1979), 283–309; R.D.Gregory, The semi-infinite strip $x \geq 0, -1 \leq y \leq 1$; completeness of the Papkovich-Fadle eigenfunctions when $\phi_{xx}(0, y), \phi_{yy}(0, y)$ are prescribed, J.Elasticity, Vol. 10 (1980), 57–80; R.D.Gregory, The traction boundary-value problem for the elastostatic semi-infinite strip; existence of solution and completeness of the Papkovich-Fadle eigenfunctions, J.Elasticity, Vol. 10 (1980), 295-327. These papers also include extensive references to earlier investigations of the problem.

It is worth noting that, as in many related problems, the eigenfunctions oscillate in $y$ with increasing frequency as $\lambda_i$ increases and in fact every time we increase $i$ by 1, an extra zero appears in the function $g_i(y)$ in the range $0 < y < b$. Thus, there is a certain similarity to the process of approximating functions by Fourier series and in particular, the residual error in case of truncation will always cross zero once more than the last eigenfunction included.

This is also helpful in that it enables us to estimate the decay rate of the first excluded term. We see from equation (6.10) that the distance between zeros in $y$ in any of the separate terms would be $(\pi/\lambda_R)$, where $\lambda_R$ is the real part of $\lambda$. It follows that over a corresponding distance in the $x$ direction, the field would decay by the factor $\exp(-\pi) = 0.0432$. This suggests that we might estimate the decay rate of the end field by noting the distance between zeros in the corresponding tractions[2]. For example, the traction of equation (6.1) has zeros at $y = 0, 3b/5$, corresponding to $\lambda_R = (5\pi/3b) = 5.23/b$.

An alternative way of estimating the decay rate is to note that the decay rate for the various terms in (6.21) increases with $i$ and hence as $x$ increases, the leading term will tend to predominate. The tractions of (6.1) are antisymmetric and hence the leading term corresponds to the real part of the first eigenvalue of equation (6.20), which is found numerically to be $\lambda_R b = 3.7$.

Either way, we can conclude that the error associated with the end tractions in the approximate solution of §5.2.2 has decayed to around $e^{-4}$, i.e. to about 2% of the values at the end, within a distance $b$ of the end. Thus the region affected by the end condition — the shaded region in Figure 6.1 — is very small.

For problems which are *symmetric* in $y$, the leading self-equilibrated term is likely to have the form of Figure 6.3($a$), which has a longer wavelength than the corresponding antisymmetric form, 6.3($b$). The end effects in symmetric problems therefore decay more slowly and this is confirmed by the fact that the real part of the first eigenvalue of the symmetric equation (6.19) is only $\lambda_R b = 2.4$

## 6.3 Other Saint-Venant problems

The general strategy used in this chapter can be used in other curvilinear coördinate systems to correct the errors incurred by imposing the weak boundary conditions on appropriate edges. The essential steps are:-

1. to define a coördinate system $(\xi, \eta)$such that the boundaries on which the strong conditions are applied are of the form, $\eta$ =constant;

2. to find a class of separated variable biharmonic functions containing a parameter ($\lambda$ in the above case);

[2]This assumes that the wavelength of the tractions is the same as that of $\phi$, which of course is an approximation, since neither function is purely sinusoidal.

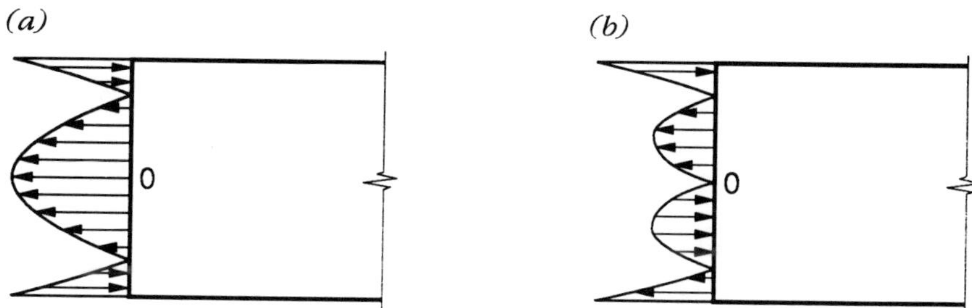


Figure 6.3: Leading term in self-equilibrated tractions for $(a)$ symmetric loading and $(b)$ antisymmetric loading

3. to set up a system of four homogeneous equations for the coefficients of each function, based on the four traction-free boundary conditions for the corrective solution on the edges $\eta$ =constant;

4. to find the eigenvalues of the parameter for which the system has a non-trivial solution and the corresponding eigenfunctions, which are then used as the terms in an eigenfunction expansion to define a general form for the corrective field;

5. to determine the coefficients in the eigenfunction expansion from the prescribed inhomogeneous boundary conditions on the end $\xi$ =constant.

# Chapter 7

# BODY FORCES

A *body force* is defined as one which acts directly on the interior particles of the body, rather than on the boundary. Since the interior of the body is not accessible, it follows necessarily that body forces can only be produced by some kind of physical process which acts 'at a distance'. The commonest examples are forces due to gravity and magnetic attraction. In addition, we can formulate elastic problems for moving bodies in terms of body forces, using D'Alembert's principle (see §7.2.2 below)

## 7.1 Stress function formulation

We noted in §4.4 that the Airy stress function formulation satisfies the equilibrium equations if and only if the body forces are identically zero, but the method can be extended to the case of non-zero body forces provided the latter can be expressed as the gradient of a scalar potential, $V$.

We adopt the new definitions

$$\sigma_{xx} = \frac{\partial^2\phi}{\partial y^2} + V \; ; \tag{7.1}$$

$$\sigma_{yy} = \frac{\partial^2\phi}{\partial x^2} + V \; ; \tag{7.2}$$

$$\sigma_{xy} = = -\frac{\partial^2\phi}{\partial x \partial y} \, , \tag{7.3}$$

in which case the two-dimensional equilibrium equations will be satisfied provided

$$p_x = -\frac{\partial V}{\partial x} \; ; \; p_y = -\frac{\partial V}{\partial y} \, , \tag{7.4}$$

i.e.

$$\boldsymbol{p} = -\boldsymbol{\nabla} V \, . \tag{7.5}$$

We also note that the definitions of equations (7.1–7.3) transform correctly into other coördinate systems, since all we have done is to add a biaxial hydrostatic tension of magnitude $V$, which is of course invariant under coördinate transformation (see §1.1.3).

### 7.1.1 Conservative vector fields

If a force field is capable of being represented as the gradient of a scalar potential, as in equation (7.5), it is referred to as *conservative.* This terminology arises from gravitational theory, since if we move a particle around in a gravitational field and if the force on the particle varies with position, the principle of conservation of energy demands that the work done in moving the particle from point $A$ to point $B$ should be path-independent — or equivalently, the work done in moving it around a closed path should be zero. If this were not the case, we could choose a direction for the particle to move around the path which would release energy and hence have an inexhaustible source of energy.

If all such integrals *are* path-independent, we can use the work done in bringing a particle from infinity to a given point as the definition of a unique local potential. Then, by equating the work done in an infinitesimal motion to the corresponding change in potential energy, we can show that the local force is proportional to the gradient of the potential, thus demonstrating that a conservative force field must be capable of a representation like (7.5). Conversely, if a given force field can be represented in this form, we can show by integration that the work done in moving a particle from $A$ to $B$ is proportional to $V(A) - V(B)$ and is therefore path-independent.

Not all body force fields are conservative and hence the formulation of §7.1 is not sufficiently general for all problems. However, we shall show below that most of the important problems involving body forces can be so treated.

We can develop a condition for a vector field to be conservative in the same way as we developed the compatibility conditions for strains. We argue that the two independent body force components $p_x, p_y$ are defined in terms of a single scalar potential $V$ and hence we can obtain a constraint equation on $p_x, p_y$ by eliminating $V$ between equations (7.4) with the result

$$\frac{\partial p_y}{\partial x} - \frac{\partial p_x}{\partial y} = 0 \ . \tag{7.6}$$

In three dimensions there are three equations like (7.6), from which we conclude that a vector field $\boldsymbol{p}$ is conservative if and only if curl $\boldsymbol{p} = 0$. Another name for such fields is *irrotational*, since we note that if we replace $\boldsymbol{p}$ by $\boldsymbol{u}$, the conditions like (7.6) are equivalent to the statement that the rotation $\boldsymbol{\omega}$ is identically zero (*cf* equation (1.31)).

If the body force field satisfies equation (7.6), the corresponding potential can be recovered by partial integration. We shall illustrate this procedure in §7.2 below.

### 7.1.2 The compatibility condition

We demonstrated in §4.4.3 above that, in the absence of body forces, the compatibility condition reduces to the requirement that the Airy stress function $\phi$ be biharmonic. This condition is modified when body forces are present.

We follow the same procedure as in §4.4.3, but use equations (7.1–7.3) in place of (4.1). Substituting into the compatibility equation in terms of stresses (4.4), we obtain

$$\frac{\partial^4\phi}{\partial y^4}+\frac{\partial^2 V}{\partial y^2}-\nu\frac{\partial^4\phi}{\partial x^2\partial y^2}-\nu\frac{\partial^2 V}{\partial y^2}+2(1+\nu)\frac{\partial^4\phi}{\partial x^2\partial y^2}$$
$$+\frac{\partial^4\phi}{\partial x^4}+\frac{\partial^2 V}{\partial x^2}-\nu\frac{\partial^4\phi}{\partial x^2\partial y^2}-\nu\frac{\partial^2 V}{\partial x^2} = 0\;, \tag{7.7}$$

i.e.

$$\nabla^4\phi=-(1-\nu)\nabla^2 V\;. \tag{7.8}$$

Methods of obtaining suitable functions which satisfy this equation will be discussed in §7.3 below.

## 7.2 Particular cases

It is worth noting that the vast majority of mechanical engineering components are loaded principally by boundary tractions rather than body forces. Of course, most components are subject to gravity loading, but the boundary loads are generally so much larger that gravity can be neglected. This is less true for civil engineering structures such as buildings, where the self-weight of the structure may be much larger than the weight of the contents or wind loads, but even in this case it is important to distinguish between the gravity loading on the individual component and that transmitted to the component by way of boundary tractions.

It might be instructive at this point for the reader to draw free-body diagrams for a few common engineering components and identify the sources and relative magnitudes of the forces acting upon them. There are really comparatively few ways of applying a load to a body. By far the commonest is to push against it with another body — in other words to apply the load by *contact*. This is why contact problems occupy a central place in elasticity theory[1]. Significant loads may also be applied by fluid pressure as in the case of turbine blades or aircraft wings. Notice that it is fairly easy to apply a compressive normal traction to a boundary, but much harder to apply tension or shear.

This preamble might be taken as a justification for not studying the subject of body forces at all, but there are a few applications in which they are of critical importance, most notably those dynamic problems in which large accelerations occur.

[1]See Chapters 12, 21.

We shall develop expressions for the body force potential for some important cases in the following sections.

### 7.2.1 Gravitational loading

The simplest type of body force loading is that due to a gravitational field. If the problem is to remain two-dimensional, the direction of the gravitational force must lie in the $xy$-plane and we can choose it to be in the negative $y$-direction without loss of generality. The magnitude of the force will be $\rho g$ per unit volume, where $\rho$ is the density of the material, so in the notation of §2.1 we have

$$p_x = 0 \ ; \ p_y = -\rho g \ . \tag{7.9}$$

This force field clearly satisfies condition (7.6) and is therefore conservative and by inspection we note that it can be derived from the body force potential

$$V = \rho g y \ . \tag{7.10}$$

It also follows that $\nabla^2 V = 0$ and hence that the stress function, $\phi$ for problems involving gravitational loading is biharmonic.

### 7.2.2 Inertia forces

It might be argued that D'Alembert achieved immortality simply by moving a term from one side of an equation to the other, since *D'Alembert's principle* consists merely of writing Newton's second law of motion in the form $\boldsymbol{F} - m\boldsymbol{a} = 0$ and treating the term $-m\boldsymbol{a}$ as a fictitious force in order to reduce the dynamic problem to one in statics.

This simple process enables us to formulate elasticity problems for moving bodies as body force problems, the corresponding body forces being

$$p_x = -\rho a_x \ ; \ p_y = -\rho a_y \ , \tag{7.11}$$

where $a_x, a_y$ are the local components of acceleration, which may of course vary with position through the body.

### 7.2.3 Quasi-static problems

If the body were rigid, the accelerations of equation (7.11) would be restricted to those associated with rigid body translation and rotation, but in a deformable body, the distance between two points can change, giving rise to additional, stress-dependent terms in the accelerations.

These two effects give qualitatively distinct behaviour, both mathematically and physically. In the former case, the accelerations will generally be defined *a priori* from the kinematics of the problem, which therefore reduces to an elasticity problem with body forces. We shall refer to such problems as *quasi-static.*

By contrast, when the accelerations associated with deformations are important, they are not known *a priori*, since the stresses producing the deformations are themselves part of the solution. In this case the kinematic and elastic problems are coupled and must be solved together. The resulting equations are those governing the propagation of elastic waves through a solid body and their study is the that of *Elastodynamics.*

In this chapter, we shall restrict attention to quasi-static problems. As a practical point, we note that the characteristic time scale of elastodynamic problems is very short. For example it generally takes only a very short time for an elastic wave to traverse a solid. If the applied loads are applied gradually in comparison with this time scale, the quasi-static assumption generally gives good results. A measure of the success of this approximation is that it works quite well even for the case of elastic impact between bodies, which may have a duration of the order of a few milliseconds[2].

### 7.2.4 Rigid-body kinematics

The most general acceleration for a rigid body in the plane involves arbitrary translation and rotation. We choose a coördinate system fixed in the body and suppose that, at some instant, the origin has velocity $\boldsymbol{v}_\circ$ and acceleration $\boldsymbol{a}_\circ$ and the body is rotating in the clockwise sense with absolute angular velocity $\Omega$ and angular acceleration $\dot{\Omega}$. The instantaneous acceleration of the point $(r,\theta)$ *relative to the origin* can then be written

$$a_r = -\Omega^2 r \ ; \ \ a_\theta = -\dot{\Omega} r \ . \tag{7.12}$$

Transforming these results into the $x,y$-coördinate system and adding the acceleration of the origin, we obtain the components of acceleration of the point $(x,y)$ as

$$a_x = a_{0x} - \Omega^2 x + \dot{\Omega} y \ ; \tag{7.13}$$

$$a_y = a_{0y} - \Omega^2 y - \dot{\Omega} x \tag{7.14}$$

and hence the corresponding body force field is

$$p_x = -\rho(a_{0x} - \Omega^2 x + \dot{\Omega} y) \ ; \tag{7.15}$$

$$p_y = -\rho(a_{0y} - \Omega^2 y - \dot{\Omega} x) \ . \tag{7.16}$$

The astute reader will notice that the case of gravitational loading can be recovered as a special case of these results by writing $a_{0y} = g$ and setting all the other terms to

[2] For more information about elastodynamic problems, the reader is referred to the classical texts of J.D.Achenbach, *Wave Propagation in Elastic Solids*, North Holland, Amsterdam, (1973) and A.C.Eringen and E.S.Şuhubi, *Elastodynamics*, Academic Press, New York, (1975). For a more detailed discussion of the impact of elastic bodies, see K.L.Johnson, *Contact Mechanics*, Cambridge University Press, Cambridge, (1985), §11.4.

zero. In fact, a reasonable interpretation of the gravitational force is as a D'Alembert force consequent on resisting the gravitational acceleration. Notice that if a body is in free fall — i.e. it is accelerating freely in a gravitational field and not rotating — there is no body force and hence no internal stress unless the boundaries are loaded.

Substitution of (7.15, 7.16) into (7.6) shows that the inertia forces due to rigid-body accelerations are conservative if and only if $\dot{\Omega} = 0$ — i.e. if the angular velocity is constant. We shall determine the body force potential for this special case. Methods of treating the problem with non-zero angular acceleration are discussed in §7.4 below.

From equations (7.4, 7.15, 7.16) with $\dot{\Omega} = 0$ we have

$$\frac{\partial V}{\partial x} = \rho(a_{0x} - \Omega^2 x) \; ; \tag{7.17}$$

$$\frac{\partial V}{\partial y} = \rho(a_{0y} - \Omega^2 y) \; , \tag{7.18}$$

and hence, on partial integration of (7.17)

$$V = \rho\left(a_{0x}x - \frac{1}{2}\Omega^2 x^2\right) + h(y) \; , \tag{7.19}$$

where $h(y)$ is an arbitrary function of $y$ only. Substituting this result into (7.18) we obtain the ordinary differential equation

$$\frac{dh}{dy} = \rho(a_{0y} - \Omega^2 y) \tag{7.20}$$

for $h(y)$, which has the general solution

$$h(y) = \rho\left(a_{0_y}y - \frac{1}{2}\Omega^2 y^2\right) + C \; , \tag{7.21}$$

where $C$ is an arbitrary constant which can be taken to be zero without loss of generality, since we are only seeking a *particular* potential function $V$.

The final expression for $V$ is therefore

$$V = \rho\left(a_{0x}x + a_{0y}y - \frac{1}{2}\Omega^2(x^2 + y^2)\right) . \tag{7.22}$$

The reader might like to try this procedure on a set of body forces which *do not* satisfy the condition (7.6). It will be found that the right hand side of the ordinary differential equation like (7.20) then contains terms which depend on $x$ and hence this equation cannot be solved for $h(y)$.

## 7.3 Solution for the stress function

Once the body force potential $V$ has been determined, the next step is to find a suitable function $\phi$, which satisfies the compatibility condition (7.8) and which defines

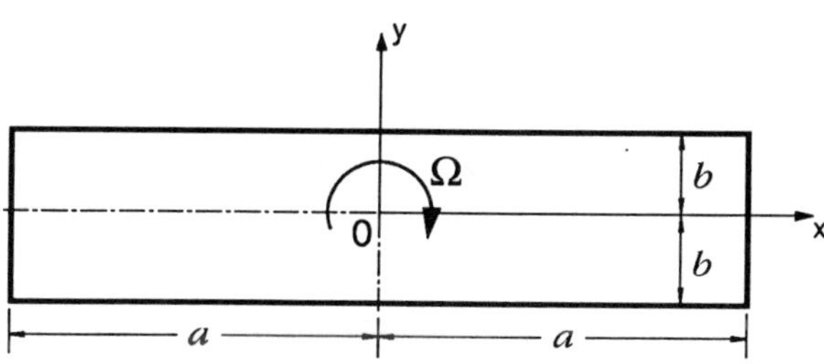


Figure 7.1: The rotating rectangular bar

stresses through equations (7.1–7.3) which satisfy the boundary conditions of the problem. There are broadly speaking two ways of doing this. One is to choose some suitable form (such as a polynomial) without regard to equation (7.8) and then satisfy the constraint conditions resulting from (7.8) in the same step as those arising from the boundary conditions: the other is to seek a general solution of the inhomogeneous equation (7.8) and then determine the resulting arbitrary constants from the boundary conditions.

### 7.3.1 The rotating rectangular beam

As an illustration of the first method, we consider the problem of the rectangular beam $-a < x < a$, $-b < y < b$, rotating about the origin at constant angular velocity $\Omega$, all the boundaries being traction-free (see Figure 7.1).

The body force potential for this problem is obtained from equation (7.22) as

$$V = -\frac{1}{2}\rho\Omega^2(x^2+y^2) , \tag{7.23}$$

and hence the stress function must satisfy the equation

$$\nabla^4\phi = 2\rho(1-\nu)\Omega^2 , \tag{7.24}$$

from (7.8, 7.23).

The geometry suggests a formulation in Cartesian coördinates and equation (7.24) leads us to expect a polynomial of degree 4 in $x, y$. We also note that $V$ is even in both $x$ and $y$ and that the boundary conditions are homogeneous, so we propose the candidate stress function

$$\phi = A_1x^4 + A_2x^2y^2 + A_3y^4 + A_4x^2 + A_5y^2 , \tag{7.25}$$

which contains all the terms of degree 4 and below with the required symmetry.

The constants $A_1, \ldots, A_5$ will be determined from equation (7.24) and from the boundary conditions

$$\sigma_{yx} = 0 \; ; \; y = \pm b \; ; \tag{7.26}$$

$$\sigma_{yy} = 0 \; ; \; y = \pm b \; ; \tag{7.27}$$

$$\int_{-b}^{b} \sigma_{xx} dy = 0 \; ; \; x = \pm a \; , \tag{7.28}$$

where we have applied the weak boundary conditions only on $x = \pm a$. Note also that the other two weak boundary conditions — that there should be no moment and no shear force on the ends — are satisfied identically in view of the symmetry of the problem about $y = 0$.

As in the problem of §5.2.2, it is algebraically simpler to start by satisfying the strong boundary conditions (7.26, 7.27). Substituting (7.25) into (7.1–7.3), we obtain

$$\sigma_{xx} = 2A_2x^2 + 12A_3y^2 + 2A_5 - \frac{1}{2}\rho\Omega^2(x^2 + y^2) \; ; \tag{7.29}$$

$$\sigma_{yy} = 12A_1x^2 + 2A_2y^2 + 2A_4 - \frac{1}{2}\rho\Omega^2(x^2 + y^2) \; , \tag{7.30}$$

$$\sigma_{yx} = -4A_2xy \; . \tag{7.31}$$

It follows that conditions (7.26, 7.27) will be satisfied for all $x$ if and only if

$$A_1 = \rho\Omega^2/24 \; ; \; A_2 = 0 \; ; \; A_4 = \rho\Omega^2 b^2/4 \; . \tag{7.32}$$

The constant $A_3$ can now be determined by substituting (7.25) into (7.24), with the result

$$24(A_1 + A_3) = 2\rho\Omega^2(1 - \nu) \tag{7.33}$$

which is the inhomogeneous equivalent of the constraint equations (see §5.1) and hence

$$A_3 = \rho\Omega^2(1 - 2\nu)/24 \; , \tag{7.34}$$

from (7.32).

Finally, we determine the remaining constant $A_5$ by substituting (7.25) into the boundary condition (7.28), with the result

$$A_5 = \rho\Omega^2 \left( \frac{\nu b^2}{6} + \frac{a^2}{4} \right) \; . \tag{7.35}$$

The final stress field is therefore

$$\sigma_{xx} = \rho\Omega^2 \left( \frac{(a^2 - x^2)}{2} + \frac{\nu(b^2 - 3y^2)}{3} \right) \; ; \tag{7.36}$$

$$\sigma_{yy} = \frac{\rho\Omega^2}{2}(b^2 - y^2) \; ; \tag{7.37}$$

$$\sigma_{yx} = 0 \; , \tag{7.38}$$

from equations(7.29–7.31).

Notice that the boundary conditions on the ends $x = \pm a$ agree with those of the physical problem except for the second term in $\sigma_{xx}$, which represents a symmetric self-equilibrated traction. From §6.2.2, we anticipate that the error due to this disagreement will be confined to regions near the ends of length comparable with the half-length, $b$.

### 7.3.2 Solution of the governing equation

Equation (7.24) is an inhomogeneous partial differential equation — i.e. it has a known function of $x, y$ on the right hand side — and it can be solved in the same way as an inhomogeneous *ordinary* differential equation, by first finding a particular integral of the equation and then superposing the general solution of the corresponding homogeneous equation.

In this context, a particular solution is *any* function $\phi$ that satisfies (7.24). It contains no arbitrary constants. The generality in the general solution comes from arbitrary constants in the *homogeneous* solution. Furthermore, the homogeneous solution is the solution of equation (7.24) modified to make the right hand side zero — i.e.

$$\nabla^4 \phi = 0 \tag{7.39}$$

Thus, the homogeneous solution is a general biharmonic function and we can summarize the solution method as containing the three steps:-

1. finding *any* function $\phi$ which satisfies (7.24);

2. superposing a sufficiently general biharmonic function (which therefore contains several arbitrary constants);

3. choosing the arbitrary constants to satisfy the boundary conditions.

In the problem of the preceding section, the particular solution would be any fourth degree polynomial satisfying (7.24) and the homogeneous solution, the most general fourth degree *biharmonic* function with the appropriate symmetry.

Notice that the particular solution (i) is itself a solution of a different physical problem — one in which the body forces are correctly represented, but the correct boundary conditions are not (usually) satisfied. Thus, the function of the homogeneous solution is to introduce additional degrees of freedom to enable us to satisfy the boundary conditions.

#### Physical superposition

It is often helpful to think of this superposition process in a physical rather than a mathematical sense. In other words, we devise a related problem in which the body forces are the same as in the real problem, but where the boundary conditions are simpler. For example, if a beam is subjected to gravitational loading, a simple physical 'particular' solution would correspond to the problem in which the beam is resting on a rigid foundation and hence the stress field is one-dimensional. To complete the real problem, we would then have to superpose the solution of a corrective problem with tractions equal and opposite to those provided by the foundation, but with no body force, since this has already been taken into account in the particular solution.

One advantage of this way of thinking is that it is not restricted to problems in which the body force can be represented by a potential. We can therefore use it to solve the problem of the rotationally accelerating beam in the next section.

## 7.4 Rotational acceleration

We saw in §7.2.4 that the body force potential cannot be used in problems where the angular acceleration is non-zero. In this section, we shall generate a particular solution for this problem and then generalize it, using the results without body force from Chapter 5.

### 7.4.1 The circular disk

Consider the rotationally symmetric problem of Figure 7.2. A solid circular disk, radius $a$, is initially at rest ($\Omega = 0$) and at time $t = 0$ it is caused to accelerate in a clockwise direction with angular acceleration $\dot{\Omega}$ by tractions uniformly distributed around the edge $r = a$. Note that we could determine the magnitude of these tractions by writing the equation of motion for the disk, but it will not be necessary to do this — the result will emerge from the analysis of the stress field.

At any given time, the body forces and hence the resulting stress field will be the sum of two parts, one due to the instantaneous angular velocity and the other to the angular acceleration. The former can be obtained using the body force potential and we therefore concentrate here on the contribution due to the angular acceleration. This is the *only* body force at the beginning of the process, so the following solution can also be regarded as the solution for the instant, $t = 0$.

If the tractions in Figure 7.2 were reversed, the disk would spin the other way, but the resulting problem would be simply the mirror image of that illustrated. Now suppose some point in the disk had a non-zero outward radial displacement, $u_r$. Reversing the tractions would change the sign of this displacement — i.e. make it be directed inwards — but this is impossible if the new problem is to be a mirror image of the old. We can therefore conclude from symmetry that $u_r = 0$ throughout the disk. In the same way, we can conclude that the stress components $\sigma_{rr}, \sigma_{\theta\theta}$ are zero everywhere. Incidentally, we also note that if the problem is conceived as being one of plane strain, symmetry demands that $\sigma_{zz}$ be everywhere zero. It follows that this

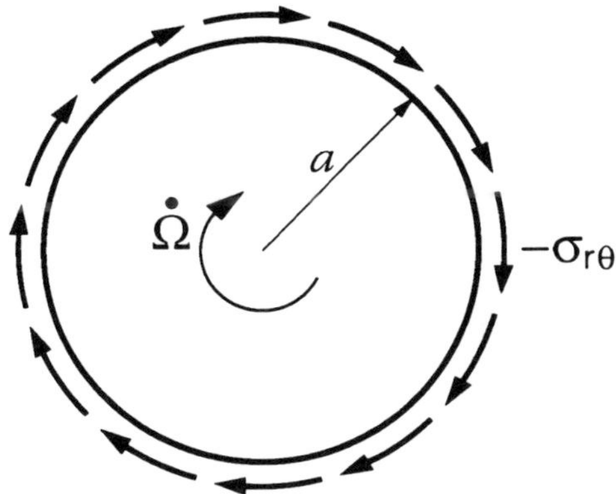


Figure 7.2: The disk with rotational acceleration

is one of those lucky problems in which plane stress and plane strain are the same and hence the plane stress assumption involves no approximation.

We conclude that there is only one non-zero displacement component, $u_\theta$, and one non-zero strain component, $e_{r\theta}$. Thus, the number of strain and displacement components is equal and no non-trivial compatibility conditions can be obtained by eliminating the displacement components. (An alternative statement is that all the compatibility equations are satisfied identically, as can be verified by substitution — the compatibility equations in cylindrical polar coördinates are given by Saada[3]).

It follows that the only non-zero stress component, $\sigma_{r\theta}$ can be determined from equilibrium considerations alone. Considering the equilibrium of a small element due to forces in the $\theta$-direction and dropping terms which are zero due to the symmetry of the system, we obtain

$$\frac{\partial \sigma_{r\theta}}{\partial r} + \frac{2\sigma_{r\theta}}{r} = -\rho r \dot{\Omega} \; , \tag{7.40}$$

which has the general solution

$$\sigma_{r\theta} = -\frac{\rho \dot{\Omega} r^2}{4} + \frac{A}{r^2} \; , \tag{7.41}$$

where $A$ is an arbitrary constant which must be set to zero to retain continuity at the origin.

[3]A.S.Saada, *Elasticity*, Pergamon Press, New York, (1973), §6.9.

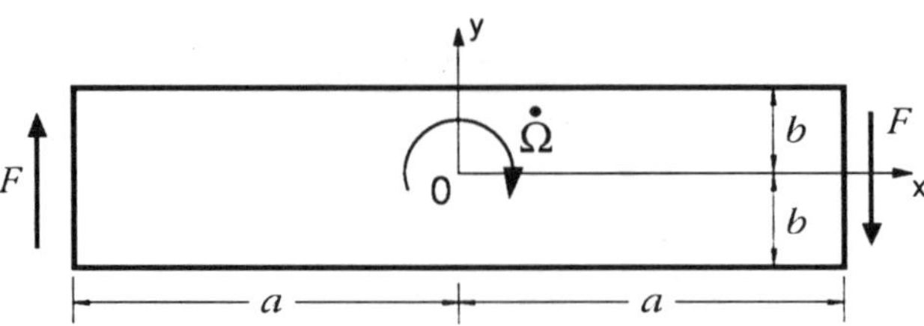


Figure 7.3: The rectangular bar with rotational acceleration

### 7.4.2 The rectangular bar

We can use the above result as a particular solution for determining the stresses in a body of arbitrary shape due to angular acceleration. One way to think of this is to imagine cutting out the real body from an imaginary disk of sufficiently large radius. The stresses in the cut-out body will be the same as those in the disk, provided we arrange to apply tractions to the boundaries of the body that are equal to the stress components on that surface before the cut was made. These will not generally be the correct boundary tractions for the real problem, but we can correct the boundary tractions by superposing a homogeneous solution — i.e. a corrective solution for the actual body with prescribed boundary tractions (equal to the difference between those applied and those obtained in the disk solution) but no body forces (since these have already been taken care of in the disk solution.)

As an illustration, we consider the rectangular bar, $-a < x < a$, $-b < y < b$, accelerated by two equal shear forces, $F$, applied at the ends $x = \pm a$ as shown in Figure 7.3, the other boundaries, $y = \pm b$ being traction free.

We first transform the disk solution into rectangular coördinates, using the transformation relations (1.7–1.9), obtaining

$$\sigma_{xx} = -2\sigma_{r\theta}\sin\theta\cos\theta = \frac{\rho\dot{\Omega}xy}{2}\,; \tag{7.42}$$

$$\sigma_{yy} = 2\sigma_{r\theta}\sin\theta\cos\theta = -\frac{\rho\dot{\Omega}xy}{2}\,; \tag{7.43}$$

$$\sigma_{xy} = \sigma_{r\theta}(\cos^2\theta - \sin^2\theta) = -\frac{\rho\dot{\Omega}(x^2-y^2)}{4}\,. \tag{7.44}$$

This stress field is clearly odd in both $x$ and $y$ and involves normal tractions on $y = \pm b$ which vary linearly with $x$. The bending moment will therefore vary with

$x^2$ suggesting a stress function $\phi$ with a leading term $x^5y$ — i.e. a sixth degree polynomial.

The most general polynomial with the appropriate symmetry is

$$\phi = A_1x^5y + A_2x^3y^3 + A_3xy^5 + A_4x^3y + A_5xy^3 + A_6xy \, . \tag{7.45}$$

The stress components are the sum of those obtained from the homogeneous stress function (7.45) *using the definitions (5.1–5.3)* — remember the homogeneous solution here is one without body force — and those from the disk problem given by equations (7.42–7.44). We find

$$\sigma_{xx} = \frac{1}{2}\rho\dot{\Omega}xy + 6A_2x^3y + 20A_3xy^3 + 6A_5xy \; ; \tag{7.46}$$

$$\sigma_{yy} = -\frac{1}{2}\rho\dot{\Omega}xy + 20A_1x^3y + 6A_2xy^3 + 6A_4xy \; ; \tag{7.47}$$

$$\begin{aligned} \sigma_{xy} = & -\frac{1}{4}\rho\dot{\Omega}(x^2 - y^2) - 5A_1x^4 - 9A_2x^2y^2 \\ & -5A_3y^4 - 3A_4x^2 - 3A_5y^2 - A_6 \, . \end{aligned} \tag{7.48}$$

The boundary conditions are

$$\sigma_{yx} = 0 \; ; \; y = \pm b \; ; \tag{7.49}$$

$$\sigma_{yy} = 0 \; ; \; y = \pm b \; ; \tag{7.50}$$

$$\int_{-b}^{b} y\sigma_{xx}dy = 0 \; ; \; x = \pm a \; , \tag{7.51}$$

where we note that weak boundary conditions are imposed on the ends $x = \pm a$. Since the solution is odd in $y$, only the moment and shear force conditions are non-trivial and the latter need not be explicitly imposed, since they will be satisfied by global equilibrium (as in the problem of §5.2.2).

Conditions (7.49, 7.50) have to be satisfied for all $x$ and hence the corresponding coefficients of all powers of $x$ must be zero. It follows immediately that $A_1 = 0$ and the remaining conditions can be written

$$-\frac{1}{2}\rho\dot{\Omega} + 6A_2b^2 + 6A_4 = 0 \; ; \tag{7.52}$$

$$-\frac{1}{4}\rho\dot{\Omega} - 9A_2b^2 - 3A_4 = 0 \; ; \tag{7.53}$$

$$\frac{1}{4}\rho\dot{\Omega}b^2 - 5A_3b^4 - 3A_5b^2 - A_6 = 0 \, . \tag{7.54}$$

We get one additional condition from the requirement that $\phi$ (equation (7.45)) be biharmonic

$$72A_2 + 120A_3 = 0 \tag{7.55}$$

and another from the boundary condition (7.51), which with (7.46) yields

$$\frac{1}{6}\rho\dot{\Omega} + 2A_2a^2 + 4A_3b^2 + 2A_5 = 0\ . \tag{7.56}$$

The solution of these equations is routine, giving the stress function

$$\phi = -\frac{\rho\dot{\Omega}}{60b^2}(5x^3y^3 - 3xy^5 - 10b^2x^3y + 11b^2xy^3 - 5a^2xy^3 + 15a^2b^2xy - 33b^4xy)\ . \tag{7.57}$$

The complete stress field, including the particular solution terms, is

$$\sigma_{xx} = \rho\dot{\Omega}xy\left(\frac{y^2}{b^2} - \frac{3}{5} + \frac{(a^2 - x^2)}{2b^2}\right)\ ; \tag{7.58}$$

$$\sigma_{yy} = \frac{\rho\dot{\Omega}xy}{2}\left(1 - \frac{y^2}{b^2}\right)\ ; \tag{7.59}$$

$$\sigma_{xy} = \rho\dot{\Omega}(b^2 - y^2)\left(\frac{y^2 + a^2 - 3x^2}{4b^2} - \frac{11}{20}\right)\ , \tag{7.60}$$

from (7.46–7.48).

Finally, we can determine the forces $F$ on the ends (which alternatively could have been found from the rigid-body equations of motion) by integrating the shear traction, $\sigma_{xy}$ over either end as

$$F = -\int_{-b}^{b} \sigma_{xy}(a)dy = \frac{2}{3}\rho\dot{\Omega}b(a^2 + b^2)\ . \tag{7.61}$$

## PROBLEMS

1. The beam $-c<y<c$, $0<x<L$ is built-in at the edge $x = L$ and is loaded merely by its own weight, $\rho g$ per unit volume.

Find a solution for the stress field, using the weak conditions on the end $x = 0$.

2. Figure 7.4($a$) shows a triangular cantilever, defined by the boundaries $y = 0$, $y = x\tan\alpha$ and built-in at $x = a$. It is loaded by its own weight, $\rho g$ per unit volume. Find a solution for the complete stress field and compare the maximum tensile bending stress with that predicted by the Strength of Materials theory.

Would the maximum tensile stress be lower if the alternative configuration of Figure 7.4($b$) were used?

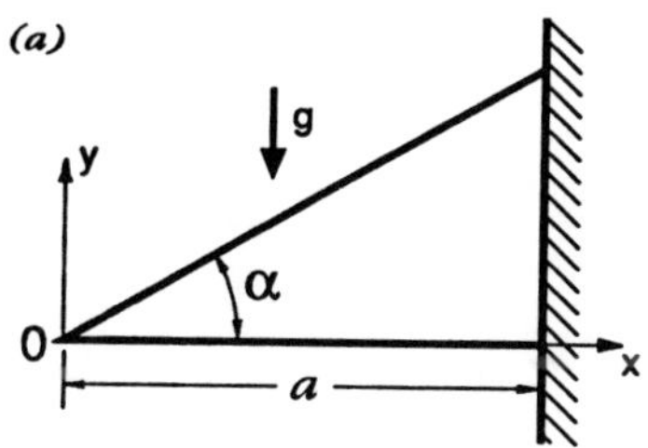


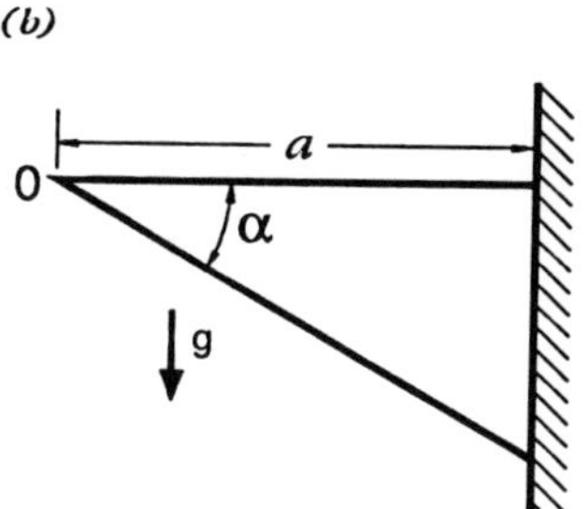


Figure 7.4: The triangular cantilever loaded by self-weight

# Chapter 8

# PROBLEMS IN POLAR COÖRDINATES

Polar coördinates $(r, \theta)$ are particularly suited to problems in which the boundaries can be expressed in terms of equations like $r = a, \theta = \alpha$. This includes the stresses in a circular disk or around a circular hole, the curved bar with circular boundaries and the wedge, all of which will be discussed in this and subsequent chapters.

## 8.1 Expressions for stress components

We first have to transform the biharmonic equation (4.7) and the expressions for stress components (4.1) into polar coördinates, using the relations

$$x = r\cos\theta \;\; ; \;\; y = r\sin\theta \; ; \tag{8.1}$$

$$r = \sqrt{x^2 + y^2} \;\; ; \;\; \theta = \arctan\left(\frac{y}{x}\right) . \tag{8.2}$$

The derivation is tedious, but routine. We first note by differentiation that

$$\frac{\partial}{\partial x} = \frac{\partial r}{\partial x}\frac{\partial}{\partial r} + \frac{\partial \theta}{\partial x}\frac{\partial}{\partial \theta} = \cos\theta\frac{\partial}{\partial r} - \frac{\sin\theta}{r}\frac{\partial}{\partial \theta} \; ; \tag{8.3}$$

$$\frac{\partial}{\partial y} = \frac{\partial r}{\partial y}\frac{\partial}{\partial r} + \frac{\partial \theta}{\partial y}\frac{\partial}{\partial \theta} = \sin\theta\frac{\partial}{\partial r} + \frac{\cos\theta}{r}\frac{\partial}{\partial \theta} \; . \tag{8.4}$$

It follows that

$$\begin{aligned}
\frac{\partial^2}{\partial x^2} &= \left(\cos\theta\frac{\partial}{\partial r} - \frac{\sin\theta}{r}\frac{\partial}{\partial \theta}\right)\left(\cos\theta\frac{\partial}{\partial r} - \frac{\sin\theta}{r}\frac{\partial}{\partial \theta}\right) \\
&= \cos^2\theta\frac{\partial^2}{\partial r^2} + \sin^2\theta\left(\frac{1}{r}\frac{\partial}{\partial r} + \frac{1}{r^2}\frac{\partial^2}{\partial \theta^2}\right) \\
&\quad +2\sin\theta\cos\theta\left(\frac{1}{r^2}\frac{\partial}{\partial \theta} - \frac{1}{r}\frac{\partial^2}{\partial r \partial \theta}\right)
\end{aligned} \tag{8.5}$$

and by a similar process, we find that

$$\begin{aligned}\frac{\partial^2}{\partial y^2} &= \sin^2\theta\frac{\partial^2}{\partial r^2}+\cos^2\theta\left(\frac{1}{r}\frac{\partial}{\partial r}+\frac{1}{r^2}\frac{\partial^2}{\partial\theta^2}\right)\\ &\quad -2\sin\theta\cos\theta\left(\frac{1}{r^2}\frac{\partial}{\partial\theta}-\frac{1}{r}\frac{\partial^2}{\partial r\partial\theta}\right)\;; \qquad (8.6)\\ \frac{\partial^2}{\partial x\partial y} &= \sin\theta\cos\theta\left(\frac{\partial^2}{\partial r^2}-\frac{1}{r}\frac{\partial}{\partial r}-\frac{1}{r^2}\frac{\partial^2}{\partial\theta^2}\right)\\ &\quad -\left(\cos^2\theta-\sin^2\theta\right)\left(\frac{1}{r^2}\frac{\partial}{\partial\theta}-\frac{1}{r}\frac{\partial^2}{\partial r\partial\theta}\right)\;. \qquad (8.7)\end{aligned}$$

Finally, we can determine the expressions for stress components, noting for example that

$$\begin{aligned}\sigma_{rr} &= \sigma_{xx}\cos^2\theta+\sigma_{yy}\sin^2\theta+2\sigma_{xy}\sin\theta\cos\theta \qquad (8.8)\\ &= \cos^2\theta\frac{\partial^2\phi}{\partial y^2}+\sin^2\theta\frac{\partial^2\phi}{\partial x^2}-2\sin\theta\cos\theta\frac{\partial^2\phi}{\partial x\partial y}\\ &= \frac{1}{r}\frac{\partial\phi}{\partial r}+\frac{1}{r^2}\frac{\partial^2\phi}{\partial\theta^2}\;, \qquad (8.9)\end{aligned}$$

after substituting for the partial derivatives from (8.5–8.7) and simplifying.

The remaining stress components, $\sigma_{\theta\theta}, \sigma_{r\theta}$ can be obtained by a similar procedure. We find

$$\sigma_{rr} = \frac{1}{r}\frac{\partial\phi}{\partial r}+\frac{1}{r^2}\frac{\partial^2\phi}{\partial\theta^2}\;;\qquad \sigma_{\theta\theta}=\frac{\partial^2\phi}{\partial r^2}\;; \qquad (8.10)$$

$$\sigma_{r\theta} = \frac{1}{r^2}\frac{\partial\phi}{\partial\theta}-\frac{1}{r}\frac{\partial^2\phi}{\partial r\partial\theta}=-\frac{\partial}{\partial r}\left(\frac{1}{r}\frac{\partial\phi}{\partial\theta}\right)\;. \qquad (8.11)$$

We also note that the Laplacian operator

$$\nabla^2\equiv\frac{\partial^2}{\partial x^2}+\frac{\partial^2}{\partial y^2}=\frac{\partial^2}{\partial r^2}+\frac{1}{r}\frac{\partial}{\partial r}+\frac{1}{r^2}\frac{\partial^2}{\partial\theta^2}\;, \qquad (8.12)$$

from equations (8.5, 8.6) and hence

$$\nabla^4\equiv\left(\frac{\partial^2}{\partial r^2}+\frac{1}{r}\frac{\partial}{\partial r}+\frac{1}{r^2}\frac{\partial^2}{\partial\theta^2}\right)^2\;. \qquad (8.13)$$

## 8.2 Strain components

A similar technique can be used to obtain the strain-displacement relations in polar coördinates. Writing

$$u_x = u_r\cos\theta - u_\theta\sin\theta \; ; \tag{8.14}$$
$$u_y = u_r\sin\theta + u_\theta\cos\theta \tag{8.15}$$

and substituting in (8.3), we find

$$e_{xx} = \frac{\partial u_x}{\partial x} = \frac{\partial u_r}{\partial r}\cos^2\theta + \left(\frac{u_\theta}{r} - \frac{\partial u_\theta}{\partial r} - \frac{1}{r}\frac{\partial u_r}{\partial \theta}\right)\sin\theta\cos\theta + \left(\frac{u_r}{r} + \frac{1}{r}\frac{\partial u_\theta}{\partial \theta}\right)\sin^2\theta \; . \tag{8.16}$$

Using the same method to obtain expressions for $e_{xy}, e_{yy}$ and substituting the results in the strain transformation relations analogous to (8.8) etc., we obtain

$$e_{rr} = \frac{\partial u_r}{\partial r} \; ; \; e_{r\theta} = \frac{1}{2}\left(\frac{1}{r}\frac{\partial u_r}{\partial \theta} + \frac{\partial u_\theta}{\partial r} - \frac{u_\theta}{r}\right) \; ; \; e_{\theta\theta} = \frac{1}{r}\frac{\partial u_\theta}{\partial \theta} + \frac{u_r}{r} \; . \tag{8.17}$$

## 8.3 Fourier series expansion

The simplest problems in polar coördinates are those in which there are no $\theta$-boundaries, the most general case being the disk with a central hole illustrated in Figure 8.1 and defined by $b<r<a$.

The stresses and displacements must be single-valued and continuous and hence they must be periodic functions of $\theta$, since $(r, \theta + 2m\pi)$ defines the same point as $(r, \theta)$, when $m$ is any integer. It is therefore natural to seek a general solution of the problem of Figure 8.1 in the form

$$\phi = \sum_{n=0}^{\infty} f_n(r)\cos n\theta + \sum_{n=1}^{\infty} g_n(r)\sin n\theta \; . \tag{8.18}$$

Substituting this expression into the biharmonic equation, using (8.13), we find that the functions $f_n, g_n$ must satisfy the ordinary differential equation

$$\left(\frac{d^2}{dr^2} + \frac{1}{r}\frac{d}{dr} - \frac{n^2}{r^2}\right)^2 f(r) = 0 \; , \tag{8.19}$$

which, if $n \neq 0, 1$ has the general solution

$$f_n(r) = A_{n1}r^{n+2} + A_{n2}r^{-n+2} + A_{n3}r^n + A_{n4}r^{-n} \; , \tag{8.20}$$

where $A_{n1}, \ldots A_{n4}$ are four arbitrary constants.

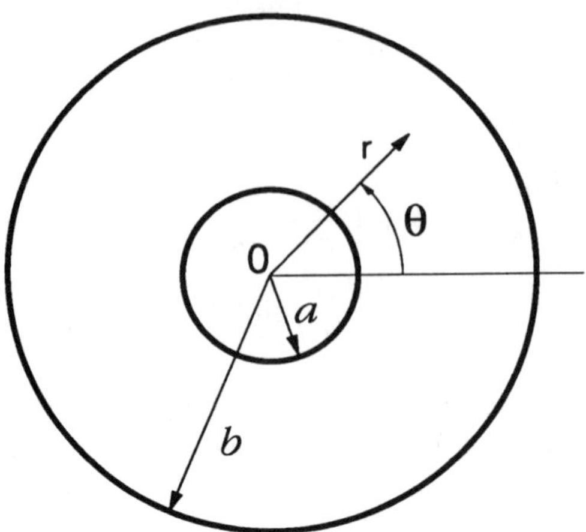


Figure 8.1: The disk with a central hole

When $n = 0, 1$, the solution (8.20) develops repeated roots and equation (8.19) has a different form of solution given by

$$f_0(r) = A_{01}r^2 + A_{02}r^2 \log r + A_{03} \log r + A_{04} \; ; \tag{8.21}$$

$$f_1(r) = A_{11}r^3 + A_{12}r \log r + A_{13}r + A_{14}r^{-1} \; . \tag{8.22}$$

In all the above equations, we note that the stress functions associated with the constants $A_{n3}, A_{n4}$ are *harmonic* and hence biharmonic *a fortiori*, whereas those associated with $A_{n1}, A_{n2}$ are biharmonic but not harmonic. We also note that 'log' will be taken to denote the natural logarithm throughout this text.

### 8.3.1 Satisfaction of boundary conditions

The boundary conditions on the surfaces $r = a, b$ will generally take the form

$$\sigma_{rr} = F_1(\theta) \; ; \; r = a \; ; \tag{8.23}$$

$$= F_2(\theta) \; ; \; r = b \; ; \tag{8.24}$$

$$\sigma_{r\theta} = F_3(\theta) \; ; \; r = a \; ; \tag{8.25}$$

$$= F_4(\theta) \; ; \; r = b \; , \tag{8.26}$$

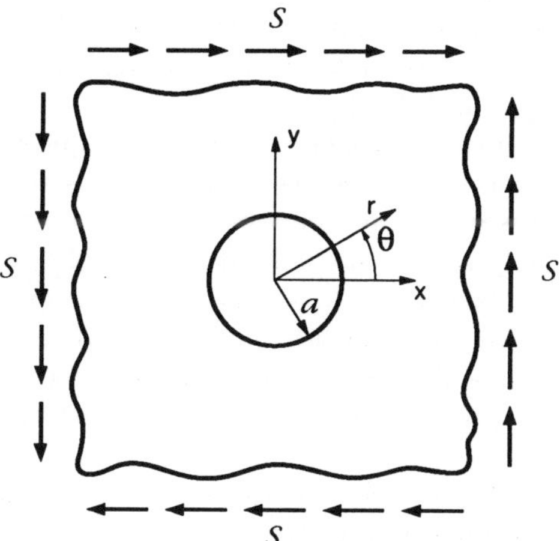


Figure 8.2: Circular hole in a shear field

each of which has to be satisfied for all values of $\theta$. This can conveniently be done by expanding the functions $F_1, \dots F_4$ as Fourier series in $\theta$. i.e.

$$F_j(\theta) = \sum_{n=0}^{\infty} a_{nj} \cos n\theta + \sum_{n=1}^{\infty} b_{nj} \sin n\theta \;\; ; \;\; j = 1, \dots, 4 \,. \tag{8.27}$$

In combination with the stress function of equation (8.18), (8.23–8.26) will then give four independent equations for each trigonometric term, $\cos n\theta, \sin n\theta$ in the series and hence serve to determine the four constants $A_{n1}, \dots, A_{n4}$. The problem of Figure 8.1 is therefore susceptible of a general solution.

### 8.3.2 Circular hole in a shear field

The general solution has some anomolous features for the special values $n = 0, 1$, but before discussing these we shall illustrate the method by solving a simple example.

Figure 8.2 shows a large plate in a state of pure shear $\sigma_{xy} = S$, perturbed by a hole of radius $a$. A formal statement of the boundary conditions for this problem is

$$\sigma_{rr} = 0 \;\; ; \;\; r = a \,; \tag{8.28}$$

$$\sigma_{r\theta} = 0 \;\; ; \;\; r = a \,; \tag{8.29}$$

$$\sigma_{xx}, \sigma_{yy} \to 0 \;\; ; \;\; r \to \infty \,; \tag{8.30}$$

$$\sigma_{xy} \to S \;\; ; \;\; r \to \infty \,. \tag{8.31}$$

Notice that we describe a body as 'large' when we mean to interpret the boundary condition at the outer boundary as being applied at $r \to \infty$. Notice also that we have stated these 'infinite' boundary conditions in $(x, y)$ coördinates, since this is the most natural way to describe a state of uniform stress. We shall see that the stress function approach makes it possible to use both rectangular and polar coördinates in the same problem without any special difficulty.

This is a typical *perturbation* problem in which a simple state of stress is perturbed by a local geometric feature (in this case a hole). It is reasonable in such cases to anticipate that the stresses distant from the hole will be unperturbed and that the effect of the hole will only be felt at moderate values of $r/a$. Perturbation problems are most naturally approached by first solving the simpler problem in which the perturbation is absent (in this case the plate *without* a hole) and then seeking a corrective solution which will describe the influence of the hole on the stress field. With such a formulation, we anticipate that the corrective solution will decay with increasing $r$.

The unperturbed field is clearly a state of uniform shear, $\sigma_{xy} = S$, and this in turn is conveniently described by the stress function

$$\phi = -Sxy = -Sr^2 \sin\theta \cos\theta = -\frac{Sr^2 \sin 2\theta}{2} , \tag{8.32}$$

from equation (4.1)

Notice that although the stress function is originally determined in rectangular coördinates, it is easily transformed into polar coördinates and then into the Fourier form of equation (8.18).

The unperturbed solution satisfies the 'infinite' boundary conditions (8.30, 8.31), but will violate the conditions at the hole surface (8.28, 8.29). However, we can correct the stress field by superposing those terms from the series (8.18) which (i) have the same Fourier dependence as (8.32) and (ii) lead to stresses which decay as $r$ increases. There will always be two such terms for any given Fourier component, permitting the two traction boundary conditions to be satisfied. In the present instance, the required terms are those derived from the constants $A_{22}, A_{24}$ in equation (8.20), giving the stress function[1]

$$\phi = -\frac{Sr^2 \sin 2\theta}{2} + A \sin 2\theta + Br^{-2} \sin 2\theta . \tag{8.33}$$

The corresponding stress components are obtained by substituting in (8.10, 8.11) with the result

$$\sigma_{rr} = \left(S - \frac{4A}{r^2} - \frac{6B}{r^4}\right) \sin 2\theta ; \tag{8.34}$$

$$\sigma_{r\theta} = \left(S + \frac{2A}{r^2} + \frac{6B}{r^4}\right) \cos 2\theta ; \tag{8.35}$$

[1] It might be thought that the term involving $A_{22}$ is inappropriate because it does not decay with $r$. However, it leads to *stresses* which decay with $r$, which is of course what we require.

$$\sigma_{\theta\theta} = \left(-S + \frac{6b}{r^4}\right) \sin 2\theta \ . \tag{8.36}$$

The two boundary conditions at the hole surface (8.28, 8.29) then yield the two equations

$$\frac{4A}{a^2} + \frac{6B}{a^4} = S \ ; \tag{8.37}$$

$$\frac{2A}{a^2} + \frac{6B}{a^4} = -S \tag{8.38}$$

for the constants $A, B$, with solution

$$A = Sa^2 \ ; \ B = -\frac{Sa^4}{2} \tag{8.39}$$

and the final stress field is

$$\sigma_{rr} = S\left(1 - 4\frac{a^2}{r^2} + 3\frac{a^4}{r^4}\right) \sin 2\theta \ ; \tag{8.40}$$

$$\sigma_{r\theta} = S\left(1 + 2\frac{a^2}{r^2} - 3\frac{a^4}{r^4}\right) \cos 2\theta \ ; \tag{8.41}$$

$$\sigma_{\theta\theta} = S\left(-1 - 3\frac{a^4}{r^4}\right) \sin 2\theta \ . \tag{8.42}$$

Notice incidentally that the maximum stress is the *hoop* stress, $\sigma_{\theta\theta} = 4S$ at the point $(a, 3\pi/4)$. At this point there is a state of uniaxial tension, so the maximum *shear* stress is $2S$. Since the unperturbed shear stress has a magnitude $S$, we say that the hole produces a *stress concentration* of 2. However, for a brittle material, we might be inclined to define the stress concentration factor as the ratio of the maximum tensile stresses in the perturbed and unperturbed solutions, which in the present problem is 4. In general, a stress concentration factor implies a measure of the severity of the stress field — usually a failure theory — which serves as a standard of comparison and the magnitude of the stress concentration factor will depend upon the measure used.

The determination of the stress concentration due to holes, notches and changes of section under various loading conditions is clearly a question of considerable practical importance. An extensive discussion of problems of this kind is given by Savin[2] and stress concentration factors for a wide range of geometries are tabulated by Peterson[3] in a form suitable for use in engineering design.

[2]G.N.Savin, *Stress Concentration around Holes*, Pergamon Press, Oxford, (1961).

[3]R.E.Peterson, *Stress Concentration Design Factors*, John Wiley, New York, (1974).

### 8.3.3 Degenerate cases

We have already remarked in §8.3 above that the solution (8.20) degenerates for $n = 0, 1$ and must be supplemented by some additional terms. In fact, even the modified stress function of (8.21, 8.22) is degenerate because the stress function

$$\phi = A + Bx + Cy = A + Br\cos\theta + Cr\sin\theta \tag{8.43}$$

defines a trivial null state of stress (see §5.1.1) and hence the constants $A_{04}, A_{13}$ in equations (8.21, 8.22) correspond to null stress fields.

This question of degeneracy arises elsewhere in Elasticity and indeed in mathematics generally, so we shall take this opportunity to develop a general technique for resolving it. As we saw in §8.3, the degeneracy often arises from the occurrence of repeated roots to an equation. Thus, in equation (8.20), the terms $A_{n2}r^{-n+2}, A_{n3}r^{n}$ degenerate to the same form when $n = 1$.

Suppose for the moment that we relax the restriction that $n$ be an integer. Clearly the degeneracy in equation (8.20) only arises exactly at the values $n = 0, 1$. There is no degeneracy for $n = 1 + \epsilon$ for any non-zero $\epsilon$, however small. We shall show therefore that we can recover the extra solution at the degenerate point by allowing the solution to tend smoothly to the limit $n \to 1$, rather than setting $n = 1$ *ab initio*.

For $n = 1 + \epsilon$, the two offending terms in (8.20) can be written

$$f(r) = Ar^{1-\epsilon} + Br^{1+\epsilon}\ , \tag{8.44}$$

where $A, B$, are two arbitrary constants.

Clearly the two terms tend to the same form as $\epsilon \to 0$. However, suppose we construct a new function from the sum and difference of these functions in the form

$$f(r) = C(r^{1+\epsilon} + r^{1-\epsilon}) + D(r^{1+\epsilon} - r^{1-\epsilon})\ . \tag{8.45}$$

In this form, the first function tends to $2Cr$ as $\epsilon \to 0$, whilst the second tends to zero. However, we can prevent the second term degenerating to zero, since the constant $D$ is arbitrary. We can therefore choose $D = E\epsilon^{-1}$ in which case the second term will tend to the limit

$$\lim_{\epsilon \to 0} E\frac{r^{1+\epsilon} - r^{1-\epsilon}}{\epsilon} = Er\log r\ , \tag{8.46}$$

where we have used L'Hôpital's rule to evaluate the limit. Notice that this result agrees with the special term included in equation (8.22) above to resolve the degeneracy in $\phi$ for $n = 1$.

This procedure can be generalized as follows: Whenever a degeneracy occurs at a denumerable set of values of a parameter (such as n in equation (8.20)), we can always make up the defect using additional terms obtained by differentiating the original form with respect to the parameter *before* allowing it to take the special value.

The pairs $r^{n+2}, r^{-n+2}$ and $r_n, r^{-n}$ both degenerate to the same form for $n = 0$, so the new functions are of the form

$$\lim_{n\to 0} \frac{d(r^{n+2})}{dn} = r^2 \log r \; ; \tag{8.47}$$

$$\lim_{n\to 0} \frac{d(r^n)}{dn} = \log r \; , \tag{8.48}$$

again agreeing with the special terms introduced in (8.21).

Sometimes the degeneracy is of a higher order, corresponding to more than two identical roots, in which case L'Hôpital's rule has to be applied more than once — i.e. we have to differentiate more than once with respect to the parameter. It is always a straightforward matter to check the resulting solutions to make sure they are of the required form.

We now turn our attention to the degeneracy implied by the triviality of the stress function of equation (8.43). Here, the stress function *itself* is not degenerate — i.e. it is a legitimate function of the required Fourier form and it is linearly independent of the other functions of the same form. The trouble is that it gives a null stress field.

We must therefore look for stress functions that are *not* of the standard Fourier form, but which give Fourier-type stress components.

Once again, the method is to approach the solution as a limit using L'Hôpital's rule, but this time we have to operate on the complete stress function — not just on the part that varies with $r$. Suppose we consider the axisymmetric degenerate term, $\phi = A$, which can be regarded as the limit of a stress function of the form

$$\phi = Ar^\epsilon \cos \epsilon\theta + Br^\epsilon \sin \epsilon\theta \; , \tag{8.49}$$

as $\epsilon \to 0$.

Differentiating this function with respect to $\epsilon$ and then letting $\epsilon$ tend to zero, we obtain the new function

$$\phi = A \log r + B\theta \; . \tag{8.50}$$

The first term is the same one that we found by operating on the function $f_n(r)$ and is of the correct Fourier form, but the second term is not appropriate to a Fourier series. However, when we substitute the second term into equations (8.10, 8.11), we obtain the stress components

$$\sigma_{rr} = \sigma_{\theta\theta} = 0 \quad ; \quad \sigma_{r\theta} = \frac{B}{r^2} \; , \tag{8.51}$$

which *are* of the required Fourier form (for $n = 0$), since the stress components do not vary with $\theta$.

In the same way, we can develop special terms to make up the deficit due to the two null terms with $n = 1$ (equation (8.43), obtaining the new stress functions

$$\phi = Br\theta \sin \theta + Cr\theta \cos \theta \; , \tag{8.52}$$

which generate the Fourier-type stress components

$$\sigma_{r\theta} = \sigma_{\theta\theta} = 0 \;\; ; \;\; \sigma_{rr} = \frac{2B\cos\theta}{r} - \frac{2C\sin\theta}{r} \,. \tag{8.53}$$

## 8.4 The Michell solution

The preceding results now permit us to write down a general solution of the elasticity problem in polar coördinates in Fourier expansion form. We have

$$\begin{aligned}\phi &= A_{01}r^2 + A_{02}r^2\log r + A_{03}\log r + A_{04}\theta \\ &\quad +(A_{11}r^3 + A_{12}r\log r + A_{14}r^{-1})\cos\theta + A_{13}r\theta\sin\theta \\ &\quad +(B_{11}r^3 + B_{12}r\log r + B_{14}r^{-1})\sin\theta + B_{13}r\theta\cos\theta \\ &\quad +\sum_{i=2}^{\infty}(A_{n1}r^{n+2} + A_{n2}r^{-n+2} + A_{n3}r^n + A_{n4}r^{-n})\cos n\theta \\ &\quad +\sum_{i=2}^{\infty}(B_{n1}r^{n+2} + B_{n2}r^{-n+2} + B_{n3}r^n + B_{n4}r^{-n})\sin n\theta \end{aligned} \tag{8.54}$$

This solution is due to Michell[4]. The corresponding stress components are easily obtained by substituting into equations (8.10, 8.11). For convenience, we give them in tabular form in Table 8.1, since we shall often wish to select a few components from the general solution in treating specific problems.

### 8.4.1 Hole in a tensile field

As a second example — and to illustrate the use of Table 8.1 — we consider the case where the the body of Figure 8.2 is subjected to uniform tension at infinity instead of shear, so that the boundary conditions become

$$\sigma_{rr} = 0 \;;\; r = a \;; \tag{8.55}$$

$$\sigma_{r\theta} = 0 \;;\; r = a \;; \tag{8.56}$$

$$\sigma_{xy}, \sigma_{yy} \to 0 \;;\; r \to \infty \;; \tag{8.57}$$

$$\sigma_{xx} \to S \;;\; r \to \infty \,. \tag{8.58}$$

The unperturbed problem in this case can clearly be described by the stress function

$$\phi = \frac{Sy^2}{2} = \frac{Sr^2\sin^2\theta}{2} = \frac{Sr^2}{4} - \frac{Sr^2\cos 2\theta}{4} \,. \tag{8.59}$$

This function contains both an axisymmetric term and a $\cos 2\theta$ term, so to complete the solution, we supplement it with those terms from Table 8.1 which have the same form and for which the stresses decay as $r \to \infty$.

[4]J.H.Michell, On the direct determination of stress in an elastic solid, with application to the theory of plates, Proc. London Math. Soc., Vol. 31 (1899), 100–124.

Table 8.1: The Michell solution — stress components

| $\phi$ | $\sigma_{rr}$ | $\sigma_{r\theta}$ | $\sigma_{\theta\theta}$ |
|---|---|---|---|
| $r^2$ | $2$ | $0$ | $2$ |
| $r^2\log r$ | $2\log r+1$ | $0$ | $2\log r+3$ |
| $\log r$ | $1/r^2$ | $0$ | $-1/r^2$ |
| $\theta$ | $0$ | $1/r^2$ | $0$ |
| $r^3\cos\theta$ | $2r\cos\theta$ | $2r\sin\theta$ | $6r\cos\theta$ |
| $r\theta\sin\theta$ | $2\cos\theta/r$ | $0$ | $0$ |
| $r\log r\cos\theta$ | $\cos\theta/r$ | $\sin\theta/r$ | $\cos\theta/r$ |
| $\cos\theta/r$ | $-2\cos\theta/r^3$ | $-2\sin\theta/r^3$ | $2\cos\theta/r^3$ |
| $r^3\sin\theta$ | $2r\sin\theta$ | $-2r\cos\theta$ | $6r\sin\theta$ |
| $r\theta\cos\theta$ | $-2\sin\theta/r$ | $0$ | $0$ |
| $r\log r\sin\theta$ | $\sin\theta/r$ | $-\cos\theta/r$ | $\sin\theta/r$ |
| $\sin\theta/r$ | $-2\sin\theta/r^3$ | $2\cos\theta/r^3$ | $2\sin\theta/r^3$ |
| $r^{n+2}\cos n\theta$ | $-(n+1)(n-2)r^n\cos n\theta$ | $n(n+1)r^n\sin n\theta$ | $(n+1)(n+2)r^n\cos n\theta$ |
| $r^{-n+2}\cos n\theta$ | $-(n+2)(n-1)r^{-n}\cos n\theta$ | $-n(n-1)r^{-n}\sin n\theta$ | $(n-1)(n-2)r^{-n}\cos n\theta$ |
| $r^n\cos n\theta$ | $-n(n-1)r^{n-2}\cos n\theta$ | $n(n-1)r^{n-2}\sin n\theta$ | $n(n-1)r^{n-2}\cos n\theta$ |
| $r^{-n}\cos n\theta$ | $-n(n+1)r^{-n-2}\cos n\theta$ | $-n(n+1)r^{-n-2}\sin n\theta$ | $n(n+1)r^{-n-2}\cos n\theta$ |
| $r^{n+2}\sin n\theta$ | $-(n+1)(n-2)r^n\sin n\theta$ | $-n(n+1)r^n\cos n\theta$ | $(n+1)(n+2)r^n\sin n\theta$ |
| $r^{-n+2}\sin n\theta$ | $-(n+2)(n-1)r^{-n}\sin n\theta$ | $n(n-1)r^{-n}\cos n\theta$ | $(n-1)(n-2)r^{-n}\sin n\theta$ |
| $r^n\sin n\theta$ | $-n(n-1)r^{n-2}\sin n\theta$ | $-n(n-1)r^{n-2}\cos n\theta$ | $n(n-1)r^{n-2}\sin n\theta$ |
| $r^{-n}\sin n\theta$ | $-n(n+1)r^{-n-2}\sin n\theta$ | $n(n+1)r^{-n-2}\cos n\theta$ | $n(n+1)r^{-n-2}\sin n\theta$ |

The resulting stress function is

$$\phi = \frac{Sr^2}{4} - \frac{Sr^2\cos 2\theta}{4} + A\log r + B\cos 2\theta + Cr^{-2}\cos 2\theta \tag{8.60}$$

and the corresponding stress components are

$$\sigma_{rr} = \frac{S}{2} + \frac{S\cos 2\theta}{2} + \frac{A}{r^2} - \frac{4B\cos 2\theta}{r^2} - \frac{6C\cos 2\theta}{r^4}\ ; \tag{8.61}$$

$$\sigma_{r\theta} = -\frac{S\sin 2\theta}{2} - \frac{2B\sin 2\theta}{r^2} - \frac{6C\sin 2\theta}{r^4}\ ; \tag{8.62}$$

$$\sigma_{\theta\theta} = \frac{S}{2} - \frac{S\cos 2\theta}{2} - \frac{A}{r^2} + \frac{6C\cos 2\theta}{r^4}\ . \tag{8.63}$$

The boundary conditions (8.55, 8.56) will be satisfied if and only if the coefficients of both Fourier terms are zero on $r = a$ and hence we obtain the equations

$$\frac{S}{2} + \frac{A}{a^2} = 0\ ; \tag{8.64}$$

$$\frac{S}{2} - \frac{4B}{a^2} - \frac{6C}{a^4} = 0\ ; \tag{8.65}$$

$$-\frac{S}{2} - \frac{2B}{a^2} - \frac{6C}{a^4} = 0\ , \tag{8.66}$$

which have the solution

$$A = -\frac{Sa^2}{2}\ ;\ B = \frac{Sa^2}{2}\ ;\ C = -\frac{Sa^4}{4}\ . \tag{8.67}$$

The final stress field is then obtained as

$$\sigma_{rr} = \frac{S}{2}\left(1 - \frac{a^2}{r^2}\right) + \frac{S\cos 2\theta}{2}\left(\frac{3a^4}{r^4} - \frac{4a^2}{r^2} + 1\right)\ ; \tag{8.68}$$

$$\sigma_{r\theta} = \frac{S\sin 2\theta}{2}\left(\frac{3a^4}{r^4} - \frac{2a^2}{r^2} - 1\right)\ ; \tag{8.69}$$

$$\sigma_{\theta\theta} = \frac{S}{2}\left(1 + \frac{a^2}{r^2}\right) - \frac{S\cos 2\theta}{2}\left(\frac{3a^4}{r^4} + 1\right)\ . \tag{8.70}$$

The maximum tensile stress is $\sigma_{\theta\theta} = 3S$ at $(a, \pi/2)$ and hence the stress concentration factor for tensile loading is 3. The maximum stress point and the unperturbed region at infinity are both in uniaxial tension and hence the stress concentration factor will be the same whatever criterion is used to measure the severity of the stress state. This contrasts with the problem of §8.3.2, where the maximum stress is uniaxial tension, but the unperturbed field is pure shear.

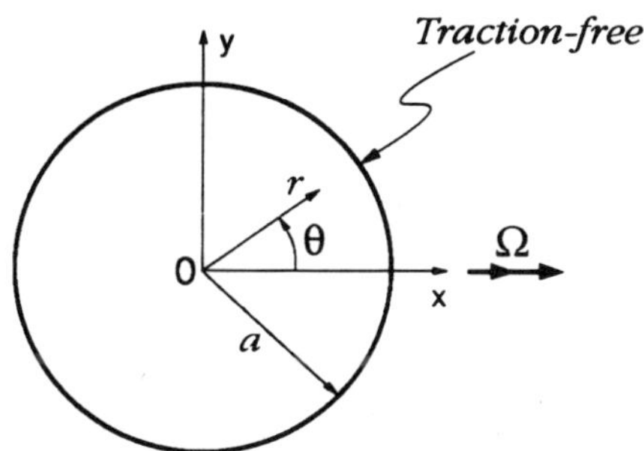


Figure 8.3: The disk rotating about a diameter

## PROBLEMS

1. Figure 8.3 shows a thin uniform circular disk, which rotates at constant speed $\Omega$ about the diametral axis $y = 0$, all the surfaces being traction-free. Determine the complete stress field in the disk.

2. A series of experiments is conducted in which a thin plate is subjected to biaxial tension/compression, $\sigma_1, \sigma_2$, the plane surface of the plate being traction-free (i.e. $\sigma_3 = 0$).

Unbeknown to the experimenter, the material contains microscopic defects which can be idealized as a sparse distribution of small holes through the thickness of the plate. Show graphically the relation which will hold at yield between the tractions $\sigma_1, \sigma_2$ applied to the defective plate, if the Tresca (maximum shear stress) criterion applies for the undamaged material.

# Chapter 9

# CALCULATION OF DISPLACEMENTS

So far, we have restricted attention to the calculation of stresses and to problems in which the boundary conditions are stated in terms of tractions or force resultants, but there are many problems in which displacements are also of interest. For example, we may wish to find the deflection of the rectangular beams considered in Chapter 5, or calculate the stress concentration factor due to a rigid circular inclusion in an elastic matrix, for which a displacement boundary condition is implied at the bonded interface.

If the stress components are known, the strains can be written down from the stress-strain relations (1.57) and these in turn can be expressed in terms of displacement gradients through (1.34). The problem is therefore reduced to the integration of these gradients to recover the displacement components.

The method is most easily demonstrated by examples, of which we shall give two — one in rectangular and one in polar coördinates.

## 9.1 The cantilever with an end load

We first consider the cantilever beam loaded by a transverse force, $F$, at the free end (Figure 5.2), for which the stress components were calculated in §5.2.1, being

$$\sigma_{xx} = \frac{3Fxy}{2b^3} \; ; \; \sigma_{xy} = \frac{3F(b^2-y^2)}{4b^3} \; ; \; \sigma_{yy} = 0 \, , \tag{9.1}$$

from (5.34–5.36).

The corresponding strain components are therefore

$$e_{xx} = \frac{\sigma_{xx}}{E} - \frac{\nu\sigma_{yy}}{E} = \frac{3Fxy}{2Eb^3} \, ; \tag{9.2}$$

$$e_{xy} = \frac{\sigma_{xy}(1+\nu)}{E} = \frac{3F(1+\nu)(b^2-y^2)}{4Eb^3} \, ; \tag{9.3}$$

$$e_{yy} = \frac{\sigma_{yy}}{E} - \frac{\nu\sigma_{xx}}{E} = -\frac{3F\nu xy}{2Eb^3} , \tag{9.4}$$

for plane stress, from (1.57).

We next make use of the strain-displacement relation to write

$$e_{xx} = \frac{\partial u_x}{\partial x} = \frac{3Fxy}{2Eb^3} , \tag{9.5}$$

which can be integrated with respect to $x$ to give

$$u_x = \frac{3Fx^2y}{4Eb^3} + f(y) , \tag{9.6}$$

where we have introduced an arbitrary function $f(y)$ of $y$, since any such function would make no contribution to the partial derivative in (9.5). A similar operation on (9.4) yields the result

$$u_y = -\frac{3F\nu xy^2}{4Eb^3} + g(x) , \tag{9.7}$$

where $g(x)$ is an arbitrary function of $x$.

To determine the two functions $f(y), g(x)$, we use the definition of shear strain and (9.3) to write

$$e_{xy} = \frac{1}{2}\left(\frac{\partial u_y}{\partial x} + \frac{\partial u_x}{\partial y}\right) = \frac{3F(1+\nu)(b^2-y^2)}{4Eb^3} . \tag{9.8}$$

Substituting for $u_x, u_y$, from (9.6, 9.7) and rearranging the terms, we obtain

$$\frac{3Fx^2}{8Eb^3} + \frac{1}{2}\frac{dg}{dx} = \frac{3F\nu y^2}{8Eb^3} - \frac{1}{2}\frac{df}{dy} + \frac{3F(1+\nu)(b^2-y^2)}{4Eb^3} . \tag{9.9}$$

Now, the left hand side of this equation is independent of $y$ and the right hand side is independent of $x$. Thus, the equation can only be satisfied for all $x, y$, if *both* sides are independent of both $x$ and $y$ — i.e. if they are equal to a constant, which we shall denote by $\frac{1}{2}C$.

Equation (9.9) can then be partitioned into the two ordinary differential equations

$$\frac{dg}{dx} = -\frac{3Fx^2}{4Eb^3} + C ; \tag{9.10}$$

$$\frac{df}{dy} = \frac{3F\nu y^2}{4Eb^3} + \frac{3F(1+\nu)(b^2-y^2)}{2Eb^3} - C , \tag{9.11}$$

which have the solution

$$g(x) = -\frac{Fx^3}{4Eb^3} + Cx + B ; \tag{9.12}$$

$$f(y) = \frac{F\nu y^3}{4Eb^3} + \frac{F(1+\nu)(3b^2y - y^3)}{2Eb^3} - Cy + A , \tag{9.13}$$

where $A, B$ are two new arbitrary constants.

The final expressions for the displacements are therefore

$$u_x = \frac{3Fx^2y}{4Eb^3} + \frac{3F(1+\nu)y}{2Eb} - \frac{F(2+\nu)y^3}{4Eb^3} + A - Cy \; ; \quad (9.14)$$

$$u_y = -\frac{3F\nu xy^2}{4Eb^3} - \frac{Fx^3}{4Eb^3} + B + Cx \; . \quad (9.15)$$

### 9.1.1 Rigid-body displacements and end conditions

The three constants $A, B, C$ define the three degrees of freedom of the cantilever as a rigid body, $A, B$ corresponding to translations in the $x$-,$y$-directions respectively and $C$ to a small anticlockwise rotation about the origin.

These rigid-body terms always arise in the integration of strains to determine displacements and they reflect the fact that a complete knowledge of the stresses and hence the strains throughout the body is sufficient to determine its deformed shape, but not its location in space.

As in Strength of Materials, the rigid-body displacements can be determined from appropriate information about the way in which the structure is supported. Since the cantilever is built-in at $x = a$, we would ideally like to specify

$$u_x = u_y = 0 \; ; \; x = a \, , \; -b < y < b \, , \quad (9.16)$$

but the three constants in (9.14, 9.15) do not give us sufficient freedom to satisfy such a strong boundary condition. We should anticipate this deficiency, since the complete solution procedure only permitted us to satisfy weak boundary conditions on the ends of the beam and (9.16), in addition to locating the beam in space, implies a pointwise traction distribution sufficient to keep the end plane and unstretched.

Many authors adapt the Strength of Materials support conditions for a cantilever by demanding that the mid-point $(a, 0)$ of the end have zero displacement and the axis of the beam ($y = 0$) have zero slope at that point. This leads to the three conditions

$$u_x = u_y = 0 \; ; \; \frac{\partial u_y}{\partial x} = 0 \; ; \; x = a \, , \; y = 0 \quad (9.17)$$

and yields the values

$$A = 0 \; ; \; B = -\frac{Fa^3}{2Eb^3} \; ; \; C = \frac{3Fa^2}{4Eb^3} \, , \quad (9.18)$$

when (9.14, 9.15) are substituted in (9.17).

However, the displacement $u_x$ at $x = a$ corresponds to the deformed shape of Figure 9.1($a$). This might be an appropriate end condition if the cantilever is supported in a horizontal groove, but other built-in support conditions might approximate more closely to the configuration of Figure 9.1($b$), where the third of conditions (9.17) is replaced by

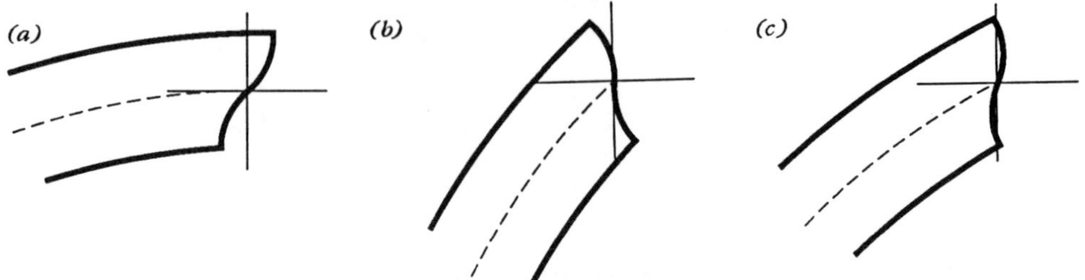


Figure 9.1: End conditions for the cantilever

$$\frac{\partial u_x}{\partial y} = 0 \ ; \ x = a \, , \, y = 0 \, . \tag{9.19}$$

This modified boundary condition leads to the values

$$A = 0 \ ; \ B = -\frac{Fa^3}{2Eb^3}\left(1 + 3(1+\nu)\frac{b^2}{a^2}\right) \ ; \ C = \frac{3Fa^2}{4Eb^3}\left(1 + 2(1+\nu)\frac{b^2}{a^2}\right) \tag{9.20}$$

for the rigid-body displacement coefficients.

The configurations 9.1($a, b$) are clearly extreme cases and we might anticipate that the best approximation to (9.16) would be obtained by an intermediate case such as Figure 9.1($c$). In fact, the displacement equivalent of the weak boundary conditions of Chapter 5 would be to prescribe

$$\int_{-b}^{b} u_x dy = 0 \ ; \ \int_{-b}^{b} u_y dy = 0 \ ; \ \int_{-b}^{b} y u_x dy = 0 \ ; \ x = a \, . \tag{9.21}$$

The substitution of (9.14, 9.15) into (9.21) gives a new set of rigid-body displacement coefficients, which are

$$A = 0 \, ; \ B = -\frac{Fa^3}{2Eb^3}\left(1 + \frac{(12+11\nu)}{5}\frac{b^2}{a^2}\right) \ ; \ C = \frac{3Fa^2}{4Eb^3}\left(1 + \frac{(8+9\nu)}{5}\frac{b^2}{a^2}\right) \, . \tag{9.22}$$

### 9.1.2 Deflection of the free end

The different end conditions corresponding to ($a, b, c$) of Figure 9.1 will clearly lead to different estimates for the deflection of the cantilever. We can investigate this effect by considering the displacement of the mid-point of the free end, $x = 0, y = 0$, for which

$$u_y(0,0) = B \, , \tag{9.23}$$

from equation (9.15).

Thus, the end deflection predicted with the boundary conditions of $(a, b, c)$ respectively of Figure 9.1 are

$$\begin{aligned} u_y(0,0) &= -\frac{Fa^3}{2Eb^3} && (a)\ ; \\ &= -\frac{Fa^3}{2Eb^3}\left(1+3(1+\nu)\frac{b^2}{a^2}\right) && (b)\ ; \\ &= -\frac{Fa^3}{2Eb^3}\left(1+\frac{(12+11\nu)}{5}\frac{b^2}{a^2}\right) && (c)\ , \end{aligned} \tag{9.24}$$

The first of these results $(a)$ is also that predicted by the elementary Strength of Materials solution and the second $(b)$ is that obtained when a correction is made for the 'shear deflection', based on the shear stress at the beam axis. Case $(c)$, which is intermediate between $(a)$ and $(b)$, but closer to $(b)$, is probably the closest approximation to the true built-in boundary condition (9.16).

All three expressions have the same leading term and the corrective term in $(b, c)$ will be small as long as $b \ll a$ — i.e. as long as the cantilever can reasonably be considered as a slender beam.

## 9.2 The circular hole

The procedure of §9.1 has some small but critical differences in polar coördinates, which we shall illustrate using the problem of §8.3.2, in which a circular hole perturbs a uniform shear field. The stresses are given by equations (8.40–8.42) and, in plane stress, the strains are therefore

$$e_{rr} = \frac{\sigma_{rr}}{E} - \frac{\nu\sigma_{\theta\theta}}{E} = \frac{S}{E}\left(1+\nu-\frac{4a^2}{r^2}+\frac{3(1+\nu)a^4}{r^4}\right)\sin 2\theta\ ; \tag{9.25}$$

$$e_{r\theta} = \frac{(1+\nu)\sigma_{r\theta}}{E} = \frac{S}{E}\left(1+\nu+\frac{2(1+\nu)a^2}{r^2}-\frac{3(1+\nu)a^4}{r^4}\right)\cos 2\theta\ ; \tag{9.26}$$

$$e_{\theta\theta} = \frac{\sigma_{\theta\theta}}{E} - \frac{\nu\sigma_{rr}}{E} = \frac{S}{E}\left(-1-\nu+\frac{4\nu a^2}{r^2}-\frac{3(1+\nu)a^4}{r^4}\right)\sin 2\theta\ . \tag{9.27}$$

Two of the three strain-displacement relations (8.17) contain both displacement components, so we start with the simplest relation

$$e_{rr} = \frac{\partial u_r}{\partial r}\ , \tag{9.28}$$

substitute for $e_{rr}$ from (9.25) and integrate, obtaining

$$u_r = \frac{S}{E}\left((1+\nu)r+\frac{4a^2}{r}-\frac{(1+\nu)a^4}{r^3}\right)\sin 2\theta + f(\theta)\ , \tag{9.29}$$

where $f(\theta)$ is an arbitrary function of $\theta$.

Writing the expression (8.17) for $e_{\theta\theta}$ in the form

$$\frac{\partial u_\theta}{\partial \theta} = re_{\theta\theta} - u_r \ , \tag{9.30}$$

we can substitute for $e_{\theta\theta}, u_r$ from (9.27, 9.29) respectively and integrate with respect to $\theta$, obtaining

$$u_\theta = \frac{S}{E}\left((1+\nu)r + \frac{2(1-\nu)a^2}{r} + (1+\nu)a^4r^3\right)\cos 2\theta - F(\theta) + g(r) \ , \tag{9.31}$$

where $g(r)$ is an arbitrary function of $r$ and we have written $F(\theta)$ for $\int f(\theta)d\theta$.

Finally, we substitute for $u_r, u_\theta, e_{r\theta}$ from (9.29, 9.31, 9.26) into (8.17) to obtain an equation for the arbitrary functions $F(\theta), g(r)$, which reduces to

$$F(\theta) + F''(\theta) = g(r) - g'(r) \ . \tag{9.32}$$

As in the previous example, this equation can only be satisfied for all $r, \theta$ if both sides are equal to a constant[1]. Solving the two resulting ordinary differential equations for $F(\theta), g(r)$ and substituting into equations (9.29, 9.31), remembering that $f(\theta) = F'(\theta)$ we obtain

$$u_r = \frac{S}{E}\left((1+\nu)r + \frac{4a^2}{r} - \frac{(1+\nu)a^4}{r^3}\right)\sin 2\theta + A\cos\theta + B\sin\theta \ ; \tag{9.33}$$

$$u_\theta = \frac{S}{E}\left((1+\nu)r + \frac{2(1-\nu)a^2}{r} + (1+\nu)a^4r^3\right)\cos 2\theta - A\sin\theta + B\cos\theta + Cr \ , \tag{9.34}$$

for the displacement components.

As before, the three arbitrary constants $A, B, C$ correspond to rigid-body motions, $A, B$ being translations in the $x$-,$y$-directions respectively, whilst $C$ describes a small anticlockwise rotation about the origin.

In the present problem, we might reasonably set $A, B, C$ to zero to preserve symmetry. At the hole surface, the radial displacement is then

$$u_r = \frac{4Sa\sin 2\theta}{E} \ , \tag{9.35}$$

which implies that the hole distorts into the ellipse shown in Figure 9.2.

The reader should note that, in both of the preceding examples, the third strain-displacement relation gives an equation for two arbitrary functions of one variable which can be partitioned into two ordinary differential equations. The success of the calculation depends upon this partition, which in turn is possible if and only if the strains satisfy the compatibility conditions. In the present examples, this is of course ensured through the derivation of the stress field from a biharmonic stress function.

[1] It turns out in polar coördinate problems that the value of this constant does not affect the final expressions for the displacement components.

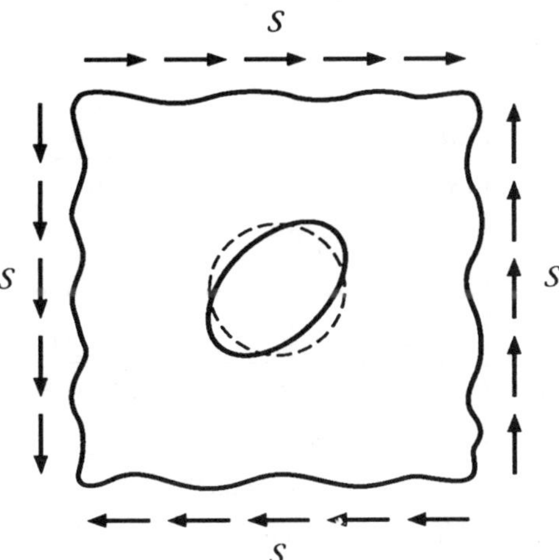


Figure 9.2: Distortion of a circular hole in a shear field

## 9.3 Displacements for the Michell solution

Displacements for all the stress functions of Table 8.1 can be obtained by the procedure of §9.2 and the results[2] are tabulated in Table 9.1. In this Table, results for plane strain can be recovered by setting $\kappa = (3-4\nu)$, whereas for plane stress, $\kappa = (3-\nu)/(1+\nu)$. We also note that it is always possible to superpose a rigid-body displacement defined by $u_r = A\cos\theta + B\sin\theta$, $u_\theta = -A\sin\theta + B\cos\theta + Cr$, where $A, B, C$ are arbitrary constants.

### 9.3.1 Equilibrium considerations

In the geometry of Figure 8.1, the boundary conditions (8.23–8.26) on the surfaces $r = a, b$ are not completely independent, since they must satisfy the condition that the body be in equilibrium. This requires that

$$\int_0^{2\pi}(F_1(\theta)\cos\theta - F_3(\theta)\sin\theta)ad\theta - \int_0^{2\pi}(F_2(\theta)\cos\theta - F_4(\theta)\sin\theta)bd\theta = 0\,; \tag{9.36}$$

$$\int_0^{2\pi}(F_1(\theta)\sin\theta + F_3(\theta)\cos\theta)ad\theta$$

[2]These results were compiled by Professor J.Dundurs of Northwestern University and his research students, and are here reprinted with his permission.

Table 9.1: The Michell solution — displacement components

| $\phi$ | $2\mu u_r$ | $2\mu u_\theta$ |
|---|---|---|
| $r^2$ | $(\kappa-1)r$ | $0$ |
| $r^2\log r$ | $(\kappa-1)r\log r - r$ | $(\kappa+1)r\theta$ |
| $\log r$ | $-1/r$ | $0$ |
| $\theta$ | $0$ | $-1/r$ |
| $r^3\cos\theta$ | $(\kappa-2)r^2\cos\theta$ | $(\kappa+2)r^2\sin\theta$ |
| $r\theta\sin\theta$ | $\frac{1}{2}[(\kappa-1)\theta\sin\theta - \cos\theta$<br>$+(\kappa+1)\log r\cos\theta]$ | $\frac{1}{2}[(\kappa-1)\theta\cos\theta - \sin\theta$<br>$-(\kappa+1)\log r\sin\theta]$ |
| $r\log r\cos\theta$ | $\frac{1}{2}[(\kappa+1)\theta\sin\theta - \cos\theta$<br>$+(\kappa-1)\log r\cos\theta]$ | $\frac{1}{2}[(\kappa+1)\theta\cos\theta - \sin\theta$<br>$-(\kappa-1)\log r\sin\theta]$ |
| $\cos\theta/r$ | $\cos\theta/r^2$ | $\sin\theta/r^2$ |
| $r^3\sin\theta$ | $(\kappa-2)r^2\sin\theta$ | $-(\kappa+2)r^2\cos\theta$ |
| $r\theta\cos\theta$ | $\frac{1}{2}[(\kappa-1)\theta\cos\theta + \sin\theta$<br>$-(\kappa+1)\log r\sin\theta]$ | $\frac{1}{2}[-(\kappa-1)\theta\sin\theta - \cos\theta$<br>$-(\kappa+1)\log r\cos\theta]$ |
| $r\log r\sin\theta$ | $\frac{1}{2}[-(\kappa+1)\theta\cos\theta - \sin\theta$<br>$+(\kappa-1)\log r\sin\theta$ | $\frac{1}{2}[(\kappa+1)\theta\sin\theta + \cos\theta$<br>$+(\kappa-1)\log r\cos\theta]$ |
| $\sin\theta/r$ | $\sin\theta/r^2$ | $-\cos\theta/r^2$ |
| $r^{n+2}\cos n\theta$ | $(\kappa-n-1)r^{n+1}\cos n\theta$ | $(\kappa+n+1)r^{n+1}\sin n\theta$ |
| $r^{-n+2}\cos n\theta$ | $(\kappa+n-1)r^{-n+1}\cos n\theta$ | $-(\kappa-n+1)r^{-n+1}\sin n\theta$ |
| $r^n\cos n\theta$ | $-nr^{n-1}\cos n\theta$ | $nr^{n-1}\sin n\theta$ |
| $r^{-n}\cos n\theta$ | $nr^{-n-1}\cos n\theta$ | $nr^{-n-1}\sin n\theta$ |
| $r^{n+2}\sin n\theta$ | $(\kappa-n-1)r^{n+1}\sin n\theta$ | $-(\kappa+n+1)r^{n+1}\cos n\theta$ |
| $r^{-n+2}\sin n\theta$ | $(\kappa+n-1)r^{-n+1}\sin n\theta$ | $(\kappa-n+1)r^{-n+1}\cos n\theta$ |
| $r^n\sin n\theta$ | $-nr^{n-1}\sin n\theta$ | $-nr^{n-1}\cos n\theta$ |
| $r^{-n}\sin n\theta$ | $nr^{-n-1}\sin n\theta$ | $-nr^{-n-1}\cos n\theta$ |

For plane strain

$$\kappa = 3 - 4\nu$$

whilst for plane stress

$$\kappa = \left(\frac{3-\nu}{1+\nu}\right)$$

$$-\int_0^{2\pi}(F_2(\theta)\sin\theta + F_4(\theta)\cos\theta)bd\theta = 0\,; \tag{9.37}$$

$$\int_0^{2\pi} F_3(\theta)a^2d\theta - \int_0^{2\pi} F_4(\theta)b^2d\theta = 0\,. \tag{9.38}$$

The orthogonality of the terms of the Fourier series ensures that these relations only concern the Fourier terms for $n = 0, 1$, but for these cases, there are only three independent algebraic equations to determine each of the corresponding sets of four constants, $A_{01}, \ldots A_{04}, A_{11}, \ldots A_{14}, B_{11}, \ldots B_{14}$.

The key to this paradox is to be seen in the displacements of Table 9.1. In the terms for $n = 0, 1$, some of the displacements include a $\theta$-multiplier. For example, we find $2\mu u_\theta = (\kappa + 1)r\theta$ for the stress function $\phi = r^2\log r$. Now the function $\theta$ is multi-valued. We can make it single-valued by defining a *principal value* — e.g., by restricting $\theta$ to the range $0<\theta<2\pi$, but we would then have a discontinuity at the line $\theta = 0, 2\pi$ which is unacceptable for the continuous body of Figure 8.1. We must therefore restrict our choice of stress functions to a set which defines a single-valued continuous displacement and it turns out that this imposes precisely one additional condition on each of the sets of four constants for $n = 0, 1$.

Notice incidentally that if the annulus were incomplete, or if it were cut along the line $\theta = 0$, the principal value of $\theta$ would then be continuous and single-valued throughout the body and the above restrictions would be removed. However, we would then also lose the equilibrium restrictions (9.36–9.38), since the body would have two new edges (e.g. $\theta = 0, 2\pi$, for the annulus with a cut on $\theta = 0$) on which there may be non-zero tractions.

We see then that the complete annulus has some complications which the incomplete annulus lacks. This arises of course because the complete annulus is *multiply-connected.* The results of this section are a direct consequence of the discussion of compatibility in multiply-connected bodies in §2.2.1. These questions and problems in which they are important will be discussed further in Chapter 13 below.

## PROBLEMS

1. A state of pure shear, $\sigma_{xy} = S$ in a large plate is perturbed by the presence of a rigid circular inclusion occupying the region $r < a$. The inclusion is perfectly bonded to the plate and is prevented from moving, so that $u_r = u_\theta = 0$ at $r = a$.

Find the complete stress field in the plate and hence determine the stress concentration factor due to the inclusion based on (i) the maximum shear stress criterion or (ii) the maximum tensile stress criterion.

Is it necessary to apply any forces or a moment to the inclusion to prevent it from moving?

2. A rigid circular inclusion of radius $a$ in a large elastic plate is subjected to a force $F$ in the $x$-direction. Find the stress field in the plate if the inclusion is perfectly bonded to the plate at $r = a$ and the stresses tend to zero as $r \to \infty$.

3. The complete annulus $b<r<a$ is subjected to a uniform internal pressure $\sigma_{rr} = -p_0$ at $r = b$, the outer surface $r = a$ being traction-free. Find the stress and displacement fields for the body and hence determine the increase in diameter of the central hole.

# Chapter 10

# CURVED BEAM PROBLEMS

If we cut the circular annulus of Figure 8.1 along two radial lines, $\theta=\alpha, \beta$, we generate a curved beam. The analysis of such beams follows that of Chapter 8, except for a few important differences — notably that (i) the ends of the beam constitute two new boundaries on which boundary conditions (usually weak boundary conditions) are to be applied and (ii) it is no longer necessary to enforce continuity of displacements (see §9.3.1), since a suitable principal value of $\theta$ can be defined which is both continuous and single-valued.

## 10.1 Loading at the ends

We first consider the case in which the curved surfaces of the beam are traction-free and the ends are loaded. As in §5.2.1, we only need to impose boundary conditions on one end — the Airy stress function formulation will ensure that the tractions on the other end have the correct force resultants to guarantee global equilibrium.

### 10.1.1 Pure bending

The simplest case is that illustrated in Figure 10.1, in which the beam $a<r<b$, $0<\theta<\alpha$ is loaded by a bending moment $M_0$, but no forces. Equilibrium considerations demand that there will be a bending moment $M_0$ and zero axial force and shear force for all $\theta$ and hence that the stress field will be independent of $\theta$. We therefore seek the solution in the axisymmetric terms in Table 8.1, using the stress function

$$\phi = Ar^2 + Br^2\log r + C\log r + D\theta \ , \tag{10.1}$$

the corresponding stress components being

$$\sigma_{rr} = 2A + B(2\log r + 1) + \frac{C}{r^2} \ ; \tag{10.2}$$

$$\sigma_{r\theta} = \frac{D}{r^2} \ ; \tag{10.3}$$

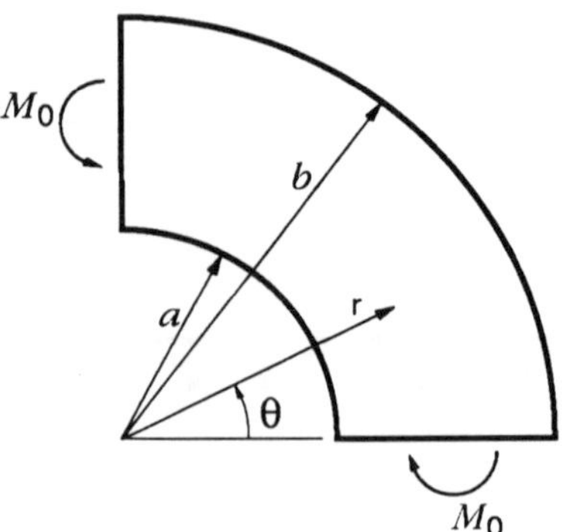


Figure 10.1: Curved beam in pure bending

$$\sigma_{\theta\theta} = 2A + B(2\log r + 3) - \frac{C}{r^2} . \quad (10.4)$$

The boundary conditions for the problem of Figure 10.1 are

$$\sigma_{rr} = 0 \ ; \ r = a, b \ ; \quad (10.5)$$
$$\sigma_{r\theta} = 0 \ ; \ r = a, b \ ; \quad (10.6)$$
$$\int_a^b \sigma_{\theta\theta} dr = 0 \ ; \ \theta = 0 \ ; \quad (10.7)$$
$$\int_a^b \sigma_{\theta r} dr = 0 \ ; \ \theta = 0 \ ; \quad (10.8)$$
$$\int_a^b \sigma_{\theta\theta} r dr = M_0 \ ; \ \theta = 0 . \quad (10.9)$$

Notice that in practice it is not necessary to impose the zero force conditions (10.7, 10.8), since if there were a non-zero force on the end $\theta = 0$, it would have to be transmitted around the beam and this would result in a non-axisymmetric stress field. Thus, our assumption of axisymmetry automatically rules out there being any force resultant.

We first impose the strong boundary conditions (10.5, 10.6), which with (10.2, 10.3) yield the four algebraic equations

$$2A + B(2\log a + 1) + \frac{C}{a^2} = 0 \ ; \quad (10.10)$$
$$2A + B(2\log b + 1) + \frac{C}{b^2} = 0 \ ; \quad (10.11)$$

$$\frac{D}{a^2} = 0 ; \quad (10.12)$$
$$\frac{D}{b^2} = 0 , \quad (10.13)$$

and an additional equation is obtained by substituting (10.4) into (10.9), giving

$$(A+B)(b^2-a^2)+B(b^2\log b-a^2\log a)-C\log(b/a)=M_0 . \quad (10.14)$$

These equations have the solution

$$A=-\frac{M}{N}(b^2-a^2+2b^2\log b-2a^2\log a) ;$$
$$B=\frac{2M}{N}(b^2-a^2) \ ; \ C=\frac{4M}{N}a^2b^2\log\frac{b}{a} \ ; \ D=0 , \quad (10.15)$$

where

$$N\equiv(b^2-a^2)^2-4a^2b^2\log^2(b/a) \quad (10.16)$$

The corresponding stress field is readily obtained by substituting these values back into equations (10.2–10.4).

The principal practical interest in this solution lies in the extent to which the classical Strength of Materials bending theory underestimates the bending stress $\sigma_{\theta\theta}$ at the inner edge, $r=a$ when $a/b$ is small. Timoshenko and Goodier give some numerical values and show that the elementary theory starts to deviate significantly from the correct result for $a/b<0.5$. The theory of curved beams — based on the application of the principle that plane sections remain plane, but allowing for the fact that the beam elements increase in length as $r$ increases due to the curvature — gives a much better agreement. However, the exact result is quite easy to compute.

Notice that an important special case is that of a 'rectangular' bend in a beam, with a small inner fillet radius (see Figure 10.2). This is reasonably modeled as the curved beam shown dotted, since there will be very little stress in the outer corner of the bend.

### 10.1.2 Force transmission

A similar type of solution can be generated for the problem in which a force is applied at the end of the beam. We consider the case illustrated in Figure 10.3, where a shear force is applied at $\theta=0$, resulting in the boundary conditions

$$\int_a^b \sigma_{\theta\theta}dr = 0 \ ; \ \theta=0 ; \quad (10.17)$$
$$\int_a^b \sigma_{\theta r}dr = F \ ; \ \theta=0 ; \quad (10.18)$$
$$\int_a^b \sigma_{\theta\theta}rdr = 0 \ ; \ \theta=0 , \quad (10.19)$$

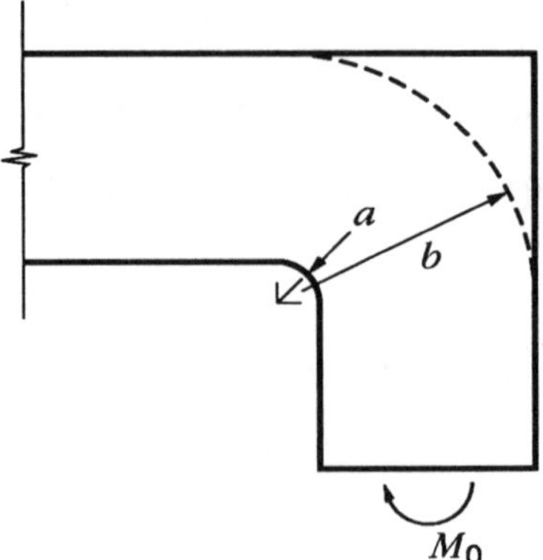


Figure 10.2: Bend in a rectangular beam with a small fillet radius

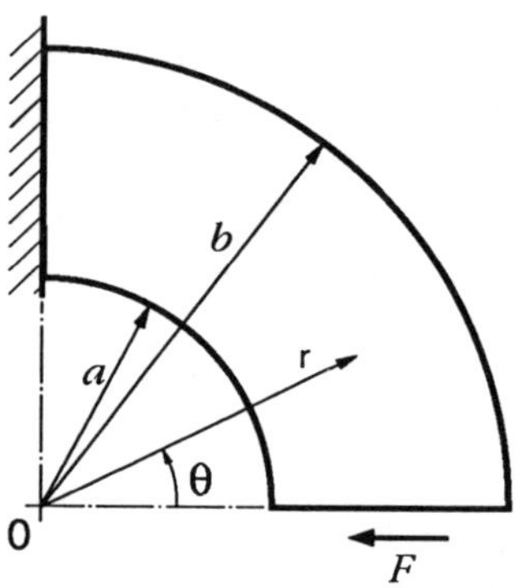


Figure 10.3: Curved beam with an end load

which replace (10.7–10.9) in the problem of §10.1.1.

Equilibrium considerations show that the shear force on any section $\theta = \alpha$ will vary as $F\cos\alpha$ and hence we seek the solution in those terms in Table 8.1 in which the shear stress $\sigma_{r\theta}$ depends on $\cos\theta$ — i.e.

$$\phi = (Ar^3 + Br^{-1} + Cr\log r)\sin\theta + Dr\theta\cos\theta \,, \tag{10.20}$$

for which the stress components are

$$\sigma_{rr} = (2Ar - 2Br^{-3} + Cr^{-1} - 2Dr^{-1})\sin\theta \,; \tag{10.21}$$
$$\sigma_{r\theta} = (-2Ar + 2Br^{-3} - Cr^{-1})\cos\theta \,; \tag{10.22}$$
$$\sigma_{\theta\theta} = (6Ar + 2Br^{-3} + Cr^{-1})\sin\theta \,. \tag{10.23}$$

The hoop stress $\sigma_{\theta\theta}$ is identically zero on $\theta = 0$ and hence the end conditions (10.17, 10.19) are satisfied identically, whilst the strong boundary conditions (10.5, 10.6) lead to the set of equations

$$2Aa - \frac{2B}{a^3} + \frac{C}{a} - \frac{2D}{a} = 0 \,; \tag{10.24}$$
$$2Ab - \frac{2B}{b^3} + \frac{C}{b} - \frac{2D}{b} = 0 \,; \tag{10.25}$$
$$2Aa - \frac{2B}{a^3} + \frac{C}{a} = 0 \,; \tag{10.26}$$
$$2Ab - \frac{2B}{b^3} + \frac{C}{b} = 0 \,. \tag{10.27}$$

The final solution that also satisfies the inhomogeneous condition (10.18) is obtained as

$$A = \frac{F}{2N_1} \;;\; B = -\frac{Fa^2b^2}{2N_1} \;;\; C = -\frac{F(a^2+b^2)}{N_1} \;;\; D = 0 \,, \tag{10.28}$$

where

$$N_1 = (a^2 - b^2) + (a^2 + b^2)\log(b/a) \,. \tag{10.29}$$

As before, the final stress field is easily obtained by back substitution into equations (10.21–10.23).

The corresponding solution for an axial force at the end $\theta = 0$ is obtained in the same way except that we interchange sine and cosine in the stress function (10.20). Alternatively, we can use the present solution and measure the angle from the end $\theta = \pi/2$ instead of from $\theta = 0$. Notice however that the axial force at $\theta = \pi/2$ must have the same line of action as the shear force at $\theta = 0$ in Figure 10.3. In other words, it must act through the origin of coördinates. If we wish to solve a problem in which an axial force is applied with some other, parallel, line of action — e.g. if the force acts through the mid-point of the beam, $r = (a+b)/2$ — we can do so by superposing an appropriate multiple of the bending solution of §10.1.1.

## 10.2 Eigenvalues and eigenfunctions

A remarkable feature of the preceding solutions is that in each case we had a set of four simultaneous homogeneous algebraic equations for four unknown constants — equations (10.10–10.13) in §10.1.1 and (10.24–10.27) in §10.1.2 — but in each we were able to obtain a non-trivial solution, because the equations were not linearly independent[1].

Suppose we were to define an *inhomogeneous* problem for the curved beam in which the curved edges $r=a,b$ were loaded by arbitrary tractions $\sigma_{rr}, \sigma_{r\theta}$. We could then decompose these tractions into appropriate Fourier series — as indicated in §8.3.1 — and use the general solution of equation (8.54) and Table 8.1. There would then be four linearly independent stress function terms for each Fourier component and hence the four boundary conditions would lead to a well-conditioned set of four algebraic equations for four arbitrary constants for each separate Fourier component.

In the special case where there are no tractions of the appropriate Fourier form on the curved surfaces, we should get four homogeneous algebraic equations and we anticipate only the trivial solution in which the corresponding four constants — and hence the stress components — are zero.

This is exactly what happens for all values of $n$ *other than* $0, 1$, but, as we have seen above, for these two special cases, there is a non-trivial solution to the homogeneous problem. In a sense, $0, 1$ are *eigenvalues* of the general Fourier problem and the corresponding stress fields are *eigenfunctions.*

## 10.3 The inhomogeneous problem

Of course, eigenvalues can be viewed from two different perspectives. They are the values of a parameter at which the homogeneous problem has a non-trivial solution — e.g. the frequency at which a dynamic system will vibrate without excitation — but they are also the values at which the *inhomogeneous* problem has an unbounded solution — excitation at the natural frequency predicts an unbounded amplitude.

In the present instance, we must therefore anticipate difficulties in the inhomogeneous problem if the boundary tractions contain Fourier terms with $n = 0$ or $1$.

### 10.3.1 Beam with sinusoidal loading

As an example, we consider the problem of Figure 10.4, in which the curved beam is subjected to a radial normal traction

$$\sigma_{rr} = S \sin\theta \;\; ; \;\; r = b \; , \tag{10.30}$$

[1]In particular, (10.12, 10.13) can both be satisfied by setting $D=0$ and (10.24–10.27) reduce to only two independent equations if $D=0$.

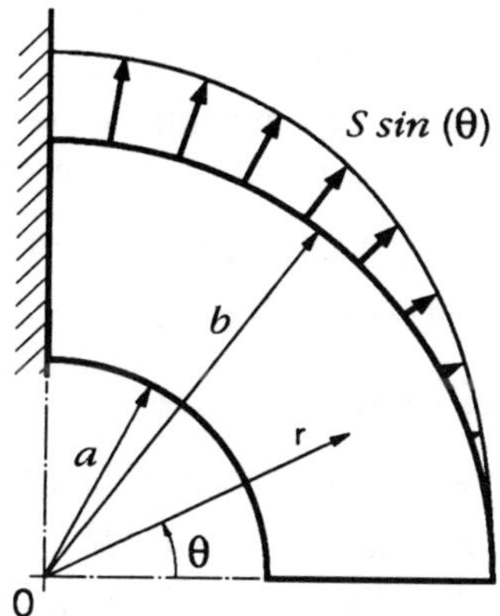


Figure 10.4: Curved beam with sinusoidal loading

the other boundary tractions being zero.

A 'naïve' approach to this problem would be to use a stress function composed of those terms in Table 8.1 for which the stress component $\sigma_{rr}$ varies with $\sin\theta$. This of course would lead to the formulation of equations (10.20–10.23) and the strong boundary conditions on the curved edges would give the four equations

$$2Aa - \frac{2B}{a^3} + \frac{C}{a} - \frac{2D}{a} = 0 \; ; \qquad (10.31)$$

$$2Ab - \frac{2B}{b^3} + \frac{C}{b} - \frac{2D}{b} = S \; ; \qquad (10.32)$$

$$2Aa - \frac{2B}{a^3} + \frac{C}{a} = 0 \; ; \qquad (10.33)$$

$$2Ab - \frac{2B}{b^3} + \frac{C}{b} = 0 \, . \qquad (10.34)$$

This looks like a well-behaved set of four equations for four unknown constants, but we know that the coefficient matrix has a zero determinant, since we successfully obtained a non-trivial solution to the corresponding homogeneous problem in §10.1.2 above.

In such cases, we must seek an additional special solution in which the dependence of the stress field on $\theta$ differs from that in the boundary conditions. Such solutions are known for a variety of geometries, but are usually presented as a *fait accompli* by the author, so it is difficult to see why they work, or how we might have been expected to determine them without guidance. In this section, we shall present the appropriate function and explain why it works, but we shall then develop a more rational approach to determining the class of solution required.

The special solution can be generated by expressing the original function (10.20) in complex variable form and then multiplying it by $\log\zeta$, where $\zeta = x + iy$ is the complex variable. Thus, (10.20) consists of a linear combination of the terms $\Im(\bar{\zeta}\zeta^2; \zeta^{-1}; \zeta\log\zeta; \bar{\zeta}\log\zeta)$, so we can generate a new biharmonic function from the terms $\Im(\bar{\zeta}\zeta^2\log\zeta; \zeta^{-1}\log\zeta; \zeta\log^2\zeta; \bar{\zeta}\log^2\zeta)$, which can be written in the form

$$\begin{aligned} \phi &= A'r^3(\log r\sin\theta + \theta\cos\theta) + B'r^{-1}(\theta\cos\theta - \log r\sin\theta) \\ & \quad C'r\log r\theta\cos\theta + D'r(\log^2 r\sin\theta - r\theta^2\sin\theta)\ . \end{aligned} \tag{10.35}$$

The corresponding stress components are

$$\begin{aligned} \sigma_{rr} &= (2A'r - 2B'r^{-3} + C'r^{-1} - 4D'r^{-1})\theta\cos\theta \\ & \quad +(2A'r\log r - A'r + 2B'r^{-3}\log r - 3B'r^{-3} \\ & \quad -2C'r^{-1}\log r + 2D'r^{-1}\log r - 2D'r^{-1})\sin\theta\ ; \end{aligned} \tag{10.36}$$

$$\begin{aligned} \sigma_{r\theta} &= (2A'r - 2B'r^{-3} + C'r^{-1})\theta\sin\theta \\ & \quad +(-2A'r\log r - 3A'r - 2B'r^{-3}\log r + 3B'r^{-3} \\ & \quad -C'r^{-1} - 2D'r^{-1}\log r)\cos\theta\ ; \end{aligned} \tag{10.37}$$

$$\begin{aligned} \sigma_{\theta\theta} &= (6A'r + 2B'r^{-3} + C'r^{-1})\theta\cos\theta \\ & \quad +(6A'r\log r + 5A'r - 2B'r^{-3}\log r + 3B'r^{-3} \\ & \quad +2D'r^{-1}\log r + 2D'r^{-1})\sin\theta\ . \end{aligned} \tag{10.38}$$

Notice in particular that these expressions contain some terms of the required form for the problem of Figure 10.4, but they also contain terms with multipliers of the form $\theta\sin\theta, \theta\cos\theta$, which are inappropriate. We therefore get four homogeneous algebraic equations for the coefficients $A', B', C', D'$ from the requirement that these inappropriate terms should vanish in the components $\sigma_{rr}, \sigma_{r\theta}$ on the boundaries $r = a, b$, i.e.

$$2A'a - \frac{2B'}{a^3} + \frac{C'}{a} - \frac{2D'}{a} = 0\ ; \tag{10.39}$$

$$2A'b - \frac{2B'}{b^3} + \frac{C'}{b} - \frac{2D'}{b} = 0\ ; \tag{10.40}$$

$$2A'a - \frac{2B'}{a^3} + \frac{C'}{a} = 0\ ; \tag{10.41}$$

$$2A'b - \frac{2B'}{b^3} + \frac{C'}{b} = 0\ . \tag{10.42}$$

Clearly these equations are identical with (10.24–10.27) and are not linearly independent. There is therefore a non-trivial solution to (10.39–10.42), which leaves us with a stress function that can supplement that of equation (10.20) to make the problem well-posed.

From here on, the solution is algebraically tedious, but routine. Adding the two stress functions (10.20, 10.35), we obtain a function with 8 unknown constants which is required to satisfy 9 boundary conditions comprising (i) equations (10.39–10.42), (ii) equations (10.31–10.34) modified to include the $\sin\theta, \cos\theta$ terms from equations (10.36, 10.37) respectively and (iii) the weak traction-free condition

$$\int_a^b \sigma_{\theta r} = 0 \ \ ; \ \ \theta = 0 \tag{10.43}$$

on the end of the beam. Since two of equations (10.39–10.42) are not independent, the system reduces to a set of 8 equations for 8 constants whose solution is

$$\begin{aligned}
A' &= \frac{Sb}{4N}; \ \ B' = -\frac{Sa^2b^3}{4N}; \ \ C' = -\frac{Sb(a^2+b^2)}{2N}; \ \ D' = 0 \, ; \\
A &= \frac{Sb}{8N^2}[2(b^2-a^2) + (3a^2+b^2)\log(b) + (3b^2+a^2)\log(a) \\
&\quad -2\{a^2\log(a) + b^2\log(b)\}\log(b/a)] \, ; \\
B &= \frac{Sa^2b^3}{8N^2}[-2(b^2-a^2) + (3b^2+a^2)\log(b) - (3a^2+b^2)\log(a) \\
&\quad -2\{a^2\log(b) + b^2\log(a)\}\log(b/a)] \, ; \\
C &= \frac{Sb}{4N^2}[2(b^4+a^4)\log(b/a) - (b^4-a^4)] \, ; \\
D &= \frac{Sb}{4N}[(b^2-a^2) + 2(b^2+a^2)\log(a)] \, .
\end{aligned} \tag{10.44}$$

Of course, it is not fortuitous that the stress function (10.35) leads to a set of equations (10.39–10.42) identical with (10.24–10.27). When we differentiate the function $f(\zeta,\bar{\zeta})\log(\zeta)$ by parts to determine the stresses, the extra multiplier $\log(\zeta)$ is only preserved in those terms where it is not differentiated and hence in which all the differential operations are performed on $f(\zeta,\bar{\zeta})$. But these operations on $f(\zeta,\bar{\zeta})$ are precisely those leading to the stresses in the original solution and hence to equations (10.24–10.27). The reader will notice a parallel here with the procedure for determining the general solution of a differential equation with repeated differential multipliers and with that for dealing with degeneracy of solutions discussed in §8.3.3.

### 10.3.2 The near-singular problem

Suppose we next consider a more general version of the problem of Figure 10.4 in which the inhomogeneous boundary condition (10.30) is replaced by

$$\sigma_{rr} = S\sin(\lambda\theta) \quad ; \quad r = b \, , \tag{10.45}$$

where $\lambda$ is a constant. In the special case where $\lambda = 1$, this problem reduces to that of §10.3.1. For all other values (excluding $\lambda = 0$), we can use the stress function

$$\phi = [Ar^{\lambda+2} + Br^{\lambda} + Cr^{-\lambda} + Dr^{-\lambda+2}]\sin(\lambda\theta) \, , \tag{10.46}$$

with stress components

$$\begin{aligned}
\sigma_{rr} &= -[A(\lambda-2)(\lambda+1)r^{\lambda}+B\lambda(\lambda-1)r^{\lambda-2} \\
&\quad +C\lambda(\lambda+1)r^{-\lambda-2}+D(\lambda+2)(\lambda-1)r^{-\lambda}]\sin(\lambda\theta)\;; \\
\sigma_{r\theta} &= [-A\lambda(\lambda+1)r^{\lambda}-B\lambda(\lambda-1)r^{\lambda-2} \\
&\quad +C\lambda(\lambda+1)r^{-\lambda-2}+D\lambda(\lambda-1)r^{-\lambda}]\cos(\lambda\theta)\;; \\
\sigma_{\theta\theta} &= [A(\lambda+1)(\lambda+2)r^{\lambda}+B\lambda(\lambda-1)r^{\lambda-2} \\
&\quad +C\lambda(\lambda+1)r^{-\lambda-2}+D(\lambda-1)(\lambda-2)r^{-\lambda}]\sin(\lambda\theta)\;.
\end{aligned} \tag{10.47}$$

The boundary conditions on the curved edges lead to the four equations

$$\begin{aligned}
A(\lambda-2)(\lambda+1)a^{\lambda}+B\lambda(\lambda-1)a^{\lambda-2}+C\lambda(\lambda+1)a^{-\lambda-2}+D(\lambda+2)(\lambda-1)a^{-\lambda} &= 0\;; \\
A(\lambda-2)(\lambda+1)b^{\lambda}+B\lambda(\lambda-1)b^{\lambda-2}+C\lambda(\lambda+1)b^{-\lambda-2}+D(\lambda+2)(\lambda-1)b^{-\lambda} &= -S\;; \\
-A\lambda(\lambda+1)a^{\lambda}-B\lambda(\lambda-1)a^{\lambda-2}+C\lambda(\lambda+1)a^{-\lambda-2}+D\lambda(\lambda-1)a^{-\lambda} &= 0\;; \\
-A\lambda(\lambda+1)b^{\lambda}-B\lambda(\lambda-1)b^{\lambda-2}+C\lambda(\lambda+1)b^{-\lambda-2}+D\lambda(\lambda-1)b^{-\lambda} &= 0\;,
\end{aligned} \tag{10.48}$$

which have the solution

$$\begin{aligned}
A &= \frac{Sb^{\lambda}}{2(\lambda+1)M}[(\lambda+1)f(\lambda)-\lambda b^2 f(\lambda-1)]\;; \\
B &= \frac{-Sb^{\lambda}}{2(\lambda-1)M}[\lambda f(\lambda+1)-(\lambda-1)b^2 f(\lambda)]\;; \\
C &= \frac{-Sa^{2\lambda}b^{\lambda+2}}{2(\lambda+1)M}[\lambda b^{2\lambda-2}f(1)+f(\lambda)]\;; \\
D &= \frac{Sa^{2\lambda-2}b^{\lambda}}{2(\lambda-1)M}[\lambda b^{2\lambda}f(1)+a^2 f(\lambda)]\;,
\end{aligned} \tag{10.49}$$

where

$$M = \lambda^2 f(\lambda-1)f(\lambda+1)-(\lambda^2-1)f(\lambda)^2\;; \tag{10.50}$$

$$f(p) = b^{2p}-a^{2p}\;. \tag{10.51}$$

On casual inspection, it seems that this problem behaves rather remarkably as $\lambda$ passes through unity. The stress field is sinusoidal for all $\lambda\neq 1$ and increases without limit as $\lambda$ approaches unity from either side, since $M=0$ when $\lambda=1$. (Notice that an additional singularity is introduced through the factor $(\lambda-1)$ in the denominator of the coefficients $B, D$.) However, when $\lambda$ is exactly equal to unity, the problem has the bounded solution derived in §10.3.1, in which the stress field is not sinusoidal.

However, we shall demonstrate that the solution is not as discontinuous as it looks. The stress field obtained by substituting (10.49) into (10.46) is not a complete solution to the problem since the end condition (10.43) is not satisfied. The solution from (10.46) will generally involve a non-zero force on the end, given by

$$\begin{aligned} F &= \int_a^b \sigma_{r\theta} dr \\ &= -\lambda \left[ Af\left(\frac{\lambda+1}{2}\right) + Bf\left(\frac{\lambda-1}{2}\right) + Cf\left(\frac{-\lambda-1}{2}\right) + Df\left(\frac{-\lambda+1}{2}\right) \right] . \end{aligned} \tag{10.52}$$

To restore the no-traction condition on the end (in the weak sense of zero force resultant) we must subtract the solution of the homogeneous problem of §10.1.2, with $F$ given by (10.52). The final solution therefore involves the superposition of the stress functions (10.20) and (10.46). As $\lambda$ approaches unity, both components of the solution increase without limit, since (10.52) contains the singular terms (10.49) but they also tend to assume the same form since, for example, $r^{2+\lambda}\sin(\lambda\theta)$ approaches $r^3\sin(\theta)$. In the limit, the corresponding term takes the form

$$\lim_{\lambda\to 1} \frac{A_1(\lambda) r^{2+\lambda}\sin(\lambda\theta) - A_2(\lambda) r^3 \sin(\theta)}{M(\lambda)} , \tag{10.53}$$

where

$$A_1(\lambda) = \frac{Sb^\lambda}{2(\lambda+1)}[(\lambda+1)f(\lambda) - \lambda b^2 f(\lambda-1)] ; \tag{10.54}$$

$$A_2(\lambda) = \frac{F(\lambda)M(\lambda)}{2N} . \tag{10.55}$$

Setting $\lambda=1$ in equations (10.54, 10.55), we find

$$A_1(1) = A_2(1) = \frac{Sb^2(b^2-a^2)}{2} . \tag{10.56}$$

The zero in $M(\lambda)$ is therefore cancelled and (10.53) has a bounded limit which can be recovered by using L'Hôpital's rule. The resulting algebra is routine but lengthy and will not be reproduced here, but it is clear that in addition to a term proportional to $r^3\sin(\theta)$, the differentiation of the numerator in (10.53) will generate a term proportional to

$$\frac{\partial}{\partial\lambda} r^{2+\lambda}\sin(\lambda\theta) = r^{2+\lambda}[\log(r)\sin(\lambda\theta) + \theta\cos(\lambda\theta)] , \tag{10.57}$$

which in the limit $\lambda=1$ reduces to $r^3[\log(r)\sin(\theta) + \theta\cos(\theta)]$. The coefficient of this term will be

$$A' = \frac{A_1(1)}{M'(1)} = \frac{Sb}{4N} , \tag{10.58}$$

agreeing with the corresponding coefficient, $A'$, in the special solution of §10.3.1 (see equations (10.44)).

A similar limiting process yields the form and coefficients of the remaining 7 stress functions in the solution of §10.3.1. In particular, we note that the coefficients $B, D$ in equations (10.49) have a second singular term in the denominator and hence L'Hôpital's rule has to be applied twice, leading to stress functions of the form of the last two terms in equation (10.35).

Thus, we find that the solution to the more general problem of equation (10.45) includes that of §10.3.1 as a limiting case and there are no values of $\lambda$ for which the stress field is singular. The limiting process also shows us a more general way of obtaining the special stress functions required for the problem of §10.3.1. We simply take the stress function (10.46), which degenerates when $\lambda = 1$, and differentiate it with respect to the parameter, $\lambda$. Since the last two terms of (10.20) are already of this differentiated form, a second differentiation is required to generate the last two terms of (10.35).

## 10.4 Some general considerations

Similar examples could be found for other geometries in the two-dimensional theory of elasticity. To fix ideas, suppose we consider the body $a < \xi < b$, $c < \eta < d$ in the general system of curvilinear coördinates, $\xi$, $\eta$. In general we might anticipate a class of separated variable solutions of the form

$$\phi = f_\lambda(\xi) g_\lambda(\eta) , \tag{10.59}$$

where $\lambda$ is a parameter. An important physical problem is that in which the boundaries $\eta = a, b$ are traction-free and the tractions on the remaining boundaries have a non-zero force or moment resultant. In such cases, we can generally use dimensional and/or equilibrium arguments to determine the form of the stress variation in the $\xi$-direction and hence the appropriate function $f_\lambda^*(\xi)$. The biharmonic equation will then reduce to a fourth order ordinary differential equation for $g_\lambda^*(\eta)$, with four linearly independent solutions. Enforcement of the traction-free boundary conditions will give a set of four homogeneous equations for the unknown multipliers of these four solutions.

If the solution is to be possible, the equations must have a non-trivial solution, implying that the matrix of coefficients is singular. It therefore follows that the corresponding *homogeneous* problem cannot be solved by the stress function of equation (10.59) if the tractions on $\eta = a, b$ vary with $f_\lambda^*(\xi)$ — i.e. in the same way as the stresses in the homogeneous (end loaded) problem. However, special stress functions appropriate to this limiting case can be obtained by differentiating (10.59) with respect to the parameter $\lambda$.

The special solution for $\lambda = \lambda^*$ is not qualitatively different from that at more general values of $\lambda$, but appears as a regular limit *once appropriate boundary conditions are imposed on the edges* $\xi = c, d$.

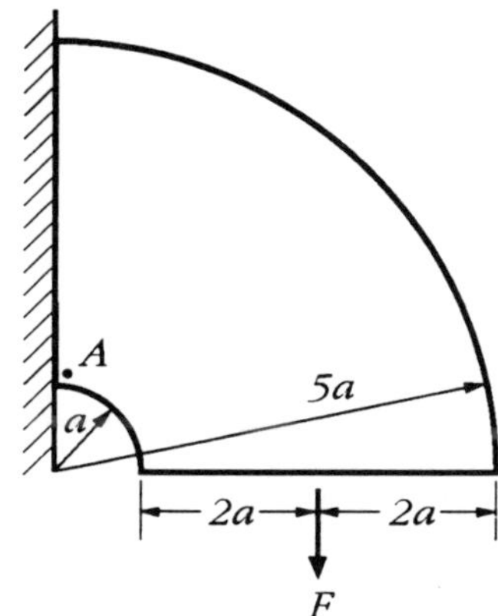


Figure 10.5: Curved beam loaded by a central force

Other simple examples in the two-dimensional theory of elasticity include the curved bar in bending (§10.1.1) and the wedge with traction-free faces (§11.2). Similar arguments can also be applied to axisymmetric problems — e.g. to problems of the cone or cylinder with traction-free curved surfaces.

### 10.4.1 Conclusions

1. Whenever we find a classical solution in which there are two traction-free boundaries, we can formulate a related inhomogeneous problem for which the equation system resulting from these boundary conditions will be singular.
2. Special solutions for these cases can be obtained by parametric differentiation of more general solutions for the same geometry.
3. Alternatively, the special case can be recovered as a limit as the parameter tends to its eigenvalue, provided that the remaining boundary conditions are satisfied in an appropriate sense (e.g. in the weak sense of force resultants) before the limit is taken.

**PROBLEM**

1. A curved beam of inner radius $a$ and outer radius $5a$ is subjected to a force $F$ at its end as shown in Figure 10.5. The line of action of the force passes through the mid-point of the beam.

By superposing the solutions for the curved beam subjected to an end force and to pure bending respectively, find the hoop stress $\sigma_{\theta\theta}$ at the point A$(a, \frac{\pi}{2})$ and compare it with the value predicted by the elementary Strength of Materials bending theory.

# Chapter 11

# WEDGE PROBLEMS

In this chapter, we shall consider a class of problems for the semi-infinite wedge defined by the lines $\alpha<\theta<\beta$, illustrated in Figure 11.1.

## 11.1 Power law tractions

We first consider the case in which the tractions on the boundaries vary with $r^n$, in which case equations (8.10, 8.11) suggest that the required stress function will be of the form

$$\phi = r^{n+2} f(\theta) \; . \tag{11.1}$$

The function $f(\theta)$ can be found by substituting (11.1) into the biharmonic equation (8.13), giving the ordinary differential equation

$$\left(\frac{d^2}{d\theta^2} + (n+2)^2\right)\left(\frac{d^2}{d\theta^2} + n^2\right) f = 0 \; . \tag{11.2}$$

For $n \neq 0, -2$, the four solutions of this equation define the stress function

$$\phi = r^{n+2}(A_1 \cos(n+2)\theta + A_2 \sin(n+2)\theta + A_3 \cos n\theta + A_4 \sin n\theta) \tag{11.3}$$

and the corresponding stress and displacement components can be taken from Tables 8.1, 9.1 respectively, with appropriate values of $n$.

### 11.1.1 Uniform tractions

Timoshenko and Goodier, in a footnote in §45, state that the uniform terms in the four boundary tractions are not independent and that only three of them may be prescribed. This is based on the argument that in a more general non-singular series solution, the uniform terms represent the stress in the corner, $r = 0$ and the stress at a point only involves three independent components in two-dimensions.

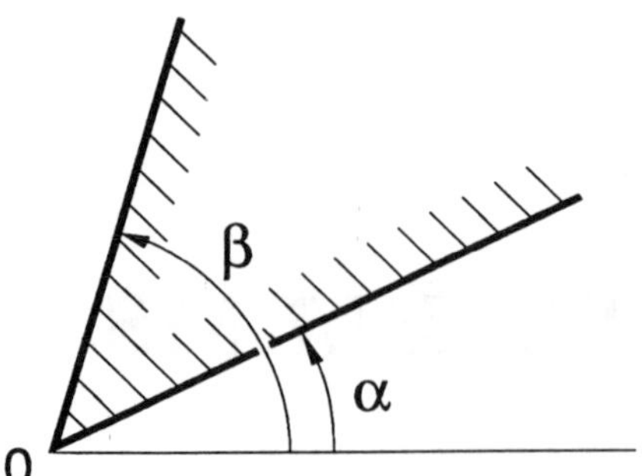


Figure 11.1: The semi-infinite wedge

However, the differential equation (11.2) has four independent non-trivial solutions for $n = 0$, corresponding to the stress function[1]

$$\phi = r^2(A_1 \cos 2\theta + A_2 \sin 2\theta + A_3 + A_4\theta) . \tag{11.4}$$

The corresponding stress components are

$$\sigma_{rr} = -2A_1 \cos 2\theta - 2A_2 \sin 2\theta + 2A_3 + 2A_4\theta ; \tag{11.5}$$
$$\sigma_{r\theta} = 2A_1 \sin 2\theta - 2A_2 \cos 2\theta - A_4 ; \tag{11.6}$$
$$\sigma_{\theta\theta} = 2A_1 \cos 2\theta + 2A_2 \sin 2\theta + 2A_3 + 2A_4\theta , \tag{11.7}$$

and, in general, these permit us to solve the problem of the wedge with any combination of four independent traction components on the faces. Timoshenko's assertion is therefore incorrect.

### Uniform shear on a right-angle wedge

To explore this paradox further, it is convenient to consider the problem of Figure 11.2, in which the right-angle corner $x>0,\ y>0$ is subjected to uniform shear on one face defined by

$$\sigma_{xy} = S \ ; \ \sigma_{xx} = 0 \ ; \ x = 0 \ ; \tag{11.8}$$
$$\sigma_{yy} = 0 \ ; \ \sigma_{yx} = 0 \ ; \ y = 0 \ . \tag{11.9}$$

[1] Another way of generating this special solution is to note that the term in (11.2) which degenerates when $n = 0$ is $A_4 \sin n\theta$. Following §8.3.3, we can find the special solution by differentiating with respect to $n$, before proceeding to the limit $n \to 0$, giving the result $A_4\theta$.

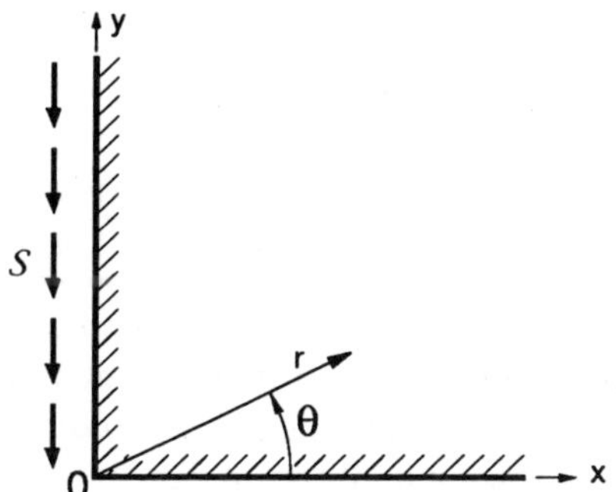


Figure 11.2: Uniform shear on a right-angle wedge

Timoshenko's footnote would seem to argue that this is not a well-posed problem, since it implies that $\sigma_{xy} \neq \sigma_{yx}$ at $x=y=0$.

Casting the problem in polar coördinates, $(r, \theta)$, the boundary conditions become

$$\sigma_{\theta r} = -S \; ; \; \sigma_{\theta\theta} = 0 \; ; \; \theta = \frac{\pi}{2} \; ; \tag{11.10}$$

$$\sigma_{\theta r} = 0 \; ; \; \sigma_{\theta\theta} = 0 \; ; \; \theta = 0 \, . \tag{11.11}$$

Using the stress field of equations (11.5–11.7), this reduces to the system of four algebraic equations

$$2A_2 - A_4 = -S \; ; \tag{11.12}$$

$$-2A_1 + 2A_3 + A_4\pi = 0 \; ; \tag{11.13}$$

$$-2A_2 - A_4 = 0 \; ; \tag{11.14}$$

$$2A_1 + 2A_3 = 0 \, , \tag{11.15}$$

with solution

$$A_1 = \frac{\pi S}{8}; \; A_2 = -\frac{S}{4}; \; A_3 = -\frac{S}{4}; \; A_4 = \frac{S}{2} \, , \tag{11.16}$$

giving the stress function

$$\phi = S\left(\frac{\pi r^2 \cos 2\theta}{8} - \frac{r^2 \sin 2\theta}{4} - \frac{\pi r^2}{8} + \frac{r^2\theta}{2}\right) \, . \tag{11.17}$$

Timoshenko's 'paradox' is associated with the apparent inconsistency in the stress components $\sigma_{xy}, \sigma_{yx}$ at $x=y=0$, so it is convenient to recast the stress function in

Cartesian coördinates as

$$\phi = S\left(\frac{\pi}{8}(x^2-y^2) - \frac{xy}{2} - \frac{\pi(x^2+y^2)}{8} + \frac{(x^2+y^2)}{2}\arctan\frac{y}{x}\right) . \tag{11.18}$$

We then determine the shear stress $\sigma_{xy}$ as

$$\sigma_{xy} = -\frac{\partial^2\phi}{\partial x\partial y} = \frac{Sy^2}{x^2+y^2} . \tag{11.19}$$

It is easily verified that this expression tends to zero for $y=0$ and to $S$ for $x=0$, except of course that it is indeterminate at the point $x=y=0$. The first three terms in the stress function (11.18) are second degree polynomials in $x, y$ and therefore define a uniform state of stress throughout the wedge, but the fourth term, resulting from $A_4r^2\theta$ in (11.4), defines stresses which are uniform along any line $\theta$ =constant, but which vary with $\theta$. Thus, any stress component $\sigma$ is a bounded function of $\theta$ only. It follows that the corner of the wedge is not a singular point, but the stress *gradients* in the $\theta$-direction $\left(\frac{1}{r}\frac{\partial\sigma}{\partial\theta}\right)$ increase with $r^{-1}$ as $r\to 0$. This arises because lines of constant $\theta$ meet at $r=0$.

### 11.1.2 More general uniform loading

More generally, there is no inconsistency in specifying four independent uniform traction components on the faces of a wedge, though values which are inconsistent with the conditions for stress at a point will give infinite stress gradients in the corner.

For the general case, it is convenient to choose a coördinate system symmetric with respect to the two faces of the wedge, which is then bounded by the lines $\theta=\pm\alpha$.

Writing the boundary conditions

$$\sigma_{\theta r} = T_1 \ ; \ \theta = \alpha \ ; \tag{11.20}$$
$$\sigma_{\theta r} = T_2 \ ; \ \theta = -\alpha \ ; \tag{11.21}$$
$$\sigma_{\theta\theta} = N_1 \ ; \ \theta = \alpha \ ; \tag{11.22}$$
$$\sigma_{\theta\theta} = N_2 \ ; \ \theta = -\alpha \ , \tag{11.23}$$

and using the solution (11.5–11.7), we obtain the four algebraic equations

$$2A_1\sin 2\alpha - 2A_2\cos 2\alpha - A_4 = T_1 \ ; \tag{11.24}$$
$$-2A_1\sin 2\alpha - 2A_2\cos 2\alpha - A_4 = T_2 \ ; \tag{11.25}$$
$$2A_1\cos 2\alpha + 2A_2\sin 2\alpha + 2A_3 + 2A_4\alpha = N_1 \ ; \tag{11.26}$$
$$2A_1\cos 2\alpha - 2A_2\sin 2\alpha + 2A_3 - 2A_4\alpha = N_2 \ . \tag{11.27}$$

As in §6.2.2, we can exploit the symmetry of the problem to partition the coefficient matrix by taking sums and differences of these equations in pairs, obtaining the simpler set

$$-4A_2\cos 2\alpha - 2A_4 = T_1 + T_2 \; ; \tag{11.28}$$

$$4A_1\sin 2\alpha = T_1 - T_2 \; ; \tag{11.29}$$

$$4A_1\cos 2\alpha + 4A_3 = N_1 + N_2 \; ; \tag{11.30}$$

$$4A_2\sin 2\alpha + 4A_4\alpha = N_1 - N_2 \; , \tag{11.31}$$

where we note that the terms involving $A_1, A_3$ correspond to a symmetric stress field and those involving $A_2, A_4$ to an antisymmetric field.

### 11.1.3 Eigenvalues for the wedge angle

The solution of these equations is routine, but we note that there are two eigenvalues for the wedge angle $2\alpha$ at which the matrix of coefficients is singular.

In particular, the solution for the symmetric terms (equations (11.29, 11.30)) is singular if

$$\sin 2\alpha = 0 \; , \tag{11.32}$$

i.e. at $2\alpha = 180^o$ or $360^o$, whilst the antisymmetric terms are singular when

$$\tan 2\alpha = 2\alpha \; , \tag{11.33}$$

which occurs at $2\alpha = 257.4^o$. The $180^o$ wedge is the half-plane $x > 0$, whilst the $257.4^o$ wedge is a re-entrant corner.

As in the problem of §10.3, special solutions are needed for the inhomogeneous problem if the wedge has one of these two special angles. As before, they can be obtained by differentiating the general solution (11.3) with respect to $n$ before setting $n = 0$, leading to the new terms, $r^2(\log r\cos 2\theta - \theta\sin 2\theta)$ and $r^2\log r$. A particular problem for which these solutions are required is that of the half-plane subjected to a uniform shear traction over half of its boundary (see Problem 11.1, below).

#### The homogeneous solution

For wedges of $180^o$ and $257.4^o$, the *homogeneous* problem, where $N_1 = N_2 = T_1 = T_2 = 0$ has a non-trivial solution. For the $180^o$ wedge (i.e. the half-plane $x > 0$), this solution can be seen by inspection to be one of uniaxial tension, $\sigma_{yy} = S$, which is non-trivial, but involves no tractions on the free surface $x = 0$. The same state of stress is also a non-trivial solution for the $360^o$ wedge, which corresponds to a semi-infinite crack (see Figure 11.6 and §11.2.3 below).

## 11.2 Williams' asymptotic method

Figure 11.3 shows a body with a notch, loaded by tractions on the remote boundaries. Intuitively we anticipate a stress concentration at the notch and we shall show in

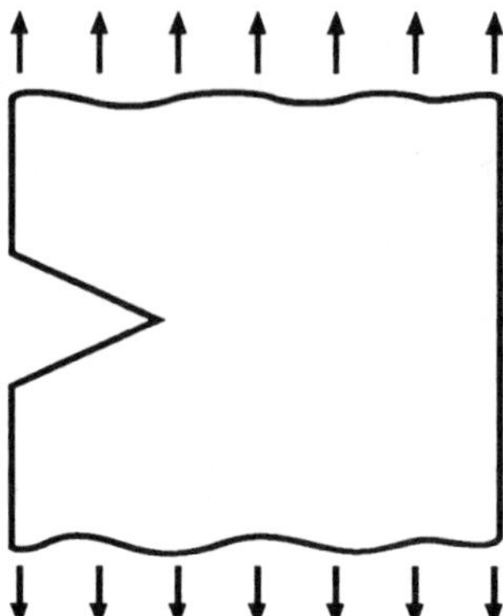

Figure 11.3: The notched bar in tension

this section that the stress field there is generally singular — i.e. that the stress components tend to infinity as we approach the sharp corner of the notch.

Williams[2] developed a method of exploring the nature of the stress field near this singularity by defining a set of polar coördinates centered on the corner and expanding the stress field as an asymptotic series in powers of $r$.

## 11.2.1 Acceptable singularities

Before developing the asymptotic solution in detail, we must first address the question as to whether singular stress fields are ever acceptable in elasticity problems.

### Engineering criteria

This question can be approached from various points of view. The engineer is usually inclined to argue that no real materials are capable of sustaining an infinite stress and hence any situation in which such a stress is predicted by an elastic analysis will in practice lead to yielding or some other kind of failure.

We also note that stress singularities are always associated with discontinuities in the geometry or the boundary conditions — for example, a sharp corner, as in the present instance, or a concentrated (delta function) load. In practice, there are no sharp corners and loads can never be perfectly concentrated. If we are to give any meaning to these solutions, they must be considered as limits of more practical

[2]M.L.Williams, Stress singularities resulting from various boundary conditions in angular corners of plates in extension, ASME J.Appl.Mech., Vol. 19 (1952), 526-528.

situations, such as a corner with a very small fillet radius or a load applied over a very small region of the boundary.

Yet another practical limitation is that of the continuum theory. Real materials are not continua – they have atomic structure and often a larger scale granular structure as well. Thus, it doesn't make much practical sense to talk about the value of quantities nearer than (say) one atomic distance to a corner, since the theory is going to break down there anyway.

Of course, we find this kind of argument whenever we try to idealize a physical system, which means essentially whenever we try to describe its behaviour in mathematical terms. Because there are so many idealizations or approximations involved in the modeling process, it is never certain which one is responsible when mathematical difficulties are encountered in a problem and often there are several ways of reformulating the problem to introduce more physical reality and avoid the difficulty.

### Mathematical criteria

A totally different approach to the question — more favoured by mathematicians — is to make choices based on questions of uniqueness, convergence and existence of solutions, rather than on the physical grounds discussed above. In terms of the functional analysis, we choose to locate the theory of elasticity in a functional space which guarantees that problems are mathematically well-posed. A functional space here denotes a class of functions in which the solution is to be sought, and generally this involves some statement about the strength of acceptable singularities in the solution.

In this context, the engineer's objections outlined above can be addressed by limiting arguments. It is not logically impossible that a material should have a yield strength of any arbitrarily large (but finite) magnitude, that the radius of a corner should be arbitrarily small, but finite, etc. Solutions of problems involving singularities are then to be conceived as the result of allowing these quantities to tend to their limits. Generally the difference between the real problem and the limiting case will be only localized.

In regions of a body or its boundary where no concentrated traction etc. is applied, we adopt the criterion that the only acceptable singularities are those for which the total strain energy in a small region surrounding the singular point vanishes as the size of that region tends to zero.

Thus, if the stresses and hence the strains vary with $r^a$ as we approach the point $r = 0$, the strain energy in a two-dimensional problem will be an integral of the form

$$U = \int_0^{2\pi} \int_0^r \sigma e r dr d\theta = C \int_0^r r^{2a+1} dr \;, \tag{11.34}$$

where $C$ is a constant which depends on the elastic constants and the nature of the stress variation with $\theta$. This integral is bounded if $a > -1$ and otherwise unbounded.

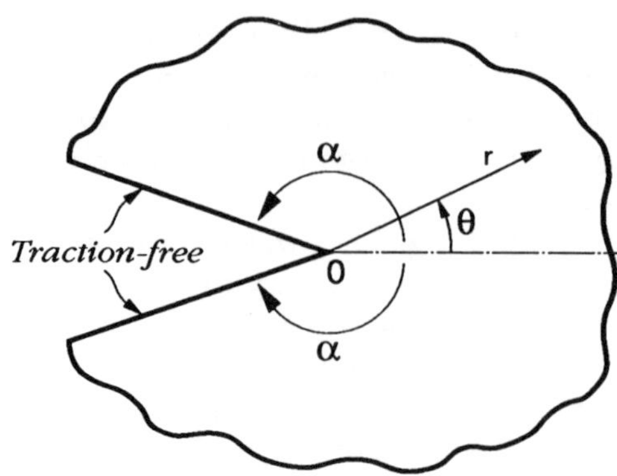


Figure 11.4: The semi-infinite notch

Thus, singular stress fields are acceptable if and only if the exponent on the stress components exceeds $-1$.

Concentrated forces and dislocations[3] involve stress singularities with exponent $a = -1$ and are therefore excluded by this criterion, since the integral in (10.34) would lead to a logarithm, which is unbounded at $r = 0$. However, these solutions can be permitted on the understanding that they are really idealizations of explicitly specified distributions. In this case, it is important to recognize that the resulting solutions are not meaningful at or near the singular point. We also use these and other singular solutions as Green's functions — i.e. as the kernels of convolution integrals to define the solution of a non-singular problem. This technique will be illustrated in Chapters 12 and 13, below.

### 11.2.2 Eigenfunction expansion

We are concerned only with the stress components in the notch at very small values of $r$ and hence we imagine looking at the corner through a strong microscope, so that we see the wedge of Figure 11.4. The magnification is so large that the other surfaces of the body, including the loaded boundaries, appear far enough away for us to treat the wedge as infinite, with 'loading at infinity'.

The stress field at the notch is of course a complicated function of $r, \theta$, but as in Chapter 6, we seek to expand it as a series of separated variable terms, each of which satisfies the traction-free boundary conditions on the wedge faces. The appropriate

[3]See Chapter 13 below.

separated variable form is of course that given by equation (11.3), except that we increase the generality of the solution by relaxing the requirement that the exponent $n$ be an integer — indeed, we shall find that most of the required solutions have *complex* exponents.

Following Williams' notation, we replace $n$ by $\lambda - 1$ in equation (11.3), obtaining the stress function

$$\phi = r^{\lambda+1}(A_1 \cos(\lambda+1)\theta + A_2 \sin(\lambda+1)\theta + A_3 \cos(\lambda-1)\theta + A_4 \sin(\lambda-1)\theta) \; , \quad (11.35)$$

with stress components

$$\begin{aligned} \sigma_{rr} &= r^{\lambda-1}(-A_1\lambda(\lambda+1)\cos(\lambda+1)\theta - A_2\lambda(\lambda+1)\sin(\lambda+1)\theta \\ &\quad -A_3\lambda(\lambda-3)\cos(\lambda-1)\theta - A_4\lambda(\lambda-3)\sin(\lambda-1)\theta) \; ; &(11.36)\\ \sigma_{r\theta} &= r^{\lambda-1}(A_1\lambda(\lambda+1)\sin(\lambda+1)\theta - A_2\lambda(\lambda+1)\cos(\lambda+1)\theta \\ &\quad +A_3\lambda(\lambda-1)\sin(\lambda-1)\theta - A_4\lambda(\lambda-1)\cos(\lambda-1)\theta) \; ; &(11.37)\\ \sigma_{\theta\theta} &= r^{\lambda-1}(A_1\lambda(\lambda+1)\cos(\lambda+1)\theta + A_2\lambda(\lambda+1)\sin(\lambda+1)\theta \\ &\quad +A_3\lambda(\lambda+1)\cos(\lambda-1)\theta + A_4\lambda(\lambda+1)\sin(\lambda-1)\theta) \; . &(11.38) \end{aligned}$$

For this solution to satisfy the traction-free boundary conditions

$$\sigma_{\theta r} = \sigma_{\theta\theta} = 0; \quad \theta = \pm\alpha \quad (11.39)$$

we require

$$\begin{aligned} &\lambda(A_1(\lambda+1)\sin(\lambda+1)\alpha - A_2(\lambda+1)\cos(\lambda+1)\alpha \\ &\quad +A_3(\lambda-1)\sin(\lambda-1)\alpha - A_4(\lambda-1)\cos(\lambda-1)\alpha) = 0 \; ; &(11.40)\\ &\lambda(-A_1(\lambda+1)\sin(\lambda+1)\alpha - A_2(\lambda+1)\cos(\lambda+1)\alpha \\ &\quad -A_3(\lambda-1)\sin(\lambda-1)\alpha - A_4(\lambda-1)\cos(\lambda-1)\alpha) = 0 \; ; &(11.41)\\ &\lambda(A_1(\lambda+1)\cos(\lambda+1)\alpha + A_2(\lambda+1)\sin(\lambda+1)\alpha \\ &\quad +A_3(\lambda+1)\cos(\lambda-1)\alpha + A_4(\lambda+1)\sin(\lambda-1)\alpha) = 0 \; ; &(11.42)\\ &\lambda(A_1(\lambda+1)\cos(\lambda+1)\alpha - A_2(\lambda+1)\sin(\lambda+1)\alpha \\ &\quad +A_3(\lambda+1)\cos(\lambda-1)\alpha - A_4(\lambda+1)\sin(\lambda-1)\alpha) = 0 \; . &(11.43) \end{aligned}$$

This is a set of four homogeneous equations for the four constants $A_1, \ldots, A_4$ and will have a non-trivial solution only for certain eigenvalues of the exponent $\lambda$. Since all the equations have a $\lambda$ multiplier, $\lambda = 0$ must be an eigenvalue for all wedge angles. We can simplify the equations by cancelling this factor and taking sums and differences in pairs to expose the symmetry of the system, with the result

$$\begin{aligned} A_1(\lambda+1)\sin(\lambda+1)\alpha + A_3(\lambda-1)\sin(\lambda-1)\alpha &= 0 \; ; &(11.44)\\ A_1(\lambda+1)\cos(\lambda+1)\alpha + A_3(\lambda+1)\cos(\lambda-1)\alpha &= 0 \; ; &(11.45)\\ A_2(\lambda+1)\cos(\lambda+1)\alpha + A_4(\lambda-1)\cos(\lambda-1)\alpha &= 0 \; ; &(11.46)\\ A_2(\lambda+1)\sin(\lambda+1)\alpha + A_4(\lambda+1)\sin(\lambda-1)\alpha &= 0 \; . &(11.47) \end{aligned}$$

Thus, the symmetric terms $A_1, A_3$ have a non-trivial solution if

$$\lambda \sin 2\alpha + \sin 2\lambda\alpha = 0 \ , \tag{11.48}$$

whilst the antisymmetric terms have a non-trivial solution if

$$\lambda \sin 2\alpha - \sin 2\lambda\alpha = 0 \ . \tag{11.49}$$

### 11.2.3 Nature of the eigenvalues

We first note from equations (11.36–11.38) that the stress components are proportional to $r^{\lambda-1}$ and hence $\lambda$ is restricted to positive values by the energy criterion of §11.2.1. Furthermore, we shall generally be most interested in the lowest possible eigenvalue[4], since this will define the strongest singularity that can occur at the notch and will therefore dominate the stress field at sufficiently small $r$.

Both equations (11.48, 11.49) are satisfied for all $\alpha$ if $\lambda = 0$ and (11.49) is also satisfied for all $\alpha$ if $\lambda = 1$.

The case $\lambda = 0$ is strictly excluded by the energy criterion, but we shall find in the next chapter that it corresponds to the important case in which a concentrated force is applied at the origin. It can legitimately be regarded as describing the stress field sufficiently distant from a load distributed on the wedge faces near the origin, but as such it is clearly not appropriate to the unloaded problem of Figure 11.3.

The solution $\lambda = 1$ of equation (11.49) is a spurious eigenvalue, since if we compute the corresponding eigenfunction, it turns out to have the null form $\phi = A_4 \sin(0)$. The correct limiting form for $\lambda = 1$ requires the modified stress function (11.4) and leads to the condition (11.33) for antisymmetric fields, which is satisfied only for $2\alpha = 257.4^o$.

Some insight into the nature of the eigenvalues for more general wedge angles can be gained from the graphical representation of Figure 11.5, where we plot the two terms in equations (11.48, 11.49) against $x = 2\lambda\alpha$. The sine wave represents the term $\sin 2\lambda\alpha$ $(= \sin x)$ and the straight lines represent various possible positions for the terms $\pm\lambda \sin 2\alpha = \pm x(\sin 2\alpha/2\alpha)$.

The simplest case is that in which $2\alpha = 2\pi$ and the wedge becomes the crack illustrated in Figure 11.6. We then have $\sin 2\alpha = 0$ and the solutions of both equations (11.48, 11.49) correspond to the points where the sine wave crosses the horizontal axis — i.e.

$$\sin 2\lambda\alpha = \frac{1}{2}, 1, \frac{3}{2}, \ldots \tag{11.50}$$

This is an important special case because of its application to fracture mechanics and we shall discuss it further in Chapter 13, below. In particular, we note that the lowest positive eigenvalue of $\lambda$ is $\frac{1}{2}$ and hence that the crack tip field is square-root singular for both symmetric and antisymmetric loading.

[4]More rigorously, a general stress field near the corner can be expressed as an eigenfunction expansion, but the term with the smallest eigenvalue will be arbitrarily larger than the next term in the expansion at sufficiently small values of $r$.

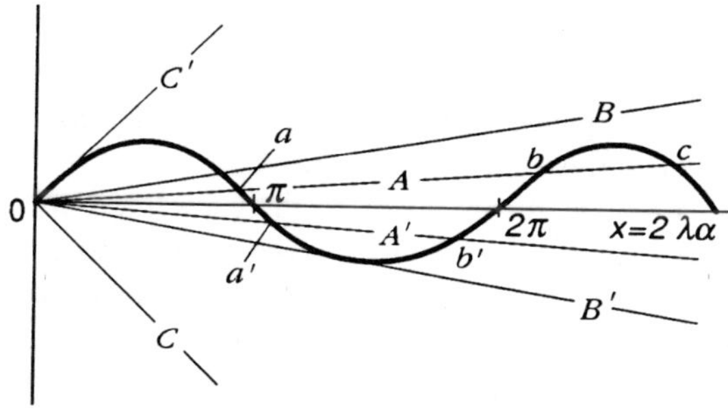


Figure 11.5: Graphical solution of equations (11.48, 11.49)

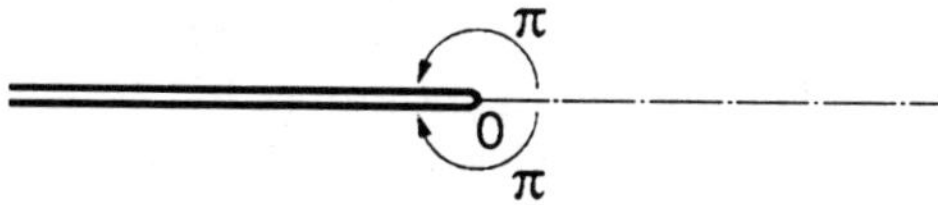


Figure 11.6: Tip of a crack considered as a 360° wedge

Suppose we now consider a wedge of somewhat less than $2\pi$, in which case equations (11.48, 11.49) are represented by the intersection between the sine wave and the lines $A, A'$ respectively in Figure 11.5.

As the wedge angle is reduced, the slope $(\sin 2\alpha/2\alpha)$ of the lines first becomes increasingly negative until a maximum is reached at $2\alpha = 257.4^o$, corresponding to the tangential line $B'$ and its mirror image $B$. Further reduction causes the slope to increase monotonically, passing through zero again at $2\alpha = \pi$ (corresponding to the case of the half-plane) and reaching the limit $C, C'$ at $2\alpha = 0$.

Clearly, if the lines $A, A'$ are even slightly inclined to the horizontal, they will only intersect a finite number of waves, corresponding to a finite set of real roots, the number of real roots increasing as the slope approaches zero. Increasing the slope causes initially equal-spaced real roots to become closer in pairs such as $b, c$ in Figure 11.5. For any given pair, there will be a critical slope at which the roots coalesce and for further increase in slope, a pair of complex conjugate roots will be developed.

In the axisymmetric case for $360^o > 2\alpha > 257.4^o$, the intersection $b'$ in Figure 11.5 corresponds to the spurious eigenvalue of (11.49), but $a'$ is a meaningful eigenvalue corresponding to a stress singularity that weakens from square-root at $2\alpha = 360^o$ to zero at $2\alpha = 257.4^o$. For smaller wedge angles, $a'$ is the spurious eigenvalue and $b'$ corresponds to a real but non-singular root.

In the symmetric case, the root $a$ gives a real eigenvalue corresponding to a singular stress field whose strength falls monotonically from square-root to zero as the wedge angle is reduced from $2\pi$ to $\pi$. The strength of the singularity $(\lambda - 1)$ for both symmetric and antisymmetric terms is plotted against total wedge angle[5] in Figure 11.7. The singularity associated with the symmetric field is always stronger than that for the antisymmetric field except in the limit $2\alpha = 2\pi$, where they are equal, and there is a range of angles $(257.4^o > 2\alpha > 180^o)$ where the symmetric field is singular, but the antisymmetric field is not.

For non-reëntrant wedges $(2\alpha < \pi)$, both fields are bounded, the dominant eigenvalue for the symmetric field becoming complex for $2\alpha < 146^o$. Since bounded solutions correspond to eigenfunctions with $r$ raised to a power with positive real part, the stress field always tends to zero in a non-reëntrant corner.

### 11.2.4 Other geometries

Williams' method can also be applied to other discontinuities in elasticity and indeed in other fields in mechanics. For example, we can extend it to determine the strength of the singularity in a composite wedge of two different materials[6]. An important special case is illustrated in Figure 11.8, where two dissimilar materials are bonded[7],

[5] Figure 11.7 is reproduced by courtesy of Dr.D.A.Hills.

[6] D.B.Bogy, Two edge-bonded elastic wedges of different materials and wedge angles under surface tractions, ASME J.Appl.Mech., Vol.38 (1971), 377–386.

[7] A problem of this kind was solved by G.G.Adams, A semi-infinite elastic strip bonded to an infinite strip, ASME J.Appl.Mech., Vol.47 (1980), 789–794.

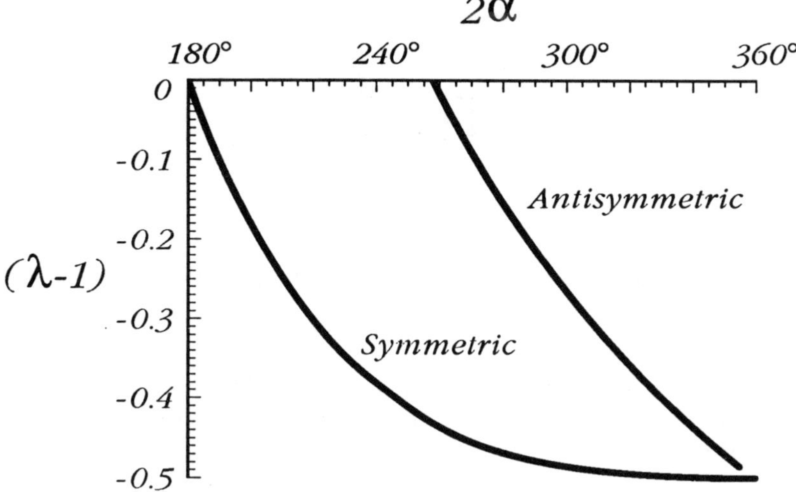


Figure 11.7: Strength of the singularity in a re-entrant corner

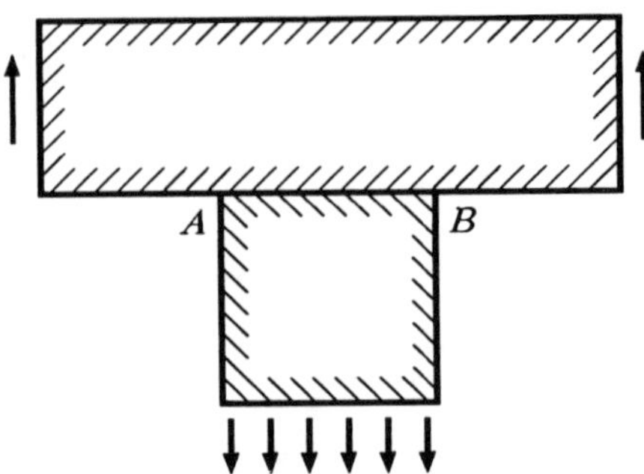


Figure 11.8: Composite T-bar, with composite notches at the re-entrant corners $A, B$.

leaving a composite wedge of $\frac{3\pi}{2}$ at the points $A, B$. Another composite wedge is the tip of a crack (or debonded zone) at the interface between two dissimilar materials (see Chapter 24).

The method has also been applied to determine the asymptotic fields in contact problems, with and without friction[8] (see Chapter 12 below).

The results of asymptotic analysis are very useful in numerical methods, since they enable us to predict the nature of the local stress field and hence devise appropriate special elements or meshes. Failure to do this in problems with singular fields will always lead to numerical inefficiency and sometimes to lack of convergence, mesh sensitivity or instability.

To look ahead briefly to three-dimensional problems, we note that when we concentrate our attention on a very small region at a corner, the resulting magnification makes all other dimensions (including radii of curvature etc) look very large. Thus, a notch in a three-dimensional body will generally have a two-dimensional asymptotic field at sufficiently small $r$. The above results and the underlying method are therefore of very general application.

---

[8]J.Dundurs and M.S.Lee, Stress concentration at a sharp edge in contact problems, J.Elasticity, Vol.2 (1972), 109–112; M.Comninou, Stress singularities at a sharp edge in contact problems with friction, Z.angew.Math.Phyz., Vol.27 (1976), 493–499.

## 11.3 General loading of the faces

If the faces of the wedge $\theta = \pm\alpha$ are subjected to tractions that can be expanded as power series in $r$, a solution can be obtained using a series of terms like (11.1). However, as with the rectangular beam, power series are of limited use in representing a general traction distribution, particularly when it is relatively localized. Instead, we can adapt the Fourier transform representation (5.64) to the wedge, by noting that the stress function (11.35) will oscillate along lines of constant $\theta$ if we choose complex values for $\lambda$. This leads to a representation as a *Mellin transform* defined as

$$\phi(r,\theta) = \frac{1}{2\pi i}\int_{c-i\infty}^{c+i\infty} f(s,\theta)r^{-s}ds \tag{11.51}$$

for which the inversion is[9]

$$f(s,\theta) = \int_0^\infty \phi(r,\theta)r^{s-1}dr \; . \tag{11.52}$$

The integrand of (11.51) is the function (11.35) with $\lambda = -s-1$. The integral itself is of course path independent in any strip of the complex plane in which the integrand is regular. If we evaluate it along the straight line $s = c + i\omega$, it can be written in the form

$$\begin{aligned} r^c\phi(r,\theta) &= \frac{1}{2\pi}\int_{-\infty}^{\infty} f(c+i\omega,\theta)r^{-i\omega}d\omega \\ &= \frac{1}{2\pi}\int_{-\infty}^{\infty} f(c+i\omega)e^{i\omega\log r}d\omega \; , \end{aligned} \tag{11.53}$$

which is readily converted to a Fourier integral[10] by the change of variable $t = \log r$. The constant $c$ has to be chosen to ensure regularity of the integrand. If there are no unacceptable singularities in the tractions and the latter are bounded at infinity, it can be shown that a suitable choice is $c = -1$.

The Mellin transform and power series methods can be applied to a wedge of any angle, but it should be remarked that, because of the singular asymptotic fields obtained in §11.3, meaningful results for the re-entrant wedge ($2\alpha > \pi$) can only be obtained if the conditions at infinity are also prescribed and satisfied.

More precisely, we could formulate the problem for the large but finite sector $-\alpha < \theta < \alpha$, $0 < r < b$, with precribed tractions on all edges[11]. The procedure

[9] I.N.Sneddon, *Fourier Transforms*, McGraw-Hill, New York, 1951.

[10] The basic theory of the Mellin transform and its relation to the Fourier transform is explained by I.N.Sneddon, *loc. cit.*. For applications to elasticity problems for the wedge, see E.Sternberg and W.T.Koiter, The wedge under a concentrated couple: A paradox in the two-dimensional theory of elasticity, ASME J.Appl.Mech., Vol. 25 (1958), 575–581, W.J.Harrington and T.W.Ting, Stress boundary-value problems for infinite wedges, J.Elasticity, Vol. 1 (1971), 65–81.

[11] A problem of this kind was considered by G.Tsamasphyros and P.S.Theocaris, On the solution of the sector problem, J.Elasticity, Vol. 9 (1979), 271–281.

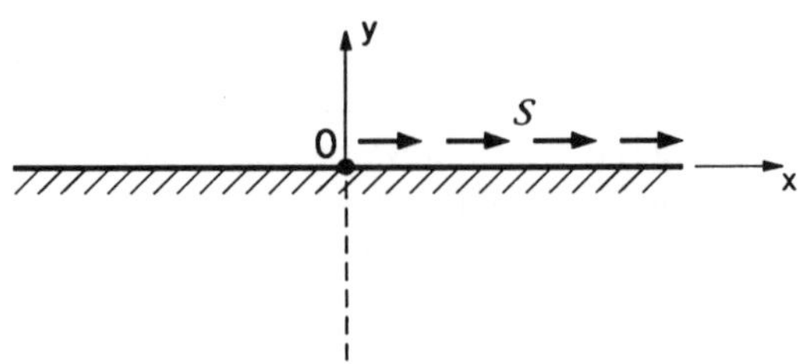


Figure 11.9: The half-plane with shear tractions

would be first to develop a particular solution for the loading on the faces $\theta = \pm\alpha$ as if the wedge were infinite and then to correct the boundary conditions at $r = b$ by superposing an infinite sequence of eigenfunctions from §11.3 with appropriate multipliers[12]. For sufficiently large $b$, the stress field near the apex of the wedge would be adequately described by the particular solution and the singular terms from the eigenfunction expansion.

## PROBLEMS

1. Figure 11.9 shows a half-plane, $y<0$, subjected to a uniform shear traction, $\sigma_{xy}=S$ on the half-line, $x>0$, $y=0$, the remaining tractions on $y=0$ being zero.

Find the complete stress field in the half-plane. **Note**: This is a problem requiring a special stress function with a logarithmic multiplier (see §11.1.3 above).

2. Show that $\phi=Ar^2\theta$ can be used as a stress function and determine the tractions which it implies on the boundaries of the region $-\frac{\pi}{2}<\theta<\frac{\pi}{2}$. Hence show that the stress function appropriate to the loading of Figure 11.10 is

$$\phi = -\frac{W}{4\pi a}(r_1^2\theta_1 + r_2^2\theta_2)$$

where $r_1, r_2, \theta_1, \theta_2$ are defined in the Figure.

Determine the principal stresses at the point B.

3. A wedge-shaped concrete dam is subjected to a hydrostatic pressure $\rho g h$ varying with depth $h$ on the vertical face $\theta=0$ as shown in Figure 11.11, the other face $\theta=\alpha$ being traction-free. The dam is also loaded by self-weight, the density of concrete being $\rho_c=2.3\rho$. Find the minimum wedge angle $\alpha$ if there is to be no tensile stress in the dam.

[12]This requires that the eigenfunction series is complete for this problem. The proof is given by R.D.Gregory, Green's functions, bi-linear forms and completeness of the eigenfunctions for the elastostatic strip and wedge, J.Elasticity, Vol. 9 (1979), 283–309.

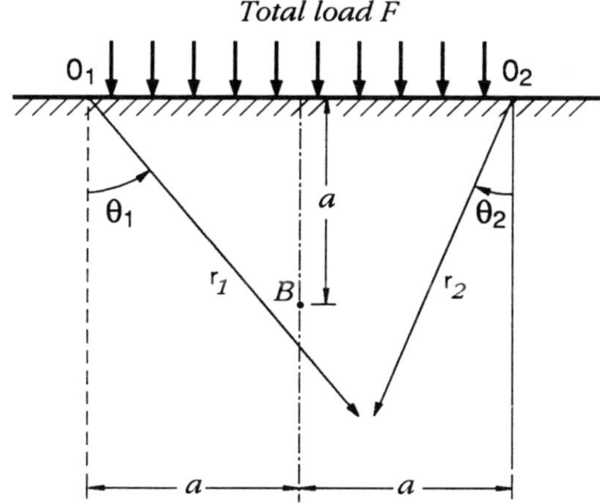


Figure 11.10: Uniform normal loading over a discrete region

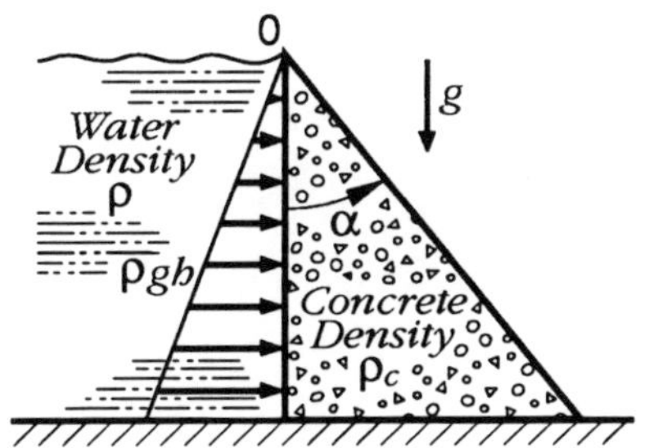


Figure 11.11: The wedge-shaped dam

# Chapter 12

# PLANE CONTACT PROBLEMS[1]

In the previous chapter, we considered problems in which the infinite wedge was loaded on its faces or solely by tractions on the infinite boundary. A related problem of considerable practical importance concerns the wedge with traction-free faces, loaded by a concentrated force $\boldsymbol{F}$ at the vertex, as shown in Figure 12.1.

## 12.1 Self-similarity

An important characteristic of this problem is that there is no inherent length scale. An enlarged photograph of the problem would look the same as the original. The solution must therefore share this characteristic and hence, for example, contours of the stress function $\phi$ must have the same geometric shape at all distances from the vertex. Problems of this type — in which the solution can be mapped into itself after a change of length scale — are described as *self-similar*.

An immediate consequence of the self-similarity is that all quantities must be capable of expression in the separated variable form

$$\sigma = f(r)g(\theta) \ . \tag{12.1}$$

Furthermore, since the tractions on the line $r=a$ must balance a constant force $\boldsymbol{F}$ for all $a$, we can deduce that $f(r)=r^{-1}$, because the area available for transmitting the force increases linearly with radius[2].

[1] For a more detailed discussion of elastic contact problems, see K.L.Johnson, *Contact Mechanics*, Cambridge University Press, (1985) and G.M.L.Gladwell, *Contact Problems in the Classical Theory of Elasticity*, Sijthoff and Noordhoff, Alphen aan den Rijn, (1980).

[2] In the same way, we can deduce that the stresses in a cone subjected to a force at the vertex must vary with $r^{-2}$.

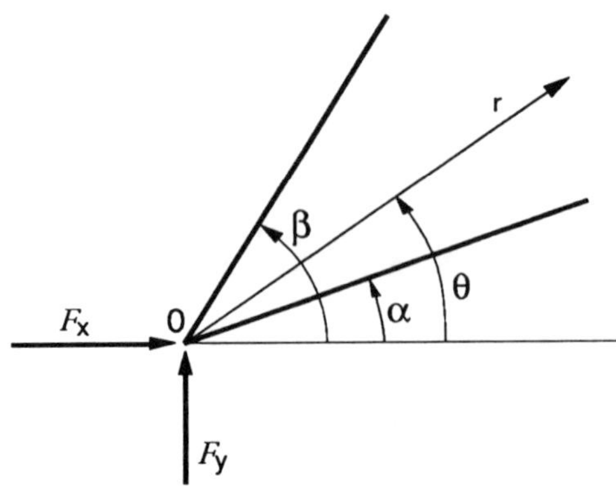


Figure 12.1: The wedge loaded by a force at the vertex

## 12.2 The Flamant Solution

Choosing those terms in Table 8.1 that give stresses proportional to $r^{-1}$, we obtain

$$\phi = C_1 r\theta \sin\theta + C_2 r\theta \cos\theta + C_3 r \log r \cos\theta + C_4 r \log r \sin\theta \ , \tag{12.2}$$

for which the stress components are

$$\begin{aligned} \sigma_{rr} &= r^{-1}(2C_1 \cos\theta - 2C_2 \sin\theta + C_3 \cos\theta + C_4 \sin\theta) \ ; \\ \sigma_{r\theta} &= r^{-1}(C_3 \sin\theta - C_4 \cos\theta) \ ; \\ \sigma_{\theta\theta} &= r^{-1}(C_3 \cos\theta + C_4 \sin\theta) \ . \end{aligned} \tag{12.3}$$

For the wedge faces to be traction-free, we require

$$\sigma_{\theta r} = \sigma_{\theta\theta} = 0 \ \ ; \ \ \theta = \alpha, \beta \ , \tag{12.4}$$

which can be satisfied by taking $C_3, C_4 = 0$.

To determine the remaining constants $C_1, C_2$, we consider the equilibrium of the region $0 < r < a$, obtaining

$$F_x + 2\int_\alpha^\beta \left(\frac{C_1 \cos\theta - C_2 \sin\theta}{a}\right) a \cos\theta d\theta \ = \ 0 \ ; \tag{12.5}$$

$$F_y + 2\int_\alpha^\beta \left(\frac{C_1 \cos\theta - C_2 \sin\theta}{a}\right) a \sin\theta d\theta \ = \ 0 \ . \tag{12.6}$$

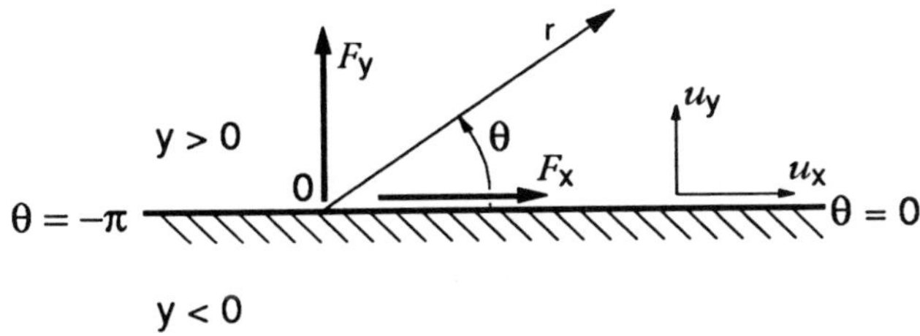


Figure 12.2: Force acting on the surface of a half-plane

These two equations permit us to choose the constants $C_1, C_2$ to give the applied force $\boldsymbol{F}$ any desired magnitude and direction. Notice that the radius $a$ cancels in equations (12.5, 12.6) and hence, if they are satisfied for any radius, they are satisfied for all.

This is known as the *Flamant solution*, or sometimes as the *simple radial distribution*, since the act of setting $C_3{=}C_4{=}0$ clears all $\theta$-surfaces of tractions, leaving

$$\sigma_{rr} = \frac{2C_1\cos\theta}{r} - \frac{2C_2\sin\theta}{r} \tag{12.7}$$

as the only non-zero stress component.

## 12.3 The half-plane

If we take $\alpha = -\pi$ and $\beta = 0$ in the above solution, we obtain the special case of a force acting at a point on the surface of the half-plane $y<0$ as shown in Figure 12.2.

For this case, performing the integrals in equations (12.5, 12.6), we find

$$F_x + 2\int_{-\pi}^{0}(C_1\cos^2\theta - C_2\sin\theta\cos\theta)d\theta = F_x + \pi C_1 = 0\,; \tag{12.8}$$

$$F_y + 2\int_{-\pi}^{0}(C_1\cos\theta\sin\theta - C_2\sin^2\theta)d\theta = F_y - \pi C_2 = 0 \tag{12.9}$$

and hence

$$C_1 = -\frac{F_x}{\pi}\ ;\ C_2 = \frac{F_y}{\pi}\,. \tag{12.10}$$

It is convenient to consider these terms separately.

### 12.3.1 The normal force $F_y$

The normal (tensile) force $F_y$ produces the stress field

$$\sigma_{rr} = -\frac{2F_y \sin\theta}{\pi r} , \qquad (12.11)$$

which reaches a maximum tensile value on the negative $y$ axis ($\theta = -\pi/2$) — i.e. directly beneath the load — and falls to zero as we approach the surface along any line other than one which passes through the point of application of the force.

The displacement field corresponding to this stress can conveniently be taken from Table 9.1 and is

$$2\mu u_r = \frac{F_y}{2\pi}[(\kappa-1)\theta\cos\theta - (\kappa+1)\log r\sin\theta + \sin\theta] ; \qquad (12.12)$$

$$2\mu u_\theta = \frac{F_y}{2\pi}[-(\kappa-1)\theta\sin\theta - (\kappa+1)\log r\cos\theta - \cos\theta] . \qquad (12.13)$$

These solutions will be used as Green's functions for problems in which a half-plane is subjected to various surface loads and hence the surface displacements are of particular interest.

On $\theta=0$, ($y=0$, $x>0$), we have

$$2\mu u_r = 2\mu u_x = 0 ; \qquad (12.14)$$

$$2\mu u_\theta = 2\mu u_y = \frac{F_y}{2\pi}[-(\kappa+1)\log r - 1] , \qquad (12.15)$$

whilst on $\theta=-\pi$, ($y=0$, $x<0$), we have

$$2\mu u_r = -2\mu u_x = \frac{F_y}{2}(\kappa-1) ; \qquad (12.16)$$

$$2\mu u_\theta = -2\mu u_y = \frac{F_y}{2\pi}[(\kappa+1)\log r + 1] . \qquad (12.17)$$

It is convenient to impose symmetry on the solution by superposing a rigid-body displacement

$$2\mu u_x = \frac{F_y(\kappa-1)}{4} \; ; \; 2\mu u_y = \frac{F_y}{2\pi} , \qquad (12.18)$$

after which equations (12.14–12.17) can be summarised in the form

$$u_x = \frac{F_y(\kappa-1)\mathrm{sgn}x}{8\mu} ; \qquad (12.19)$$

$$u_y = -\frac{F_y(\kappa+1)\log|x|}{4\pi\mu} , \qquad (12.20)$$

where the function sgn($x$) is defined to be $+1$ for $x>0$ and $-1$ for $x<0$.

Note incidentally that $r=|x|$ on $y=0$.

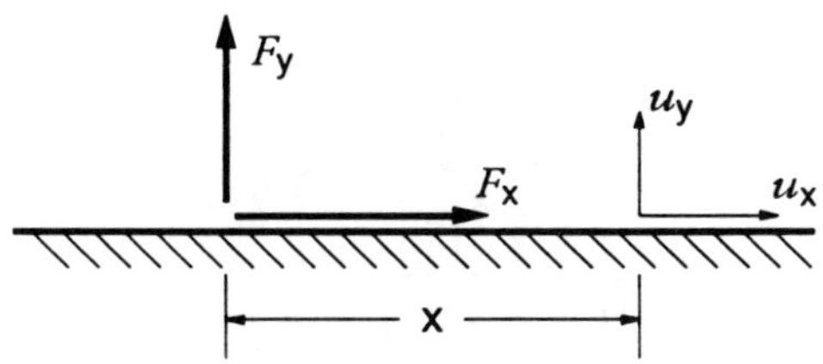


Figure 12.3: Surface displacements due to a surface force

### 12.3.2 The tangential force $F_x$

The tangential force $F_x$ produces the stress field

$$\sigma_{rr} = -\frac{2F_x \cos\theta}{\pi r}, \tag{12.21}$$

which is compressive ahead of the force ($\theta=0$) and tensile behind it ($\theta=-\pi$) as we might expect.

The corresponding displacement field is found from Table 9.1 as before and after superposing an appropriate rigid-body displacement as in §12.3.1, the surface displacements can be written

$$u_x = -\frac{F_x(\kappa+1)\log|x|}{4\pi\mu}; \tag{12.22}$$

$$u_y = -\frac{F_x(\kappa-1)\text{sgn}x}{8\mu}. \tag{12.23}$$

### 12.3.3 Summary

We summarize the results of §§12.3.1, 12.3.2 in the equations

$$u_x = -\frac{F_x(\kappa+1)\log|x|}{4\pi\mu} + \frac{F_y(\kappa-1)\text{sgn}x}{8\mu}; \tag{12.24}$$

$$u_y = -\frac{F_x(\kappa-1)\text{sgn}x}{8\mu} - \frac{F_y(\kappa+1)\log|x|}{4\pi\mu}, \tag{12.25}$$

see Figure 12.3.

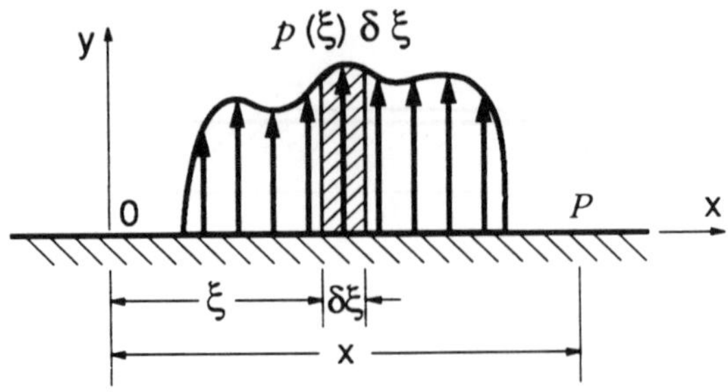


Figure 12.4: Half-plane subjected to a distributed traction

## 12.4 Distributed normal tractions

Now suppose that the surface of the half-plane is subjected to a distributed normal load $p(\xi)$ per unit length as shown in Figure 12.4.

The stress and displacement field can be found by superposition, using the Flamant solution as a Green's function — i.e. treating the distributed load as the limit of a set of point loads of magnitude $p(\xi)\delta\xi$.

Of particular interest is the distortion of the surface, defined by the normal displacement $u_y$. At the point $P(x, 0)$, this is given by

$$u_y = -\frac{\kappa+1}{4\pi\mu}\int_A p(\xi)\log|x-\xi|d\xi \ , \tag{12.26}$$

from equation (12.20), where $A$ is the region over which the load acts.

The displacement defined by equation (12.26) is logarithmically unbounded as $x$ tends to infinity. In general, if a finite region of the surface of the half-plane is subjected to a non-self-equilibrated system of loads, the stress and displacement fields at a distance $r \gg A$ will approximate those due to the force resultants applied as concentrated forces — i.e. to the expressions of equations (12.11–12.13) for normal loading. In other words, the stresses will decay with $1/r$ and the displacements (being integrals of the strains) will vary logarithmically.

This means that the half-plane solution can be used for the stresses in a finite body loaded on a region of the surface which is small compared with the linear dimensions of the body, since $1/r$ is arbitrarily small for sufficiently large $r$. However, the logarithm is not bounded at infinity and hence the rigid-body displacement of

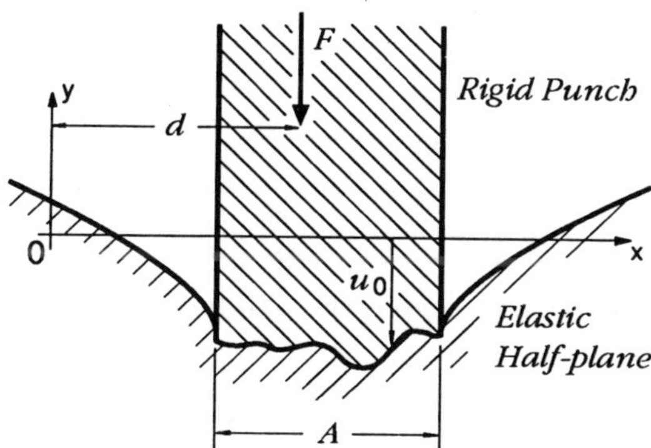


Figure 12.5: The elastic half-plane indented by a frictionless rigid punch

the loaded region with respect to distant parts of the body cannot be found without a more exact treatment taking into account the finite dimensions of the body. For this reason, contact problems for the half-plane are often formulated in terms of the displacement *gradient*.

$$\frac{du_y}{dx} = -\frac{(\kappa+1)}{4\pi\mu}\int_A \frac{p(\xi)d\xi}{x-\xi}, \tag{12.27}$$

thereby avoiding questions of rigid-body translation.

When the point $x = \xi$ lies within $A$, the integral in (12.27) is interpreted as a Cauchy principal value.

## 12.5 Frictionless contact problems

The results of the last section permit us to develop a solution to the problem of a half-plane indented by a frictionless rigid punch of known profile (see Figure 12.5).

We suppose that the load $F$ is sufficient to establish contact over the whole of the contact area $A$, in which case the displacement of the half-plane must satisfy the equation

$$u_y = -u_0(x) + C_1x + C_0 \tag{12.28}$$

in $A$, where $u_0$ is a known function of $x$ which describes the profile of the punch.

The constants $C_0, C_1$ define an unknown rigid body translation and rotation of the punch respectively, which can generally be assigned in such a way as to give a pressure distribution $p(\xi)$ which is statically equivalent to the force $F$. i.e.

$$\int_A p(\xi)d\xi = -F ; \qquad (12.29)$$

$$\int_A p(\xi)\xi d\xi = -Fd , \qquad (12.30)$$

where $d$ defines the line of action of $F$ as shown in Figure 12.5 and the negative signs are a consequence of the tensile positive convention for $p(\xi)$. These two equations are sufficient to determine the constants $C_0, C_1$.

If the contact region, $A$, is connected, we can define a coördinate system with origin at the mid-point and denote the half-length of the contact by $a$. Equations (12.27, 12.28) then give

$$-\frac{du_0}{dx} + C_1 = -\frac{(\kappa+1)}{4\pi\mu}\int_{-a}^{a}\frac{p(\xi)d\xi}{x-\xi} \; ; \; -a < x < a , \qquad (12.31)$$

which is a Cauchy singular integral equation for the unknown pressure $p(\xi)$.

### 12.5.1 Method of solution

Integral equations of this kind arise naturally in the application of the complex variable method to two-dimensional potential problems and they are extensively discussed in a classical text by Muskhelishvili[3]. However, for our purposes, a simpler solution will suffice, based on the change of variable

$$x = a\cos\phi \; ; \; \xi = a\cos\theta \qquad (12.32)$$

and expansion of the two sides of the equation as Fourier series.

Substituting (12.32) into (12.31), we find

$$\frac{1}{a\sin\phi}\frac{du_0}{d\phi} + C_1 = -\frac{(\kappa+1)}{4\pi\mu}\int_0^{\pi}\frac{p(\theta)\sin\theta d\theta}{(\cos\phi-\cos\theta)} \; ; \; 0 < \phi < \pi , \qquad (12.33)$$

which can be simplified using the result

$$\int_0^{\pi}\frac{\cos n\theta d\theta}{(\cos\phi-\cos\theta)} = -\frac{\pi\sin n\phi}{\sin\phi} . \qquad (12.34)$$

Writing

$$p(\theta) = \sum_{n=0}^{\infty}\frac{p_n\cos n\theta}{\sin\theta} ; \qquad (12.35)$$

$$\frac{du_0}{d\phi} = \sum_{n=1}^{\infty} u_n\sin n\phi , \qquad (12.36)$$

[3] N.I.Muskhelishvili, *Singular Integral Equations*, (English translation by J.R.M.Radok, Noordhoff, Groningen, (1953)). For applications to elasticity problems, including the above contact problem, see N.I.Muskhelishvili, *Some Basic Problems of the Mathematical Theory of Elasticity*, (English translation by J.R.M.Radok, Noordhoff, Groningen, (1953)). For a simpler discussion of the solution of contact problems involving singular integral equations, see K.L.Johnson, *loc. cit*, §2.7.

we find

$$\begin{aligned}\sum_{n=1}^{\infty} u_n \sin n\phi + C_1 a \sin\phi &= \frac{(\kappa+1)a\sin\phi}{4\pi\mu}\sum_{n=1}^{\infty}\frac{\pi p_n \sin n\phi}{\sin\phi} \\ &= \frac{(\kappa+1)a}{4\mu}\sum_{n=1}^{\infty} p_n \sin n\phi \qquad (12.37)\end{aligned}$$

and hence

$$p_n = \frac{4\mu u_n}{(\kappa+1)a} \; ; \; n>1 \; ; \qquad (12.38)$$

$$p_1 = \frac{4\mu(C_1 a + u_1)}{(\kappa+1)a} \; . \qquad (12.39)$$

Thus, if the shape of the punch $u_0$ is expanded as in equation (12.36), the corresponding coefficients in the pressure series (12.35) can be written down using (12.38), except for $n=0,1$.

The two coefficients $p_0, p_1$ are found by substituting (12.35) into the equilibrium conditions (12.29, 12.30). From (12.29), we have

$$\begin{aligned}-F &= \int_{-a}^{a} p(x')dx' = \int_0^{\pi} p(\theta) a \sin\theta d\theta \\ &= a\sum_{n=0}^{\infty} p_n \int_0^{\pi}\cos n\theta d\theta = \pi a p_0 \qquad (12.40)\end{aligned}$$

and from (12.30),

$$\begin{aligned}-Fd &= \int_{-a}^{a} p(x')x'dx' = \int_0^{\pi} p(\theta)a^2 \sin\theta\cos\theta d\theta \\ &= a^2\sum_{n=0}^{\infty} p_n\int_0^{\pi}\cos\theta\cos n\theta d\theta = \frac{\pi a^2 p_1}{2} \; . \qquad (12.41)\end{aligned}$$

Hence,

$$p_0 = -\frac{F}{\pi a} \; ; \; p_1 = -\frac{2Fd}{\pi a^2} \; . \qquad (12.42)$$

### 12.5.2 The flat punch

We first consider the case of a flat punch with a symmetric load as shown in Figure 12.6.

The profile of the punch is described by the equation

$$u_0 = C \; ; \; \frac{du_0}{dx} = 0 \qquad (12.43)$$

and hence $p_n = 0$ for $n>1$ from equation (12.38). Furthermore, since the load is symmetric ($d=0$), $p_1$ is also zero (equation (12.42)) and $p_0$ is given by (12.42).

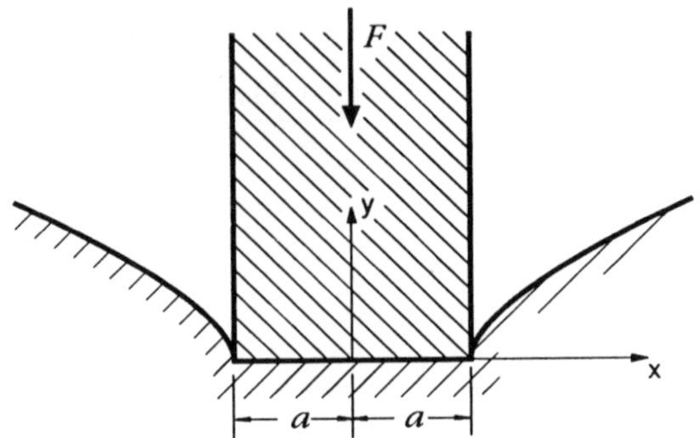


Figure 12.6: Half-plane indented by a flat rigid punch

The traction distribution under the punch is therefore

$$\begin{aligned} p(x) &= \frac{p_0}{\sin\phi} = -\frac{F}{\pi a\sqrt{1-x^2/a^2}} \\ &= -\frac{F}{\pi\sqrt{a^2-x^2}} \ . \end{aligned} \tag{12.44}$$

We note that the sharp corners of the punch cause a square-root singularity in traction at the edges, $x=\pm a$. This is typical of indentation problems where the punch has a sharp corner. The pressure distribution is illustrated in Figure 12.7.

### 12.5.3 The cylindrical punch (Hertz problem)

We next consider the indentation of a half-plane by a frictionless rigid cylinder (Figure 12.8), which is a special case of the contact problem associated with the name of Hertz.

This problem differs from the previous example in that the contact area semi-width $a$ depends on the load $F$ and cannot be determined *a priori*. Strictly, $a$ must be determined from the pair of inequalities

$$p(x) \leq 0 \ ; \ -a < x < a \ ; \tag{12.45}$$

$$u_y(x) \geq u_0(x) \ ; \ |x| > a \ , \tag{12.46}$$

which express the physical requirements that the contact traction should not be tensile (12.45) and that there should be no interference between the bodies outside the contact area (12.46). However, it can be shown that these conditions are in most

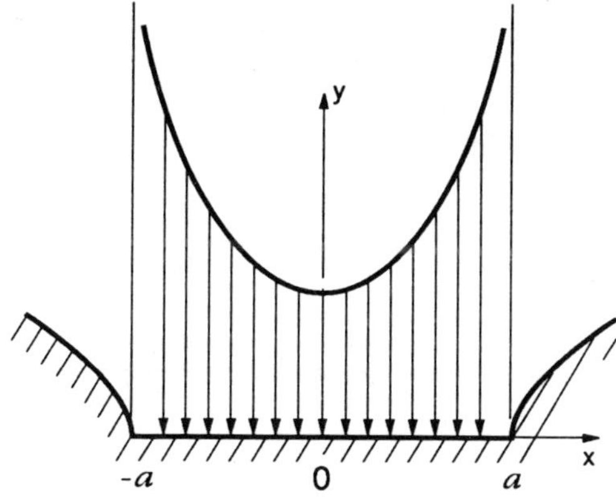


Figure 12.7: Pressure distribution under the flat rigid punch

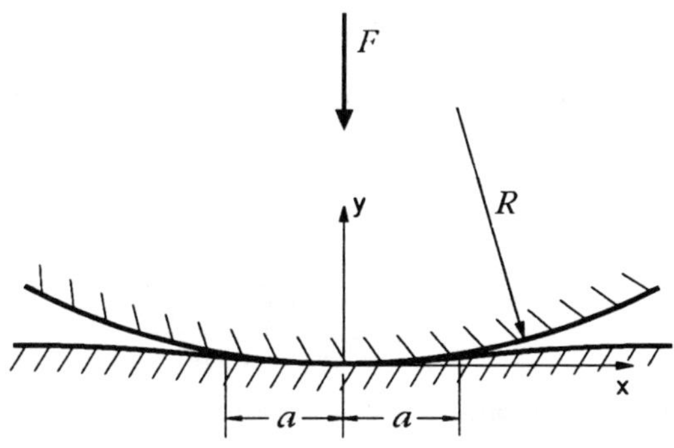


Figure 12.8: The Hertzian contact problem

cases formally equivalent to the requirement that the contact traction should not be singular at $x=\pm a$. This question will be discussed more rigorously in connection with three-dimensional contact problems in Chapter 21. Here it is sufficient to remark that a traction singularity such as that obtained at the sharp corner of the flat punch (equation (12.44)) causes a discontinuity in the displacement gradient $du_y/dx$, which will cause a violation of (12.46) if the punch is smooth.

If the radius of the cylinder is $R$, we have

$$\frac{d^2u_0}{dx^2} = -\frac{1}{R} \tag{12.47}$$

and hence

$$u_0 = C_0 - \frac{x^2}{2R} = C_0 - \frac{a^2\cos 2\phi}{4R} - \frac{a^2}{4R} \,. \tag{12.48}$$

Thus

$$u_2 = \frac{a^2}{2R} \,, \tag{12.49}$$

(see equation (12.36)) and hence

$$p_2 = \frac{2\mu a}{R(\kappa+1)} \,, \tag{12.50}$$

from (12.35).

For the cylinder, the load must be symmetrical to retain equilibrium and hence $p_1=0$. As before, $p_0$ is given by (12.42) and hence

$$p(\theta) = \left(-\frac{F}{\pi a} + \frac{2\mu a}{R(\kappa+1)}\cos 2\theta\right) / \sin\theta \,. \tag{12.51}$$

Now this expression will be singular at $\theta=0,\pi$ ($x=\pm a$) unless we choose $a$ such that

$$\frac{F}{\pi a} = \frac{2\mu a}{R(\kappa+1)} \,,$$

i.e.

$$a = \sqrt{\frac{F(\kappa+1)R}{2\pi\mu}} \,. \tag{12.52}$$

Note that for plane strain, $\kappa=3-4\nu$ and

$$\frac{2\mu}{(\kappa+1)} = \frac{E}{4(1-\nu^2)} \,, \tag{12.53}$$

whilst for plane stress, $\kappa=(3-\nu)/(1+\nu)$ and

$$\frac{2\mu}{(\kappa+1)} = \frac{E}{4} \,. \tag{12.54}$$

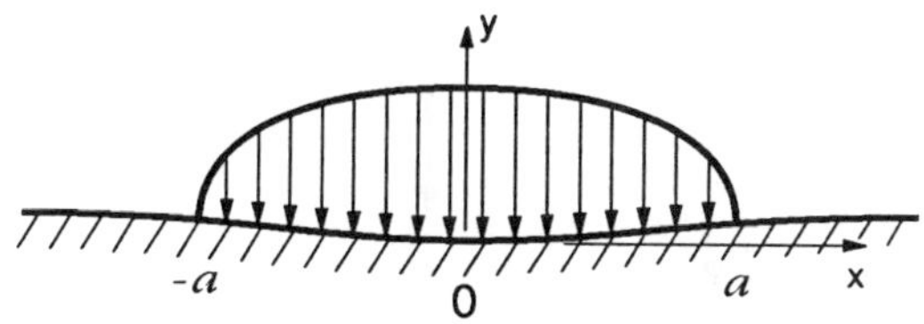


Figure 12.9: The Hertzian pressure distribution

With the value of $a$ from (12.52), we find

$$p(\theta) = -\frac{F(1-\cos 2\theta)}{\pi a \sin\theta} = -\frac{2F\sin\theta}{\pi a} \; ; \tag{12.55}$$

i.e.

$$p(x) = -\frac{2F\sqrt{a^2-x^2}}{\pi a^2} \; . \tag{12.56}$$

This pressure distribution is illustrated in Figure 12.9.

## 12.6 Problems with two deformable bodies

The same method can be used to treat problems involving the contact of two deformable bodies, provided that they have sufficiently large radii in the vicinity of the contact area to be approximated by half-planes — i.e. $R_1, R_2 \gg a$, where $R_1, R_2$ are the local radii[4] (see Figure 12.10). We shall also take the opportunity to generalize the formulation to the case where, in addition to the normal tractions $p_y$, there are tangential tractions $p_x$ at the interface due to friction.

We first note that, by Newton's third law, the tractions must be equal and opposite on the two surfaces, so the appropriate Green's function corresponds to the force pairs $F_x, F_y$ of Figure 12.11.

Suppose that the bodies are placed lightly in contact, as shown in Figure 12.12, and that the initial gap between their surfaces is a known function $g_0(x)$. We now give the upper body a vertical rigid-body translation $C_0$ and a small clockwise rotation $C_1$ such that, in the absence of deformation, the gap would become

$$g(x) = g_0(x) - C_0 - C_1 x \; . \tag{12.57}$$

[4]In fact the same condition must be satisfied even for the case where the curved body is rigid, since otherwise the small strain assumption of linear elasticity ($e_{ij} \ll 1$) will be violated near the contact area.

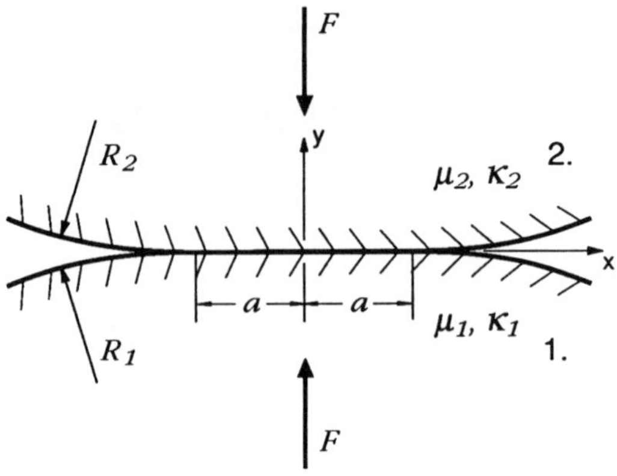


Figure 12.10: Contact of two curved deformable bodies

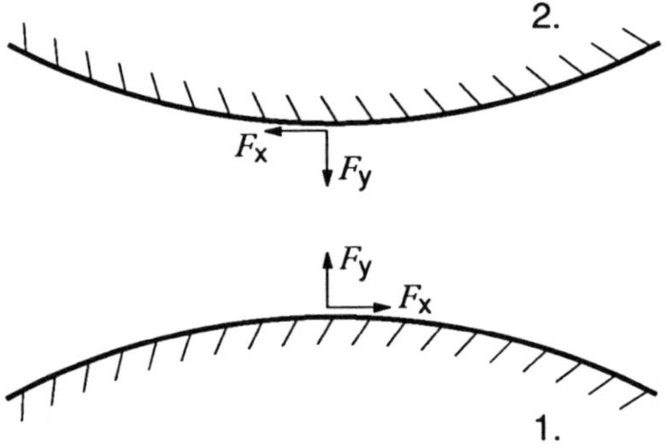


Figure 12.11: Green's function for contact problems

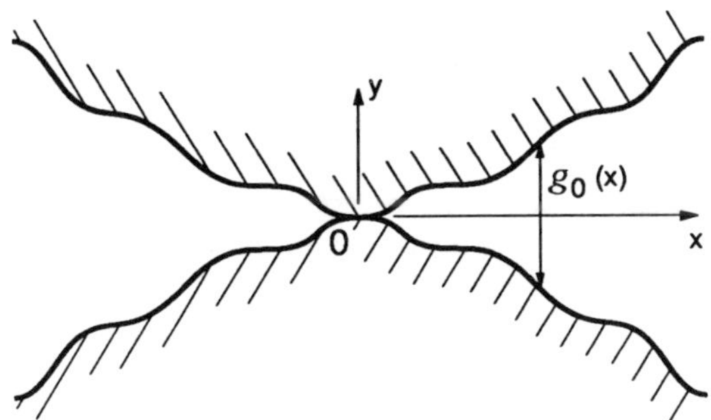


Figure 12.12: Initial gap between two unloaded contacting bodies

In addition, we assume that the contact tractions will cause some elastic deformation, represented by the displacements $\boldsymbol{u}_1,\boldsymbol{u}_2$ in bodies 1,2 respectively.

As a result of these operations, the gap will be modified to

$$g(x) = g_0(x) - u_{y1}(x,0) + u_{y2}(x,0) - C_0 - C_1 x \tag{12.58}$$

and it follows that the contact condition analogous to (12.28) can be written

$$u_{y1}(x,0) - u_{y2}(x,0) = g_0(x) - C_0 - C_1 x \;\; ; \;\; \text{in } A \, , \tag{12.59}$$

since, in a contact region, the gap is by definition zero.

The surface displacement $u_{y1}(x,0)$ of body 1 is simply (12.26) generalized to include the effect of the tangential traction $p_x$ (see equation (12.25)) — i.e.

$$\begin{aligned} u_{y1}(x,0) &= -\frac{(\kappa+1)}{4\pi\mu}\int_{-a}^{a} p_y(\xi)\log|x-\xi|d\xi \\ &\quad -\frac{(\kappa-1)}{8\mu}\int_{-a}^{a} p_x(\xi)\text{sgn}(x-\xi)d\xi \end{aligned} \tag{12.60}$$

and hence[5]

$$\frac{du_{y1}}{dx} = -\frac{(\kappa+1)}{4\pi\mu}\int_{-a}^{a}\frac{p_y(\xi)d\xi}{(x-\xi)} - \frac{(\kappa-1)}{4\mu}p_x(x) \, . \tag{12.61}$$

[5]Notice that an alternative representation of sgn($x$) is $2H(x)-1$, where $H(x)$ is the Heaviside step function. It follows that the derivative of sgn($x$) is $2\delta(x)$ and that the derivative of the integral in the second term of (12.60) is simply twice the value of $p_x(\xi)$ at the point $\xi = x$.

We also record the corresponding expression for the tangential displacement $u_{x1}$, which is

$$\frac{du_{x1}}{dx} = -\frac{(\kappa+1)}{4\pi\mu}\int_{-a}^{a}\frac{p_x(\xi)d\xi}{(x-\xi)} + \frac{(\kappa-1)}{4\mu}p_y(x) , \tag{12.62}$$

from (12.22).

Comparing Figure 12.11 with Figure 12.3, we see that equations (12.61, 12.62) can be used for the displacements $u_x, u_y$ of body 1, but for the corresponding displacements of body 2 we have to take account of the fact that the tractions $p_x, p_y$ are reversed and that the $y$-axis is now directed *into* the body. It is easily verified that this can be achieved by changing the signs in the expressions involving $u_y, p_x$, whilst leaving those expressions with $u_x, p_y$ unchanged. It then follows that

$$\frac{d}{dx}(u_{y1}-u_{y2}) = -\frac{A}{4\pi}\int_{-a}^{a}\frac{p_y(\xi)d\xi}{(x-\xi)} - \frac{B}{4}p_x(x) ; \tag{12.63}$$

$$\frac{d}{dx}(u_{x1}-u_{x2}) = -\frac{A}{4\pi}\int_{-a}^{a}\frac{p_x(\xi)d\xi}{(x-\xi)} + \frac{B}{4}p_y(x) , \tag{12.64}$$

where

$$A = \frac{(\kappa_1+1)}{\mu_1} + \frac{(\kappa_2+1)}{\mu_2} ; \tag{12.65}$$

$$B = \frac{(\kappa_1-1)}{\mu_1} - \frac{(\kappa_2-1)}{\mu_2} . \tag{12.66}$$

## 12.7 Uncoupled problems

We can now substitute (12.63) into the derivative of the contact condition (12.59), obtaining

$$g_0'(x) - C_1 = -\frac{A}{4\pi}\int_{-a}^{a}\frac{p_y(\xi)d\xi}{(x-\xi)} - \frac{B}{4}p_x(x) \; ; \; -a < x < a . \tag{12.67}$$

This equation is similar in form to (12.31), except for the presence of the term involving $p_x$. It can therefore be solved in the same way if for any reason this term is identically zero. Four important cases where this condition is satisfied are:-

1. The contact is frictionless, so $p_x=0$.

2. The materials are similar ($\kappa_1=\kappa_2,\ \mu_1=\mu_2$) and hence $B=0$.

3. Both materials are incompressible ($\nu_1=\nu_2=0.5,\ \kappa_1=\kappa_2=1,\ \mu_1\neq\mu_2$).

4. One body is rigid ($\mu_1=\infty$) and the other incompressible ($\kappa_2=1$).

Of course, no real materials are even approximately rigid, but the coupling terms in equations (12.63, 12.64) — i.e. those connecting $u_y, p_x$ and $u_x, p_y$ —can reasonably be neglected provided that the ratio

$$\beta = \frac{B}{A} = \left(\frac{(\kappa_1 - 1)}{\mu_1} - \frac{(\kappa_2 - 1)}{\mu_2}\right) \bigg/ \left(\frac{(\kappa_1 + 1)}{\mu_1} + \frac{(\kappa_2 + 1)}{\mu_2}\right) \ll 1 \ . \tag{12.68}$$

This will be true if $\mu_1 \gg \mu_2$ and $(\kappa_2 - 1) \ll 1$ and it is a reasonable approximation to the practically important case of rubber in contact with steel. The dimensionless parameter $\beta$ is one of Dundurs' bimaterial parameters (see §4.4.5).

In the rest of this chapter, we shall restrict attention to problems in which the coupling terms can be neglected for one of the reasons given above.

### 12.7.1 Contact of cylinders

If the two bodies are cylinders with radii $R_1, R_2$ (see Figure 12.10), we have

$$g_0''(x) = \frac{1}{R_1} + \frac{1}{R_2} = \frac{R_1 + R_2}{R_1 R_2} \equiv \frac{1}{R} \ . \tag{12.69}$$

Hence, comparing equations (12.69, 12.47) and (12.67, 12.31), we find that the solution can be written down from equations (12.52, 12.56) by replacing $R$ by $R_1 R_2/(R_1 + R_2)$ and $(\kappa + 1)/\mu$ by $A$ of equation (12.65). i.e.

$$a = \sqrt{\frac{F R_1 R_2}{2\pi(R_1 + R_2)} \left(\frac{(\kappa_1 + 1)}{\mu_1} + \frac{(\kappa_2 + 1)}{\mu_2}\right)} \ ; \tag{12.70}$$

$$p_y(x) = -\frac{2F\sqrt{a^2 - x^2}}{\pi a^2} \ . \tag{12.71}$$

## 12.8 Combined normal and tangential loading

Tangential tractions can be transmitted between contacting bodies only by means of friction and the complete specification of the problem then requires an assumption about the friction law relating the tangential traction to the relative tangential motion at the interface.

As in the normal contact analysis, tangential relative displacement or *shift*, $h(x)$, can result from rigid-body motion, $C$, and/or elastic deformation, being given by

$$h(x) = u_{x2}(x, 0) - u_{x1}(x, 0) + C \ , \tag{12.72}$$

where a positive shift corresponds to displacement of the upper body to the right relative to the lower body.

We shall define a state of *stick* as one in which the time derivative of the shift, $\dot{h}(x) = 0$. A state with $\dot{h}(x) \neq 0$ will be referred to as *positive* or *negative slip*, depending on the sign of $\dot{h}(x)$.

The simplest frictional assumption is that usually referred to as Coulomb's law, which, in terms of the above notation, can be defined as

$$\dot{h}(x) = \dot{u}_{x2} - \dot{u}_{x1} + \dot{C} = 0 \ ; \ \ fp_y(x) < p_x(x) < -fp_y(x) \tag{12.73}$$

in stick regions and

$$p_x(x) = -fp_y(x)\text{sgn}(\dot{h}(x)) \tag{12.74}$$

in slip regions, where $f$ is a constant known as the *coefficient of friction*[6]. We make no distinction between dynamic and static friction coefficients. In interpreting these equations, the reader should recall that the normal traction $p_y(x)$ is always compressive and hence negative in contact problems.

The function $\text{sgn}(\dot{h}(x))$ in (12.74) ensures that the frictional traction opposes the relative motion and hence dissipates energy. This condition and the inequality in (12.73) serve to determine the division of the contact region into positive slip, negative slip and stick zones in much the same way as the inequalities (12.45, 12.46) determine the contact area in the normal contact problem.

Notice that both of equations (12.73, 12.74) involve time derivatives. Thus, frictional contact problems are *incremental* in nature. It is not generally sufficient to know the final loading condition — we also need to know how that condition was reached. In other words, frictional contact problems are *history-dependent*. They share this and other properties with problems involving another well-known dissipative mechanism — plastic deformation.

### 12.8.1 Mindlin's problem

Consider the problem of two elastic cylinders which are first pressed together by a normal compressive force $F$ and then subjected to a monotonically increasing tangential force $T$ as shown in Figure 12.13. We restrict attention to the uncoupled case, $\beta = 0$.

The first phase of the loading is described by the analysis of §12.7.1, the contact semi-width $a$ and the normal tractions $p_y(x)$ being defined by equations (12.70, 12.71). The condition $\beta = 0$ ensures that these normal tractions produce no tendency for slip (see equations (12.64, 12.68)) and it follows that no tangential tractions are induced and that the whole contact area remains in a state of stick as the normal force $F$ is applied.

The absence of coupling also ensures that the contact area and the normal tractions remain constant during the tangential loading phase. Suppose we first assume that stick prevails everywhere during this phase as well.

[6] The more usual symbol $\mu$ for the coefficient of friction would lead to confusion with Lamé's constant.

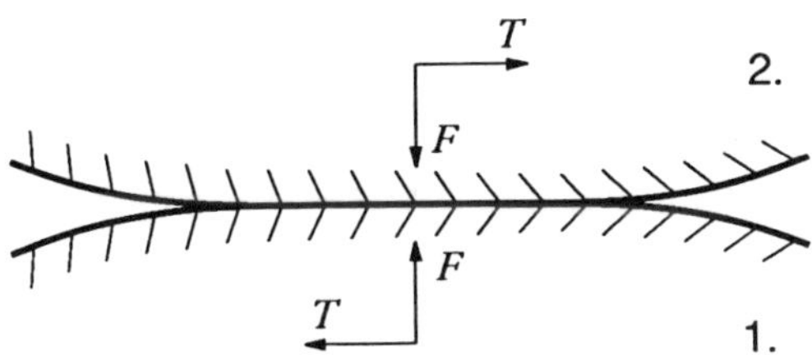


Figure 12.13: Loading for Mindlin's problem

Differentiating (12.73) with respect to $x$ and substituting for the displacement derivatives from (12.64) with $B=0$, we obtain

$$-\frac{A}{4\pi}\int_{-a}^{a}\frac{\dot{p}_x(\xi)d\xi}{(x-\xi)}=0 \;\; ; \;\; -a<x<a\,. \tag{12.75}$$

This equation is identical in form to (12.31) and is solved in the same way. The solution is easily shown to be

$$\dot{p}_x(x)=\frac{\dot{T}}{\pi\sqrt{a^2-x^2}}\,, \tag{12.76}$$

by analogy with (12.44) and hence

$$p_x(x)=\frac{T}{\pi\sqrt{a^2-x^2}}\,,. \tag{12.77}$$

since $a$ is independent of time during the tangential loading phase.

This result shows that the assumption of stick throughout the contact area $-a<x<a$ leads to a singularity in $p_x$ at the edges $x=\pm a$ and hence the frictional inequality (12.73) must be violated there for any $f$, since $p_y$ is bounded. We deduce that some slip will occur near the edges of the contact region for any non-zero $T$, however small.

The problem with slip zones was first solved apparently independently by Cattaneo[7] and Mindlin[8]. In the slip zones, the tractions satisfy the condition $p_x=-fp_y$. We therefore consider the solution for the tangential tractions as the sum of two parts:-

[7]C.Cattaneo, Sul contatto di corpi elastici, Acad. dei Lincei, Rediconti, Ser 6, Vol. 27 (1938), 342–348, 433–436, 474–478.

[8]R.D.Mindlin, Compliance of elastic bodies in contact, J.Appl.Mech., Vol. 17 (1949), 259–268.

1. a shear traction

$$p_x = -fp_y = \frac{2fF\sqrt{a^2-x^2}}{\pi a^2}\ , \tag{12.78}$$

from equation (12.71) *throughout* the contact area $-a<x<a$.

2. a *corrective* shear traction $p_x^*$, which must be zero in the slip zones and which is sufficient to restore the condition (12.73) in the stick zone.

As a first step towards finding the corrective traction $p_x^*$, we find the shift due to the traction distribution (12.78), which is defined by

$$\begin{aligned} h'(x) = \frac{d}{dx}(u_{x1}-u_{x2}) &= \frac{A}{4\pi}\int_{-a}^{a}\frac{2fF\sqrt{a^2-\xi^2}d\xi}{\pi a^2(x-\xi)} = -\frac{fFA}{2\pi a^2}\int_0^{\pi}\frac{\sin^2\theta d\theta}{(\cos\phi-\cos\theta)} \\ &= -\frac{fFA}{2\pi a}\cos\phi = -\frac{fFAx}{2\pi a^2}\ . \end{aligned} \tag{12.79}$$

Now, in the *stick* zone, we require the shift to be independent of $x$ and hence we seek corrective shear tractions in the stick zone that will cancel the right hand side of equation (12.81). By analogy with equations (12.78, 12.79), it is clear that this cancellation can be achieved by the distribution[9]

$$p_x^* = -\frac{2fF\sqrt{c^2-x^2}}{\pi a^2}\ \ ;\ \ -c<x<c\ , \tag{12.80}$$

i.e. a traction similar in form to equation (12.78), but distributed over a smaller centrally located slip zone of width $2c$.

Thus, the complete shear traction distribution is

$$p_x = \frac{2fF}{\pi a^2}[\sqrt{a^2-x^2} - H(c^2-x^2)\sqrt{c^2-x^2}]\ \ ;\ \ -a<x<a\ , \tag{12.81}$$

which is illustrated in Figure 12.14.

The stick zone semi-width $c$ can be determined by requiring

$$T = \int_{-a}^{a} p_x dx = fF - fF\left(\frac{c}{a}\right)^2\ , \tag{12.82}$$

[9]Notice that the assumption is that all points in $-c<x<c$ are in a state of stick ***throughout the loading process.*** It is therefore possible to integrate (12.73) and write the boundary condition in terms of $h(x)$ instead of $\dot{h}(x)$. In general, this is possible as long as no point passes from a state of slip to one of stick during the loading. In particular, the stick zone must not ***advance*** into the slip zone during loading. For an exhaustive study of the effect of loading history in frictional contact problems, see J.Dundurs and M.Comninou, An educational elasticity problem with friction: Part 1, Loading and unloading paths for weak friction, ASME J.Appl.Mech., Vol. 48 (1981), 841–845; Part 2: Unloading for strong friction and reloading, ***ibid.***, Vol. 49 (1982), 47–51; Part 3: General load paths, ***ibid***, Vol. 50 (1983), 77–84.

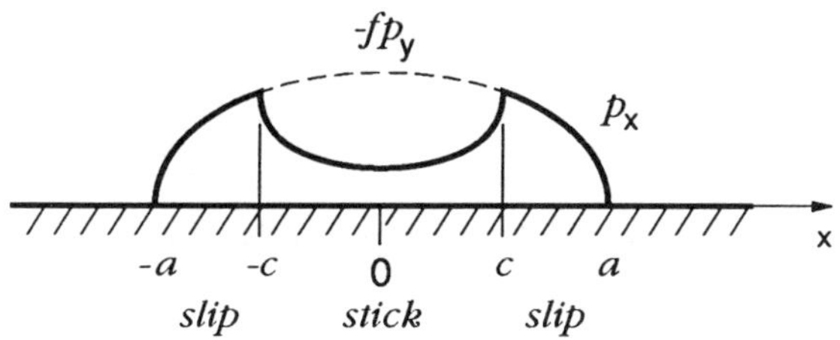


Figure 12.14: Shear traction distribution for Mindlin's problem

from (12.81) and hence

$$c = a\sqrt{1 - \frac{T}{fF}} . \tag{12.83}$$

As we would expect, the stick zone shrinks to zero as the applied tangential force $T$ approaches $fF$, after which gross slip (i.e. large scale rigid-body motion) occurs.

Finally, we note that it can be shown that the signs of the displacement derivatives satisfy the condition (12.74), provided that $T$ increases monotonically in time (i.e. $\dot{T} > 0$ for all $t$)[10].

The related solution for two spheres in contact (also due to Mindlin) has been verified experimentally by observing the damaged regions produced by cyclic microslip between tangentially loaded spheres[11].

### 12.8.2 Steady rolling: Carter's solution

A final example of considerable practical importance is that in which two cylinders roll over each other whilst transmitting a constant tangential force, $T$. If we assume that the rolling velocity is $V$, the solution will tend to a steady-state which is invariant with respect to the moving coördinate system

$$\xi = x - Vt \tag{12.84}$$

[10]The effect of non-monotonic loading in the related problem of two contacting spheres was considered by R.D.Mindlin and H.Deresiewicz, Elastic spheres in contact under varying oblique forces, J.Appl.Mech., Vol. 21 (1953), 327-344. The history-dependence of the friction law leads to quite complex arrangements of slip and stick zones and consequent variation in the load-compliance relation. These results also find application in the analysis of oblique impact, where neither normal nor tangential loading is monotonic (see N.Maw, J.R.Barber and J.N.Fawcett, The oblique impact of elastic spheres, Wear, Vol. 38 (1976), 101-114).

[11]K.L.Johnson, Energy dissipation at spherical surfaces in contact transmitting oscillating forces, J.Mech.Eng.Sci., Vol. 3 (1961), 362–368.

In this system, we can write the 'stick' condition (12.73) as

$$\dot{h}(x,t) = \frac{d}{dt}[u_{x2}(x-Vt) - u_{x1}(x-Vt) + C] = V\frac{d}{d\xi}(u_{x1} - u_{x2}) + \dot{C} = 0 \; , \quad (12.85)$$

where $\dot{C}$ is an arbitrary but constant rigid-body slip (or creep) velocity. Similarly, the slip condition (12.74) becomes

$$p_x = -fp_y \text{sgn}\left(V\frac{d}{d\xi}(u_{x1} - u_{x2})\right) + \dot{C} \; . \quad (12.86)$$

At first sight, we might think that the Mindlin traction distribution (11.79) satisfies this condition, since it gives

$$\frac{d}{d\xi}(u_{x1} - u_{x2}) = 0 \quad (12.87)$$

in the central stick zone and $\dot{C}$ can be chosen arbitrarily. However, if we substitute the resulting displacements into the slip condition (12.86), we find a sign error in the leading slip zone. This can be explained as follows: In the Mindlin problem, as $T$ is increased, positive slip (i.e. $\dot{h}(x) = \dot{u}_{x2} - \dot{u}_{x1} + \dot{C} > 0$) occurs in both slip zones and the magnitude of $h(x)$ increases from zero at the stick-slip boundary to a maximum at $x = \pm a$. It follows that $\frac{d}{d\xi}(u_{x1} - u_{x2})$ is negative in the right slip zone and positive in the left slip zone. Thus, if this solution is used for the steady rolling problem, a violation of (12.86) will occur in the right zone if $V$ is positive and in the left zone if $V$ is negative. In each case there is a violation in the *leading* slip zone — i.e. in that zone next to the edge where contact is being established.

Now in frictional problems, when we make an assumption that a given region slips and then find that it leads to a sign violation, it is usually an indication that we made the wrong assumption and that the region in question should be in a state of stick. Thus, in the rolling problem, there is no leading slip zone[12]. Carter[13] has shown that the same kind of superposition can be used for the rolling problem as for Mindlin's problem, except that the corrective traction is displaced to a zone adjoining the leading edge. A corrective traction

$$p_x^* = \frac{2fF}{\pi a^2}\sqrt{\left(\frac{a-c}{2}\right)^2 - \left(\xi - \frac{a+c}{2}\right)^2} = \frac{2fF}{\pi a^2}\sqrt{(a-\xi)(\xi-c)} \quad (12.88)$$

[12]Note that this applies to the uncoupled problem ($\beta = 0$) only. With dissimilar materials, there is generally a leading slip zone and there can also be an additional slip zone contained within the stick zone. This problem is treated by R.H.Bentall and K.L.Johnson, Slip in the rolling contact of dissimilar rollers, Int.J.Mech.Sci., Vol. 9 (1967), 389–404.

[13]F.W.Carter, On the action of a locomotive driving wheel, Proc.Roy.Soc. (London), Vol. A112 (1926), 151–157.

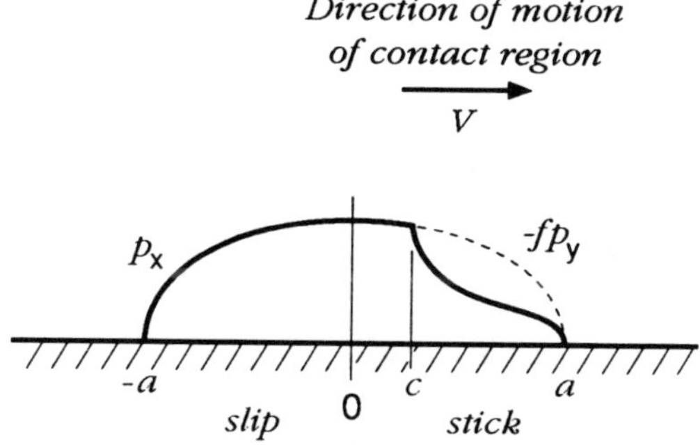


Figure 12.15: Shear traction distribution for Carter's problem.

produces a displacement distribution

$$\frac{d}{d\xi}(u_{x1}-u_{x2})=\frac{fFA}{2\pi a^2}\left(\xi-\frac{a+c}{2}\right)\ \ ;\ \ c<\xi<a\ , \tag{12.89}$$

which cancels the $x$-varying term in (12.79), leaving only an admissable constant $\dot{C}$ (see equation (12.85)). Hence, the traction distribution for positive $V$ is

$$p_x=\frac{2fF}{\pi a^2}[\sqrt{a^2-\xi^2}-H(\xi-c)\sqrt{(a-\xi)(\xi-c)}]\ \ ;\ \ -a<\xi<a\ , \tag{12.90}$$

which is illustrated in Figure 12.15.

Stick occurs in the leading zone $c<\xi<a$ and positive slip in the trailing zone $-a<\xi<c$.

The corresponding total tangential load is

$$T=fF\left[1-\left(\frac{a-c}{2a}\right)^2\right] \tag{12.91}$$

and hence the stick-slip boundary $\xi=c$ is given by

$$c=a\left(1-2\sqrt{1-\frac{T}{fF}}\right)\ . \tag{12.92}$$

An interesting feature of this solution is that the creep velocity $\dot{C}$ is not zero — i.e. there is a small steady-state relative tangential velocity between the two bodies. This has the effect of making the driven roller rotate slightly more slowly than a rigid-body kinematic analysis would lead us to expect. This in turn means that more energy is

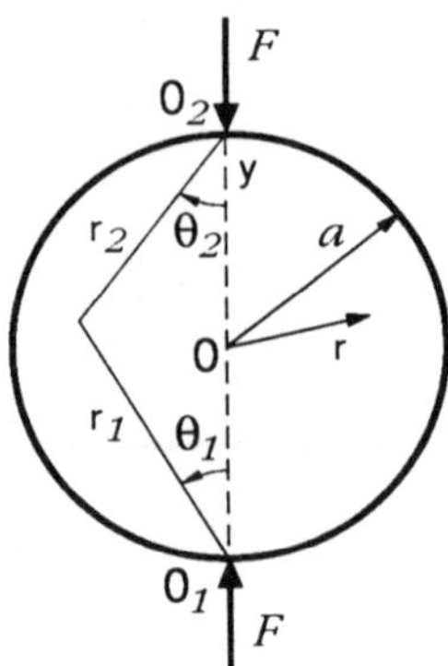


Figure 12.16: Disk loaded by concentrated forces

provided to the driving roller than is recovered from the driven roller, the balance of course being dissipated in the microslip regions in the form of heat. The creep velocity can be calculated by substituting the superposition of (12.79) and (12.89) into (12.85), with the result

$$\dot{C} = -\frac{fVa}{R}\left(1 - \sqrt{1 - \frac{T}{fF}}\right) , \tag{12.93}$$

where $R$ is given by (12.69). Notice that $\dot{C}$ increases without limit as $T \to fF$.

It is interesting to note that Carter's solution was published 20 years before Mindlin's, but there is no evidence that either Mindlin or Cattaneo was aware of it, despite the similarity of the techniques used[14].

## PROBLEMS

1. Figure 12.16 shows a disk of radius $a$ subjected to two equal and opposite forces $F$ at the points $A, B$, the rest of the boundary $r = a$ being traction-free.

The stress function

$$\phi = -\frac{F}{\pi}(r_1\theta_1 \sin\theta_1 + r_2\theta_2 \sin\theta_2)$$

is proposed to account for the localized effect of the forces. Find the stress field due to this function and, in particular, find the tractions implied upon the boundary $r = a$.

[14]For a more extensive discussion of frictional problems of this type, see K.L.Johnson, *Contact Mechanics*, Cambridge University Press, (1985), Chapters 5,7,8.

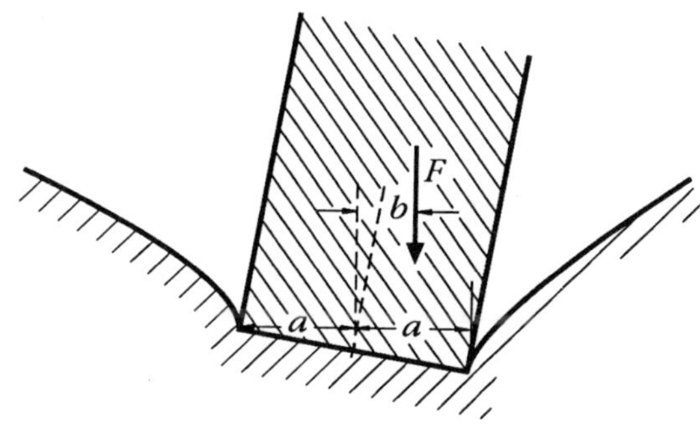


Figure 12.17: Punch with an eccentric load

Then complete the solution by superposing approriate stress functions from Table 8.1, so as to satisfy the traction-free boundary condition.

2. A rigid flat punch of width $2a$ is pressed into an elastic half-plane by a force $F$ whose line of action is displaced a distance $b$ from the centreline, as shown in Figure 12.17.

1. Assuming that the punch makes contact over the entire face $-a < x < a$, find the pressure distribution $p(x)$ and the angle of tilt ot the punch. Assume the half-plane is prevented from rotating at infinity.

2. Hence find the maximum value of $b$ for which there is contact throughout $-a < x < a$.

3. Re-solve the problem, assuming that $b$ is larger than the critical value found in (ii). Contact will now occur in the range $c < x < a$ and the unknown left hand end of the contact region ($x = c$) must be found from a smoothness condition on $p(x)$. Express $c$ and $p(x)$ as functions of $F, x$ and $b$.

# Chapter 13

# FORCES, DISLOCATIONS AND CRACKS

In this chapter, we shall discuss the applications of two solutions which are singular at an *interior* point of a body, and which can be combined to give the stress field due to a concentrated force (the *Kelvin solution*) and a dislocation. Both solutions involve a singularity in stress with exponent $-1$ and are therefore inadmissable according to the criterion of §11.2.1. However, like the Flamant solution considered in Chapter 12, they can be used as Green's functions to describe distributions, resulting in convolution integrals in which the singularity is integrated out. The Kelvin solution is also useful for describing the far field (i.e. the field a long way away from the loaded region) due to a force distributed over a small region.

## 13.1 The Kelvin solution

We consider the problem in which a concentrated force, $F$ acts in the $x$-direction at the origin in an infinite body. This is not a perturbation problem like the stress field due to a hole in an otherwise uniform stress field, since the force has a non-zero resultant. Thus, no matter how far distant we make the boundary of the body, there will have to be some traction to oppose the force. In fact, self-similarity arguments like those used in §§12.1, 12.2 show that the stress field must decay with $r^{-1}$.

The Flamant solution has this behaviour and it corresponds to a concentrated force (see §12.2), but it cannot be used at an interior point in the body, since the corresponding displacements (equations (12.12, 12.13)) are multivalued[1]. However, we can construct a solution with the same character and with single-valued displacements from the more general stress function (12.2) by choosing the coefficients in such a way that the multivalued terms cancel.

[1]This was not a problem for the surface loading problem, since the wedge of Figure 12.1 only occupies a part of the $\theta$-domain and hence a suitable principal value of $\theta$ can be chosen to be both single-valued and continuous.

In view of the symmetry of the problem about $\theta=0$, we restrict attention to the symmetric terms

$$\phi = C_1 r\theta \sin\theta + C_3 r \log r \cos\theta \tag{13.1}$$

of (12.2), for which the stress components are

$$\sigma_{rr} = \frac{2C_1\cos\theta}{r} + \frac{C_3\cos\theta}{r} ; \tag{13.2}$$
$$\sigma_{r\theta} = \frac{C_3\sin\theta}{r} ; \tag{13.3}$$
$$\sigma_{\theta\theta} = \frac{C_3\cos\theta}{r} , \tag{13.4}$$

from Table 8.1 and the displacement components are

$$2\mu u_r = \frac{C_1}{2}[(\kappa-1)\theta\sin\theta - \cos\theta + (\kappa+1)\log r\cos\theta] + \frac{C_3}{2}[(\kappa+1)\theta\sin\theta - \cos\theta + (\kappa-1)\log r\cos\theta] ; \tag{13.5}$$
$$2\mu u_\theta = \frac{C_1}{2}[(\kappa-1)\theta\cos\theta - \sin\theta - (\kappa+1)\log r\sin\theta] + \frac{C_3}{2}[(\kappa+1)\theta\cos\theta - \sin\theta - (\kappa-1)\log r\sin\theta] , \tag{13.6}$$

from Table 9.1.

Suppose we make an imaginary cut in the plane at $\theta=0, 2\pi$ and define a principal value of $\theta$ such that $0 \leq \theta < 2\pi$. This makes $\theta$ discontinuous at $\theta = 2\pi$, but the trigonometric functions of course remain continuous. The only potential difficulty is associated with the expressions $\theta\sin\theta, \theta\cos\theta$.

Now $\theta\sin\theta=0$ at $\theta=2\pi$ and hence this expression has the same value at the two sides of the cut and is continuous. We can therefore make the whole displacement field continuous by choosing $C_1, C_3$ such that the terms $\theta\cos\theta$ in $u_\theta$ cancel — i.e. by setting[2]

$$C_1(\kappa-1) + C_3(\kappa+1) = 0 , \tag{13.7}$$

which for plane stress is equivalent to

$$(1-\nu)C_1 + 2C_3 = 0 . \tag{13.8}$$

This leaves us with one degree of freedom (one free constant) to satisfy the condition that the force at the origin is equal to $F$. Considering the equilibrium of a small circle of radius $r$ surrounding the origin, we have

$$F + \int_0^{2\pi} (\sigma_{rr}\cos\theta - \sigma_{r\theta}\sin\theta) r d\theta = 0 . \tag{13.9}$$

[2]Notice that this choice also has the effect of cancelling the $\theta\sin\theta$ terms in (13.5), so that the complete displacement field, like the stress field, depends on $\theta$ only through sine and cosine terms.

We now substitute for the stress components from equations (13.2, 13.3), obtaining

$$C_1 = -\frac{F}{2\pi} , \tag{13.10}$$

after which we recover the constant $C_3$ from equation (13.8) as

$$C_3 = \frac{(1-\nu)F}{4\pi} . \tag{13.11}$$

Finally, we substitute these constants back into (13.2–13.4) to obtain

$$\sigma_{rr} = \frac{(3+\nu)F\cos\theta}{4\pi r} ; \tag{13.12}$$

$$\sigma_{r\theta} = \frac{(1-\nu)F\sin\theta}{4\pi r} ; \tag{13.13}$$

$$\sigma_{\theta\theta} = \frac{(1-\nu)F\cos\theta}{4\pi r} . \tag{13.14}$$

These are the stress components for Kelvin's problem, where the force acts in the $x$-direction. The corresponding results for a force in the $y$-direction can be obtained in the same way, using the antisymmetric terms in (12.2). Alternatively, we can simply rotate the axis system in the above solution by redefining $\theta \rightarrow (\theta - \pi/2)$.

### 13.1.1 Body force problems

Kelvin's problem is a special case of a body force problem — that in which the body force is a delta function at the origin. The solution can also be used to solve more general body force problems by convolution. We consider the body force $p_x \delta x \delta y$ acting on the element $\delta x \delta y$ as a concentrated point force and use the above solution to determine its effect on the stress components at an arbitrary point. Treating the component $p_y$ in the same way and summing over all the elements of the body then gives a double integral representation of the stress field.

This method is not restricted to the infinite body, since we only seek a particular solution of the body force problem. We can therefore use the convolution method to develop a solution for the stresses in an infinite body with the appropriate body force distribution, after which we 'cut out' the shape of the real body and correct the boundary conditions as required, using an appropriate homogeneous solution (i.e. a solution without body forces).

Any body force distribution can be treated this way — the method is not restricted to conservative vector fields. It is generally more algebraically tedious than the methods developed in Chapter 7, but it lends itself naturally to numerical implementation. For example, it can be used to extend the boundary integral method to body force problems.

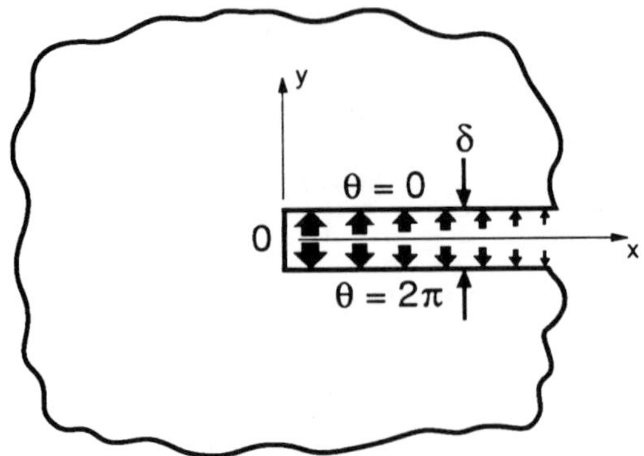


Figure 13.1: The climb dislocation solution

## 13.2 Dislocations

The term *dislocation* has related, but slightly different meanings in Elasticity and in Materials Science.

In Elasticity, the material is assumed to be a continuum — i.e. to be infinitely divisible. Suppose we take an infinite continuous body and make a cut along the half-plane $x>0$, $y=0$. We then apply equal and opposite tractions to the two surfaces of the cut such as to open up a gap of constant thickness, as illustrated in Figure 12.1. We then slip a thin slice of the same material into the space to keep the surface apart and weld the system up, leaving a new continuous body which will now be in a state of residual stress.

This is the stress field referred to as the dislocation solution. It can be obtained from the stress function of equation (13.1) by requiring that there be no net force at the origin, with the result $C_1=0$ (see equation 13.10). We therefore have

$$\phi = C_3 r \log r \cos\theta \; . \tag{13.15}$$

The strength of the dislocation can be defined in terms of the discontinuity in the displacement $u_\theta$ on $\theta=0, 2\pi$, which is also the thickness of the slice of extra material which must be inserted to restore continuity of material. This thickness is

$$\delta = u_\theta(0) - u_\theta(2\pi) = -\frac{\pi(\kappa+1)C_3}{2\mu} \tag{13.16}$$

$$= -\frac{2\pi C_3}{\mu(1+\nu)} \; , \tag{13.17}$$

for the case of plane stress.

Thus, we can define a dislocation of strength $B_y$ as one which opens a gap $\delta = B_y$, corresponding to $C_3 = -B_y\mu(1+\nu)/2\pi$ and the stress field

$$\sigma_{rr} = \sigma_{\theta\theta} = -\frac{\mu(1+\nu)B_y \cos\theta}{2\pi r} ; \tag{13.18}$$

$$\sigma_{r\theta} = -\frac{\mu(1+\nu)B_y \sin\theta}{2\pi r} . \tag{13.19}$$

We also record the stress components at $y=0$, — i.e. $\theta = 0, \pi$ — in rectangular coördinates, which are

$$\sigma_{xx} = \sigma_{yy} = -\frac{\mu(1+\nu)B_y}{2\pi x} ; \tag{13.20}$$

$$\sigma_{yx} = 0 . \tag{13.21}$$

This solution is called a *climb dislocation*, because it opens a gap on the cut at $\theta = 0, 2\pi$. A corresponding solution can be obtained from the stress function $\phi = r \log r \sin\theta$ which is discontinuous in the displacement component $u_r$ — i.e. for which the two surfaces of the cut experience a relative tangential displacement. This is called a *glide dislocation*. The solutions actually differ only in orientation. The climb dislocation becomes a glide dislocation if we choose to make the cut along the line $\theta = -3\pi/2, \pi/2$ (i.e. along the $y$-axis instead of the $x$-axis[3]). We distinguish the two orientations by using the symbol $B_x$ to denote the strength of a glide dislocation and $B_y$ that of a climb dislocation. We can regard $B_x, B_y$ as the components of a vector, known as the *Burgers vector*.

### 13.2.1 Dislocations in Materials Science

Of course, real materials have a discrete atomic or molecular structure. However, we can follow a similar procedure by imagining cleaving the solid between two sheets of molecules up to the line $x = y = 0$ and inserting *one extra layer* of molecules. When the system is released, there will be some motion of the molecules, mostly concentrated at the end of the added layer, resulting in an imperfection in the regular molecular array. This is what is meant by a dislocation in Materials Science.

Considerable insight into the role of dislocations in material behaviour was gained by the work of L.Bragg[4] with bubble models. Bragg devised a method of generating a two-dimensional collection of identical size bubbles. Attractive forces between the bubbles ensured that they adopted a regular array wherever possible, but dislocations are identifiable as can be seen in Figure 13.2. When forces are applied to the edges of

[3]In fact, the cut can be made along any (not necessarily straight) line from the origin to infinity. The solution of equations (13.18, 13.19) will then exhibit a discontinuity in $u_y$ of magnitude $B_y$ at all points along the cut, as long as the principal value of $\theta$ is appropriately defined.

[4]L.Bragg and J.F.Nye, A dynamical model of a crystal structure, Proc. Roy. Soc., Ser.A, Vol.190 (1947), 474–481.

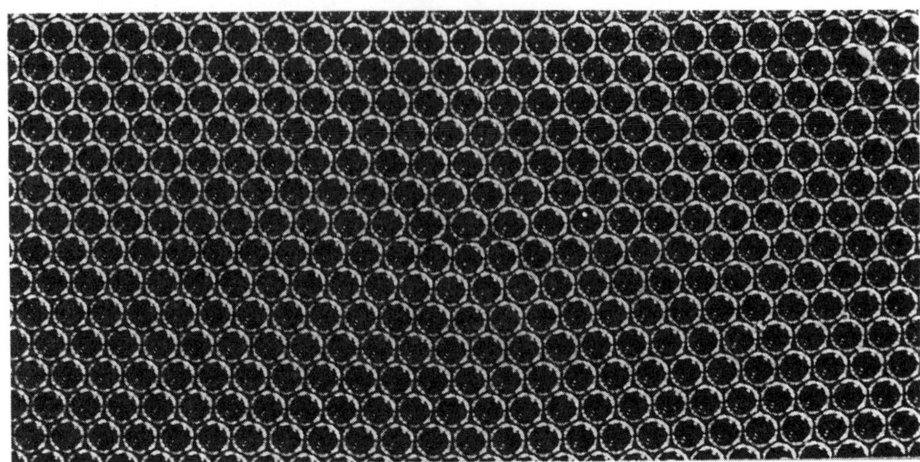

Figure 13.2: The bubble model — a dislocation

the bubble assembly, the dislocations are found to move in such a way as to permit the boundaries to move. This dislocation motion is believed to be the principal mechanism of plastic deformation in ductile materials. Larger bubble assemblies exhibit discrete regions in which the arrays are differently aligned as in Figure 13.3. These are analogous with grains in multigranular materials. When the structure is deformed, the dislocations typically move until they reach a grain boundary, but the misalignment prohibits further motion and the stiffness of the assembly increases. This pile-up of dislocations at grain boundaries is responsible for work-hardening in ductile materials. Also, the accumulated dislocations coalesce into larger disturbances in the crystal structure such as voids or cracks, which function as initiation points for failure by fatigue or fracture.

### 13.2.2 Similarities and differences

It is tempting to deduce that the elastic solution of §13.2 describes the stresses due to the molecular structure dislocation, if we multiply by a constant defining the thickness of a single layer of molecules. However, the concept of stress is rather vague over dimensions comparable with interatomic distances. In fact, this is pre-eminently a case where the apparent singularity of the mathematical solution is moderated in reality by the discrete structure of the material.

Notice also that the continuum dislocation of §13.2 can have any strength, cor-

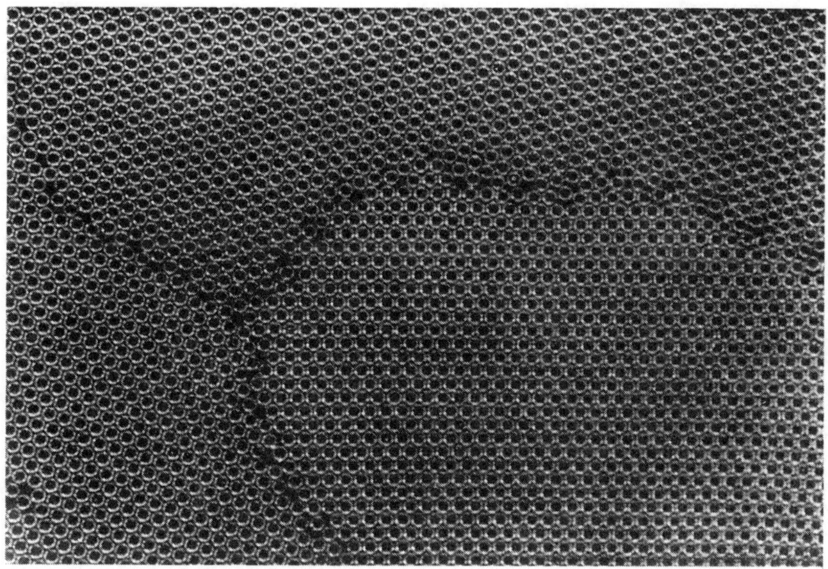

Figure 13.3: The bubble model — grain boundaries

responding to the fact that $B_x, B_y$ are arbitrary real constants. By contrast, the thickness of the inserted layer is restricted to one layer of molecules in the discrete theory. This thickness is sufficiently small to ensure that the 'stresses' due to a material dislocation are very small in comparison with typical engineering magnitudes, except in the immediate vicinity of the defect. Furthermore, although a stress-relieved metallic component will generally contain numerous dislocations, they will be randomly oriented, leaving the components essentially stress-free on the macroscopic scale.

However, if the component is plastically deformed, the dislocation motion will lead to a more systematic structure, which can result in residual stress. Indeed, one way to represent the stress field during plastic deformation is as a distribution of mathematical dislocations. Notice, however, that during plastic deformation, dislocations move, but are not created or destroyed. This can be represented mathematically by introducing a dislocation pair — i.e. a negative dislocation at the original location and an equal positive dislocation at the final location. This leads to a *closure condition* as in §§13.2.3, 13.2.4 below.

### 13.2.3 Dislocations as Green's functions

In the Theory of Elasticity, the principal use of the dislocation solution is as a Green's function to represent localized processes. Suppose we place a distribution of dislocations in some interior domain $\Omega$ of the body, which is completely surrounded by elastic material. The resulting stress field obtained by integration will satisfy the conditions of equilibrium everywhere (since the dislocation solution involves no force) and will satisfy the compatibility condition everywhere *except* in $\Omega$. There is of course the possibility that the displacement may be multiple-valued outside $\Omega$, but we can prevent this by enforcing the two *closure conditions*

$$\int_\Omega B_x(x,y)dxdy = 0 \;\; ; \;\; \int_\Omega B_y(x,y)dxdy = 0 \; , \tag{13.22}$$

which state that the total strength of the dislocations in $\Omega$ is zero[5].

### 13.2.4 Stress concentrations

A special case of some importance is that in which the enclosed domain $\Omega$ represents a hole which perturbs the stress field in an elastic body.

It may seem strange to place dislocations in a region which is strictly not a part of the body. However, we might start with an infinite body with no hole, place dislocations in $\Omega$ generating a stress field and then make a cut along the boundary of $\Omega$ producing the body with a hole. The stress field will be unchanged by the cut provided we place tractions on its boundary equal to those which were transmitted across the same surface in the original continuous body. In particular, if we choose the dislocation distribution so as to make the boundaries of $\Omega$ traction-free, we can cut out the hole without changing the stress distribution, which is therefore the solution of the original problem for the body with a hole.

The general idea of developing perturbation solutions by placing singularities in a region where the governing equations (here the compatibility condition) are not required to be enforced is well-known in many branches of Applied Mechanics. For example, the solution for the flow of a fluid around a rigid obstruction can be developed in many cases by placing an appropriate distribution of sources and sinks in the region occupied by the body, the distribution being chosen so as to make the velocity component normal to the body surface be everywhere zero.

The closure conditions (13.22) ensure that acceptable dislocation distributions can be represented in terms of dislocation pairs — i.e. as matched pairs of dislocations of equal magnitude and opposite direction. It follows that acceptable distributions can be represented as distributions of *dislocation derivatives*, since, for example, a dislocation at $P$ and an equal negative dislocation at $Q$ is equivalent to a uniform distribution of derivatives on the straight line joining $P$ and $Q$[6].

Now the stress components in the dislocation solution (equations (13.18, 13.19)) decay with $r^{-1}$ and hence those in the *dislocation derivative* solution will decay with

[5]This does *not* mean that the stress field is null, since the various self-cancelling dislocations have different locations.

[6]To see this, think of the derivative as the limit of a pair of equal and opposite dislocations separated by a distance $\delta S$ whose magnitude is proportional to $1/\delta S$.

$r^{-2}$. It therefore follows that the dominant term in the perturbation (or corrective) solution due to a hole will decay at large $r$ with $r^{-2}$. We see this in the particular case of the circular hole in a uniform stress field (§§8.3.2, 8.4.1). The same conclusion follows for the perturbation in the stress field due to an inclusion — i.e. a localized region whose properties differ from those of the bulk material.

## 13.3 Crack problems

Crack problems are particularly important in Elasticity because of their relevance to the subject of Fracture Mechanics, which broadly speaking is the study of the stress conditions under which cracks grow. For our purposes, a crack will be defined as the limiting case of a hole whose volume (at least in the unloaded case) has shrunk to zero, so that opposite faces touch. It might also be thought of as an interior surface in the body which is incapable of transmitting tension.

In practice, cracks will generally have some small thickness, but if it is small, the crack will behave unilaterally with respect to tension and compression — i.e. it will open if we try to transmit tension, but close in compression, transmitting the tractions by means of contact. For this reason, a cracked body will appear stiffer in compression than it does in tension. Also, the crack acts as a stress concentration in tension, but not in compression, so cracks do not generally propagate in compressive stress fields.

### 13.3.1 Linear Elastic Fracture Mechanics

We saw in §11.2.3 that the asymptotic stress field at the tip of a crack has a square-root singularity and we shall find this exemplified in the particular solutions that follow. The simplest and most prevalent theory of brittle fracture — that due to Griffith — states in essence that crack propagation will occur when the scalar multiplier on this singular stress field exceeds a certain critical value. More precisely, Griffith proposed the thesis that a crack would propagate when propagation caused a reduction in the total energy of the system. Crack propagation causes a reduction in strain energy in the body, but also generates new surfaces which have *surface energy*. Surface energy is related to the force known as surface tension in fluids and follows from the fact that to cleave a solid body along a plane involves doing work against the interatomic forces across the plane. When this criterion is applied to the stress field in particular cases, it turns out that for a small change in crack length, propagation is predicted when the multiplier on the singular term, known as the *stress intensity factor*, exceeds a certain critical value, which is a constant for the material known as the *fracture toughness.*

It may seem paradoxical to found a theory of real material behaviour on properties of a singular elastic field, which clearly cannot accurately represent conditions in the precise region where the failure is actually to occur. However, if the material is brittle, non-linear effects will be concentrated in a relatively small *process zone* surrounding the crack tip. Furthermore, the undoubtedly very complicated conditions in this process zone can only be influenced by the surrounding elastic material and hence the conditions for failure must be expressible in terms of the characteristics of the much simpler surrounding elastic field. As long as the process zone is small compared with the other linear dimensions of the body (notably the crack length), it will have only a very localized effect on the surrounding elastic field, which will therefore be adequately characterized by the dominant singular term in the linear elastic solution, whose multiplier (the stress intensity factor) then determines the conditions for crack propagation.

It is notable that this argument requires no assumption about or knowledge of the actual mechanism of failure in the process zone and, by the same token, the success of Linear Elastic Fracture Mechanics (LEFM) as a predictor of the strength of brittle components provides no evidence for or against any particular failure theory[7].

### 13.3.2 Plane crack in a tensile field

Two-dimensional crack problems are very conveniently formulated using the methods outlined in §§13.2.3, 13.2.4. Thus, we seek a distribution of dislocations on the plane of the crack (which appears as a line in two-dimensions), which, when superposed on the unperturbed stress field, will make the surfaces of the crack traction-free. We shall illustrate the method for the simple case of a plane crack in a tensile stress field.

Figure 13.4 shows a plane crack of width $2a$ occupying the region $-a<x<a$; $y=0$ in a two-dimensional body subjected to uniform tension $\sigma_{yy}=S$ at its remote boundaries.

Assuming that the crack opens, the boundary conditions for this problem can be stated in the form

$$\sigma_{yx}=\sigma_{yy}=0 \;\; ; \;\; -a<x<a,\; y=0 \; ; \tag{13.23}$$

$$\sigma_{yy}\rightarrow S \;\; ; \;\; \sigma_{xy},\sigma_{xx}\rightarrow 0 \;\; ; \;\; r\rightarrow\infty \; . \tag{13.24}$$

Following the procedure of §8.3.2, we represent the solution as the sum of the stress field in the corresponding body without a crack — here a uniform uniaxial tension $\sigma_{yy}=S$ — and a corrective solution, for which the boundary conditions are

$$\sigma_{yx}=0 \;\; ; \;\; \sigma_{yy}=-S \;\; ; \;\; -a<x<a,\; y=0 \; ; \tag{13.25}$$

$$\sigma_{xx},\sigma_{xy},\sigma_{yy}\rightarrow 0 \;\; ; \;\; r\rightarrow\infty \; . \tag{13.26}$$

[7] For more details of the extensive development of the field of Fracture Mechanics, the reader is reffered to the many excellent texts on the subject, such as M.F.Kanninen and C.H.Popelar, *Advanced Fracture Mechanics*, Clarendon Press, Oxford, (1985), H.Leibowitz, ed., *Fracture, An Advanced Treatise*, 7 Vols., Academic Press, New York, (1971). Stress intensity factors for a wide range of geometries are tabulated by G.C.Sih, *Handbook of Stess Intensity Factors*, Inst. of Fracture and Solid Mechanics, Lehigh University, Bethlehem, PA, (1973).

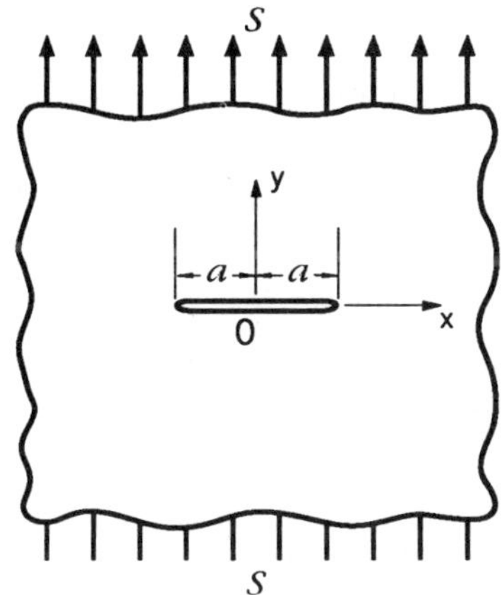


Figure 13.4: Plane crack in a tensile field

Notice that the corrective solution corresponds to the problem of a crack in an otherwise stress-free body opened by compressive normal tractions of magnitude $S$.

The most general stress field due to the crack would involve both climb and glide dislocations, but in view of the symmetry of the problem about the plane $y = 0$, we conclude that there is no relative tangential motion between the crack faces and hence that the solution can be constructed with a distribution of climb dislocations only. More precisely, we respresent the solution in terms of a distribution $B_y(x)$ of dislocations per unit length in the range $-a < x < a$; $y = 0$.

We consider first the traction $\sigma_{yy}$ at the point $(x, 0)$ due to those dislocations between $\xi$ and $\xi + \delta\xi$ on the line $y = 0$. If $\delta\xi$ is small, they can be considered as a concentrated dislocation of strength $B_y(\xi)\delta\xi$ and hence they produce a traction

$$\sigma_{yy} = -\frac{\mu(1+\nu)B_y(\xi)\delta\xi}{2\pi(x-\xi)}, \tag{13.27}$$

from equation (13.20), since the distance from $(\xi, 0)$ to $(x, 0)$ is $(x - \xi)$.

The traction due to the whole distribution of dislocations can therefore be written as the integral

$$\sigma_{yy} = -\frac{\mu(1+\nu)}{2\pi}\int_{-a}^{a}\frac{B_y(\xi)d\xi}{(x-\xi)} \tag{13.28}$$

and the boundary condition (13.25) leads to the following integral equation for $B_y(\xi)$

$$\int_{-a}^{a} \frac{B_y(\xi)d\xi}{(x-\xi)} = \frac{2\pi S}{\mu(1+\nu)} \;\; ; \;\; -a < x < a \; . \tag{13.29}$$

This is of exactly the same form as equation (12.31) and can be solved in the same way. Writing

$$x = a\cos\phi \;\; ; \;\; \xi = a\cos\theta \; , \tag{13.30}$$

we have

$$\int_{0}^{\pi} \frac{B_y(\theta)\sin\theta d\theta}{(\cos\phi - \cos\theta)} = \frac{2\pi S}{\mu(1+\nu)} \;\; ; \;\; 0 < \phi < \pi \; . \tag{13.31}$$

Now (12.34) with $n=1$ gives

$$\int_{0}^{\pi} \frac{\cos\theta d\theta}{(\cos\phi - \cos\theta)} = -\pi \;\; ; \;\; 0 < \phi < \pi \tag{13.32}$$

and hence

$$B_y(\theta) = -\frac{2S\cos\theta}{\mu(1+\nu)\sin\theta} + \frac{A}{\sin\theta} \; , \tag{13.33}$$

i.e.

$$B(\xi) = -\frac{2S\xi}{\mu(1+\nu)\sqrt{a^2-\xi^2}} + \frac{Aa}{\sqrt{a^2-\xi^2}} \; . \tag{13.34}$$

The arbitrary constant $A$ is determined from the closure condition (13.22) which here takes the form

$$\int_{-a}^{a} B_y(\xi)d\xi = 0 \tag{13.35}$$

and leads to the result $A=0$.

From a Fracture Mechanics perspective, we are particularly interested in the stress field surrounding the crack tip. For example, the stress component $\sigma_{yy}$ in $|x|>a$; $y=0$ is given by

$$\sigma_{yy} = -\frac{\mu(1+\nu)}{2\pi} \int_{-a}^{a} \frac{B_y(\xi)d\xi}{(x-\xi)} \tag{13.36}$$

$$= \frac{S}{\pi} \int_{-a}^{a} \frac{\xi d\xi}{(x-\xi)\sqrt{a^2-\xi^2}}$$

$$= S\left(-1 + \frac{|x|}{\sqrt{x^2-a^2}}\right) \;\; ; \;\; |x| > a, \; y=0 \tag{13.37}$$

using 3.228.2 of Gradshteyn and Ryzhik[8].

[8] I.S.Gradshteyn and I.M.Ryzhik, *Tables of Integrals, Series and Products*, Academic Press, New York, 1980.

Remembering that this is the *corrective* solution, we add the uniform stress field $\sigma_{yy}=S$ to obtain the complete stress field, which on the line $y=0$ gives

$$\sigma_{yy} = \frac{S|x|}{\sqrt{x^2-a^2}} \;\; ; \;\; |x|>a\ . \tag{13.38}$$

This tends to the uniform field as it should as $x\to\infty$ and is singular as $x\to a^+$ or $x\to -a^-$. We define the *stress intensity factor*, $K_I$ as

$$K_I \equiv \lim_{x\to a^+} \sigma_{yy}(x)\sqrt{x-a} \tag{13.39}$$

$$= \lim_{x\to a^+} \frac{Sx\sqrt{x-a}}{\sqrt{x^2-a^2}}$$

$$= S\sqrt{\frac{a}{2}}\ . \tag{13.40}$$

We can also calculate the crack opening displacement

$$u_y(x,0^+) - u_y(x,0^-) = \int_{-a}^{x} B_y(\xi)d\xi = \frac{2S}{\mu(1+\nu)}\sqrt{a^2-x^2}\ . \tag{13.41}$$

Thus, the crack is opened to the shape of a long narrow ellipse as a result of the tensile field.

## PROBLEMS

1. Figure 13.5 shows a large body with a circular hole of radius $a$, subjected to a concentrated force $F$, tangential to the hole. The remainder of the hole surface is traction-free and the stress field is assumed to tend to zero as $r\to 0$.

(i) Find the stress components at the point $B(a,\theta)$ on the surface of the hole, due to the candidate stress function

$$\phi = \frac{F}{\pi}R\psi\cos(\psi)$$

in polar coördinates $R,\psi$ centered on the point $A$ as shown.

(ii) Transform these stress components into the polar coördinate system $r,\theta$ centered on the centre of the hole.

**Note**: *For points on the surface of the hole* $(r=a)$, we have $R=2a\sin(\theta/2)$, $\sin(\psi)=\cos(\theta/2)$, $\cos(\psi)=-\sin(\theta/2)$.

(iii) Complete the solution by superposing stress functions (in $r,\theta$) from Tables 8.1, 9.1 with appropriate Fourier components in the tractions and determining the multiplying constants from the conditions that (a) the hole be traction-free (except at $A$) and (b) the displacements be everywhere single-valued.

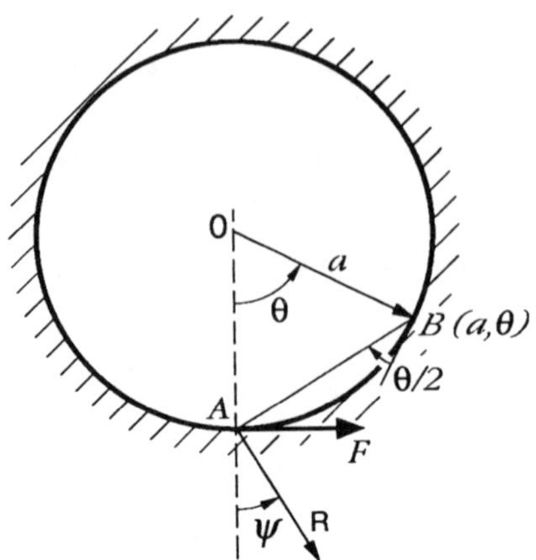


Figure 13.5: Tangential force at the surface of a hole

2. A large plate is in a state of pure bending such that $\sigma_{xx}=\sigma_{xy}=0$, $\sigma_{yy}=Cx$, where $C$ is a constant.

We now introduce a crack in the range $-a<x<a$, $y=0$.

(i) Find a suitable corrective solution which, when superposed on the simple bending field, will make the surfaces of the crack free of tractions. (**Hint**: represent the corrective solution by a distribution of climb dislocations in $-a<x<a$. Don't forget the closure condition (13.35)).

(ii) Find the corresponding crack opening displacement as a function of $x$ and show that it has an unacceptable negative value in $-a<x<0$.

(iii) Re-solve the problem assuming that there is frictionless contact in some range $-a<x<b$, where $b$ is a constant to be determined. The dislocations will now have to be distributed only in $b<x<a$ and $b$ is found from a continuity condition at $x=b$. (Move the origin to the mid-point of the range $b<x<a$.)

(iv) For the case with contact, find expressions for (a) the crack opening displacement, (b) the stress intensity factor at $x=a$, (c) the dimension $b$ and (d) the contact traction in $-a<x<b$ and hence verify that the contact inequalities are satisfied.

# Chapter 14

# THERMOELASTICITY

Most materials tend to expand if their temperature rises and, to a first approximation, the expansion is proportional to the temperature change. If the expansion is unrestrained, all dimensions will expand equally — i.e. there will be a uniform dilatation described by

$$e_{xx} = e_{yy} = e_{zz} = \alpha T \; ; \tag{14.1}$$
$$e_{xy} = e_{yz} = e_{zx} = 0 \; , \tag{14.2}$$

where $\alpha$ is the *coefficient of linear thermal expansion.* Notice that no shear strains are induced in unrestrained thermal expansion, so that a body which is heated to a uniformly higher temperature will get larger, but will retain the same shape.

Thermal strains are additive to the elastic strains due to local stresses, so that Hooke's law is modified to the form

$$e_{xx} = \frac{\sigma_{xx}}{E} - \frac{\nu\sigma_{yy}}{E} - \frac{\nu\sigma_{zz}}{E} + \alpha T \; ; \tag{14.3}$$
$$e_{xy} = \frac{\sigma_{xy}(1+\nu)}{E} \; . \tag{14.4}$$

## 14.1 The governing equation

The Airy stress function can be used for two-dimensional thermoelasticity, but the governing equation will generally include additional terms associated with the temperature field. Repeating the derivation of §4.4.3, but using (14.3) in place of (1.38), we find that the compatibility condition demands that

$$\frac{\partial^2\sigma_{xx}}{\partial y^2} - \nu\frac{\partial^2\sigma_{yy}}{\partial y^2} + E\alpha\frac{\partial^2 T}{\partial y^2} - 2(1+\nu)\frac{\partial^2\sigma_{xy}}{\partial x\partial y} + \frac{\partial^2\sigma_{yy}}{\partial x^2} - \nu\frac{\partial^2\sigma_{xx}}{\partial x^2} + E\alpha\frac{\partial^2 T}{\partial x^2} = 0 \tag{14.5}$$

and after substituting for the stress components from (4.1) and rearranging, we obtain

$$\nabla^4\phi = -E\alpha\nabla^2 T \; , \tag{14.6}$$

for plane stress.

The corresponding plane strain equations can be obtained by a similar procedure, noting that the restraint of the transverse strain $e_{zz}$ (14.1) will induce a stress $\sigma_{zz} = -E\alpha T$ and hence additional in-plane strains $\nu\alpha T$. Equation (14.6) is therefore modified to

$$\nabla^4\phi = -E\alpha(1+\nu)\nabla^2 T \ , \tag{14.7}$$

for plane strain and we can supplement the plane stress to plane strain conversions (3.18) with the relation

$$\alpha = \alpha'(1+\nu') \ . \tag{14.8}$$

Equations (14.6, 14.7) are similar in form to that obtained in the presence of body forces (7.8) and can be treated in the same way. Thus, we can seek any particular solution of (14.6) and then satisfy the boundary conditions of the problem by superposing a more general biharmonic function, since the biharmonic equation is the complementary or homogeneous equation corresponding to (14.6, 14.7).

### 14.1.1 Example

As an example, we consider the case of the thin circular disk, $r < a$, with traction-free edges, raised to the temperature

$$T = T_0 y^2 = T_0 r^2 \sin^2\theta \ , \tag{14.9}$$

where $T_0$ is a constant.

Substituting this temperature distribution into equation (14.6), we obtain

$$\nabla^4\phi = -2E\alpha T_0 \tag{14.10}$$

and a simple particular solution is

$$\phi_0 = -\frac{E\alpha T_0 r^4}{32} \ . \tag{14.11}$$

The stresses corresponding to $\phi_0$ are

$$\sigma_{rr} = -\frac{E\alpha T_0 r^2}{8} \ ; \ \sigma_{\theta\theta} = -\frac{3E\alpha T_0 r^2}{8} \ ; \ \sigma_{r\theta} = 0 \tag{14.12}$$

and the boundary $r = a$ can be made traction-free by superposing a uniform hydrostatic tension $E\alpha T_0 a^2/8$, resulting in the final stress field[1]

$$\sigma_{rr} = \frac{E\alpha T_0(a^2 - r^2)}{8} \ ; \ \sigma_{\theta\theta} = \frac{E\alpha T_0(a^2 - 3r^2)}{8} \ ; \ \sigma_{r\theta} = 0 \ . \tag{14.13}$$

[1]It is interesting to note that the stress field in this case is axisymmetric, even though the temperature field (14.9) is not.

### 14.1.2 The method of strain suppression

There is another connection between body force and thermoelastic problems. Suppose we were to constrain every particle of the body so as to prevent it from expanding — in other words, we constrain the displacement to be everywhere zero. We can solve (14.3, 14.4) for the stress field required to achieve this, but in general the stresses will not satisfy equilibrium. However, we can always find a body force distribution that will restore equilibrium — all we have to do is to substitute the non-equilibrated stresses into the equilibrium equation with body forces (2.5) and use the latter as a definition of $p_i$.

To complete the solution, we now remove the unwanted body forces by superposing the solution of a problem with no thermal strains, but with equal and opposite body forces. This can be done using the methods of Chapter 7, or by distributing point forces, using the Kelvin solution as outlined in §13.1.1. This is called the *method of strain suppression*[2].

## 14.2 Heat conduction

The temperature field might be a given quantity — for example, it might be measured using thermocouples or radiation methods — but more often it has to be calculated from thermal boundary conditions as a separate boundary-value problem. Most materials approximately satisfy the Fourier heat conduction law, according to which the heat flux is proportional to the local temperature gradient. i.e.

$$\boldsymbol{q} = -K\boldsymbol{\nabla} T \; , \tag{14.14}$$

where $K$ is the thermal conductivity of the material. The conductivity is usually assumed to be constant, though for real materials it depends upon temperature. However, the resulting non-linearity is only important when the range of temperatures under consideration is large.

We next apply the principle of conservation of energy to a small cube of material. Equation (14.14) governs the flow of heat across each face of the cube and there may also be heat generated, $Q$ per unit volume, within the cube due to some mechanism such as electrical resistive heating or nuclear reaction etc. If the sum of the heat flowing into the cube and that generated within it is positive, the temperature will rise at a rate which depends upon the *thermal capacity* of the material. Combining these arguments we find that the temperature $T$ must satisfy the equation[3]

$$\rho c \frac{\partial T}{\partial t} = K\nabla^2 T + Q \; , \tag{14.15}$$

[2]See S.P.Timoshenko and J.N.Goodier, *loc. cit.*, §153.

[3]More detail about the derivation of this equation and other information about the linear theory of heat conduction can be found in the classical text H.S.Carslaw and J.C.Jaeger, *Conduction of Heat in Solids*, 2nd.ed., Clarendon Press, Oxford (1959).

where $\rho, c$ are respectively the density and specific heat of the material, so that the product $\rho c$ is the amount of heat needed to increase the temperature of a unit volume of material by one degree.

In equation (14.15), the first term on the right hand side is the net heat flow into the element per unit volume and the second term, $Q$ is the rate of heat generated per unit volume. The algebraic sum of these terms gives the heat available for raising the temperature of the cube.

It is convenient to divide both sides of the equation by $K$, giving the more usual form of the heat conduction equation

$$\nabla^2 T = \frac{1}{\kappa}\frac{\partial T}{\partial t} - \frac{Q}{K} , \tag{14.16}$$

where

$$\kappa = \frac{K}{\rho c} \tag{14.17}$$

is the *thermal diffusivity* of the material. Thermal diffusivity has the dimensions area/time and its magnitude gives some indication of the rate at which a thermal disturbance will propagate through the body.

We can substitute (14.16) into (14.6) obtaining

$$\nabla^4 \phi = -E\alpha\left(\frac{1}{\kappa}\frac{\partial T}{\partial t} - \frac{Q}{K}\right) \tag{14.18}$$

for plane stress.

## 14.3 Steady-state problems

Equation (14.18) shows that if the temperature is independent of time and there is no internal source of heat in the body ($Q = 0$), $\phi$ will be biharmonic. It therefore follows that in the steady-state problem without heat generation, the stress field is unaffected by the temperature distribution. In particular, if the boundaries of the body are traction-free, a steady-state (and hence harmonic) temperature field will not induce any thermal stresses.

Furthermore, if there are internal heat sources — i.e. if $Q \neq 0$ — the stress field can be determined directly from equation (14.18), using $Q$, without the necessity of first solving a boundary-value problem for the temperature $T$. This also implies that the thermal boundary conditions in two-dimensional steady-state problems have no effect on the thermoelastic stress field.

All of these deductions rest on the assumption that the elastic problem is defined in terms of *tractions*. If some of the boundary conditions are stated in terms of displacements, the resulting thermal distortion will induce boundary tractions and the stress field will be affected, though it will still be the same as that which would have been produced by the same tractions if they had been applied under isothermal conditions.

Recalling the arguments of §2.2.1, we conclude that similar considerations apply to the multiply-connected body, for which there exists an implied displacement boundary condition. In other words, multiply-connected bodies will generally develop non-zero thermal stresses even under steady-state conditions with no boundary tractions. However, the resulting stress field is essentially that associated with the presence of a dislocation in the hole and hence can be characterized by relatively few parameters[4].

Appropriate conditions for multiply-connected bodies can be explicitly imposed, but in general it is simpler to obviate the need for such a condition by reverting to a displacement function representation. We shall therefore postpone discussion of thermoelastic problems for multiply-connected bodies until Chapter 17 where such a formulation is introduced.

### 14.3.1 Dundurs' Theorem

If the conditions discussed in the last section are satisfied and the temperature field therefore induces no thermal stress, the strains will be given by equations (14.1, 14.2). It therefore follows that

$$\frac{\partial^2 u_y}{\partial x^2} = -\frac{\partial^2 u_x}{\partial x \partial y} , \tag{14.19}$$

because of (14.2)

$$= -\frac{\partial e_{xx}}{\partial y} = -\alpha \frac{\partial T}{\partial y} = \frac{\alpha q_y}{K} , \tag{14.20}$$

from (14.1, 14.14). In view of (14.8), the corresponding result for plane strain can be written

$$\frac{\partial^2 u_y}{\partial x^2} = \frac{\alpha(1+\nu) q_y}{K} . \tag{14.21}$$

In this equation, the constant of proportionality $\alpha(1+\nu)/K$ is known as the *thermal distortivity* of the material and is denoted by the symbol $\delta$.

Equations (14.20, 14.21) state that the curvature of an initially straight line segment in the $x$-direction ($\partial^2 u_y/\partial x^2$) is proportional to the local heat flux across that line segment. This result was first proved by Dundurs[5] and is referred to as *Dundurs' Theorem.* It is very useful as a guide to determining the effect of thermal distortion on a structure. Figure 14.1 shows some simple traction-free bodies with various thermal boundary conditions and the resulting steady-state thermal distortion. Notice that straight boundaries that are unheated remain straight, those that are heated become convex outwards, whilst those that are cooled become concave. The angles between the edges are unaffected by the distortion, because there is no shear strain. Since the

[4]See for example J.Dundurs, Distortion of a body caused by free thermal expansion, Mech.Res.Comm., Vol. 1 (1974), 121-124.

[5]J.Dundurs *loc. cit.*

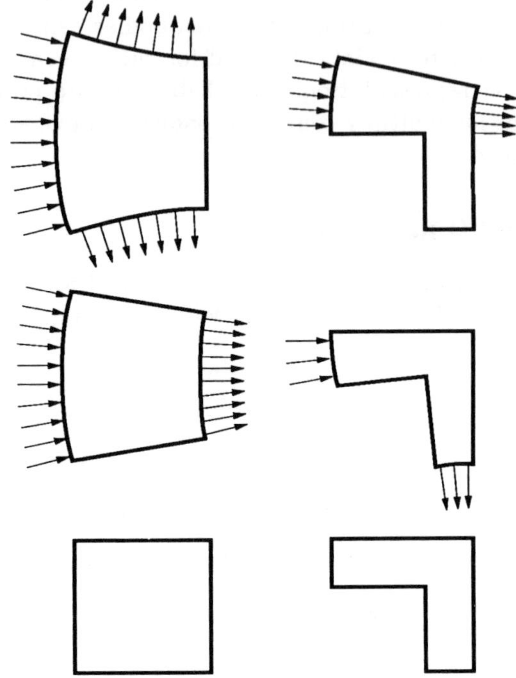

Figure 14.1: Distortion due to thermal expansion

thermal field is in the steady-state and there are no heat sources, the algebraic sum of the heat input around the boundary must be zero. Thus, although the boundary is locally rotated by the cumulative heat input from an appropriate starting point, this does not lead to incompatibility at the end of the circuit.

Many three-dimensional structures such as box-sections, tanks, rectangular hoppers etc., are fabricated from plate elements. If the unrestrained thermal distortions of these elements are considered separately, the incompatibilities of displacement developed at the junctions between elements permits the thermal stress problem to be described in terms of dislocations, the physical effects of which are more readily visualized.

Dundurs' Theorem can also be used to obtain some useful simplifications in two-dimensional contact and crack problems involving thermal distortion[6].

## PROBLEMS

1. A direct electric current $I$ flows along a conductor of rectangular cross-section $-4a<x<4a$, $-a<y<a$, all the surfaces of which are traction-free. The conductor is made of copper of electrical resistivity $\rho$, thermal conductivity $K$, Young's modulus $E$, Poisson's ratio $\nu$ and coefficient of thermal expansion $\alpha$. Assuming the current density to be uniform and neglecting electromagnetic effects, estimate the thermal stresses in the conductor when the temperature has reached a steady state.

2. A fuel element in a nuclear reactor can be regarded as a solid cylinder of radius $a$. During operation, heat is generated at a rate $Q_0(1+Ar^2/a^2)$ per unit volume, where $r$ is the distance from the axis of the cylinder and $A$ is a constant.

Assuming that the element is immersed in a fluid at pressure $p$ and that *axial* expansion is prevented, find the radial and circumferential thermal stresses produced in the steady state.

[6]For more details, see J.R.Barber, Some implications of Dundurs' Theorem for thermoelastic contact and crack problems, J.Strain Analysis, Vol. 22 (1980), 229-232.

# Part III

# THREE DIMENSIONAL PROBLEMS

# Chapter 15

# DISPLACEMENT FUNCTION SOLUTIONS

In Part II, we chose a representation for stress which satisfied the equilibrium equations identically, in which case the compatibility condition leads to a governing equation for the potential function. In three-dimensional elasticity, it is more usual to use the opposite approach — i.e. to define a potential function representation for displacement (which therefore identically satisfies the compatibility condition) and allow the equilibrium condition to define the governing equation.

A major reason for this change of method is simplicity. Stress function formulations of three-dimensional problems are generally more cumbersome than their displacement function counterparts, because of the greater complexity of the three-dimensional compatibility conditions. It is also worth noting that displacement formulations have a natural advantage in both two and three-dimensional problems when displacement boundary conditions are involved — particularly those associated with mutliply-connected bodies (see §2.2.1).

## 15.1 The strain potential

The simplest displacement function representation is that in which the displacement is set equal to the gradient of a scalar function — i.e.

$$2\mu \mathbf{u} = \nabla \phi \; . \tag{15.1}$$

The function $\phi$ must then be chosen so as to satisfy the equation of equilibrium

$$\nabla \mathrm{div}\, \boldsymbol{u} + (1-2\nu)\nabla^2 \boldsymbol{u} = 0 \tag{15.2}$$

and hence we find

$$\nabla \nabla^2 \phi = 0 \; , \tag{15.3}$$

i.e.

$$\nabla^2\phi = C \ , \tag{15.4}$$

where $C$ is an arbitrary constant.

This displacement function is generally referred to as the strain or displacement potential.

The solution of equation (15.4) is conveniently considered as the sum of a complementary function and a particular integral and, for the latter, an arbitrary state of uniform stress can be taken. Thus, the representation (15.1) can be decomposed into a uniform state of stress superposed on a solution for which $\phi$ is harmonic.

It is easy to see that this representation is not sufficiently general to represent all possible displacement fields in an elastic body, since, for example, the rotation

$$\omega_z = \left(\frac{\partial u_y}{\partial x} - \frac{\partial u_x}{\partial y}\right) = \frac{\partial^2\phi}{\partial x\partial y} - \frac{\partial^2\phi}{\partial y\partial x} \equiv 0 \ . \tag{15.5}$$

In other words, the strain potential can only be used to describe irrotational deformation fields[1].

It is associated with the name of Lamé (whose name also attaches to the Lamé constants $\lambda, \mu$) who used it to solve the problems of an elastic cylinder and sphere under axisymmetric loading. In the following work it will chiefly be useful as one term in a more general representation.

## 15.2 The Galerkin vector

The representation of the previous section is insufficiently general to describe all possible states of deformation of an elastic body and in seeking a more general form it is natural to examine representations in which the displacement is built up from second derivatives of a potential function. The most general such form is

$$2\mu\mathbf{u} = (c\nabla^2 - \boldsymbol{\nabla}\text{div })\boldsymbol{F} \ , \tag{15.6}$$

where $c$ is an arbitrary constant, which we shall later asign so as to simplify the representation. Notice that the potential function has to be a vector in this form — there are no second derivative operators which transform scalars into vectors.

The function $\boldsymbol{F}$ must be chosen to satisy the equilibrium equation (15.2) and hence

$$(c\boldsymbol{\nabla}\text{div }\nabla^2 - \boldsymbol{\nabla}\text{div }\boldsymbol{\nabla}\text{div })\boldsymbol{F} + (1-2\nu)(c\nabla^4 - \nabla^2\boldsymbol{\nabla}\text{div })\boldsymbol{F} = 0 \ . \tag{15.7}$$

[1]We encountered the same lack of generality with the body force potential of Chapter 7 (see §7.1.1).

Writing the vector operators in suffix notation, we have

$$\boldsymbol{F} = \boldsymbol{e}_i F_i \; ; \tag{15.8}$$

$$\text{div } \boldsymbol{F} = \frac{\partial F_i}{\partial x_i} \; ; \tag{15.9}$$

$$\boldsymbol{\nabla}\phi = \boldsymbol{e}_i \frac{\partial \phi}{\partial x_i} \; ; \tag{15.10}$$

$$\boldsymbol{\nabla}\text{div } \boldsymbol{F} = \boldsymbol{e}_i \frac{\partial^2 F_j}{\partial x_i \partial x_j} \; ; \tag{15.11}$$

$$\nabla^2\phi = \frac{\partial^2 \phi}{\partial x_i \partial x_i} \; , \tag{15.12}$$

where $\boldsymbol{e}_i$ is a unit vector in direction $x_i$.

It follows that

$$\boldsymbol{\nabla}\text{div } \boldsymbol{\nabla}\text{div } \boldsymbol{F} = \boldsymbol{e}_i \frac{\partial^4 F_k}{\partial x_i \partial x_k \partial x_j \partial x_j} \tag{15.13}$$

$$= \nabla^2 \boldsymbol{\nabla}\text{div } \boldsymbol{F} = \boldsymbol{\nabla}\text{div } \nabla^2 \boldsymbol{F} \; , \tag{15.14}$$

since the order of differentiation in (15.13) is immaterial. It is therefore possible to choose c so that all the terms in equation (15.4) cancel except for the biharmonic term. This will be the case if

$$c - 1 - (1 - 2\nu) = 0 \; , \tag{15.15}$$

i.e.

$$c = 2(1 - \nu) \; . \tag{15.16}$$

The representation (15.6) then takes the form

$$2\mu\mathbf{u} = 2(1 - \nu)\nabla^2 \boldsymbol{F} - \boldsymbol{\nabla}\text{div } \boldsymbol{F} \tag{15.17}$$

and the vector function $\boldsymbol{F}$ has to satisfy the biharmonic equation

$$\nabla^4 \boldsymbol{F} = 0 \; . \tag{15.18}$$

This solution is due to Galerkin and the function $\boldsymbol{F}$ is known as the Galerkin vector. A more detailed derivation is given by Westergaard[2], who gives expressions for the stress components in terms of $\boldsymbol{F}$ and who uses this representation to solve a number of classical three-dimensional problems — notably those involving concentrated forces in the infinite or semi-infinite body. We also note that the Galerkin solution was to some extent foreshadowed by Love[3], who introduced a displacement function appropriate for an axisymmetric state of stress in a body of revolution. Love's displacement function is in fact one component of the Galerkin vector $\boldsymbol{F}$.

---

[2] H.M.Westergaard, *Theory of Elasticity and Plasticity*, Dover, New York, (1964), §66.

[3] A.E.H.Love, *A Treatise on the Mathematical Theory of Elasticity*, 4th.edn., Dover, New York, (1944), §188.

## 15.3 The Papkovich-Neuber solution

A closely related solution is that developed independently by Papkovich and Neuber in terms of harmonic functions. In general, solutions in terms of harmonic functions are easier to use than those involving biharmonic functions, because the properties of harmonic functions have been extensively studied in the context of other physical theories such as electrostatics, gravitation, heat conduction and fluid mechanics.

We define the vector function

$$\boldsymbol{\psi} = -\frac{1}{2}\nabla^2 \boldsymbol{F} \tag{15.19}$$

and since the function $\boldsymbol{F}$ is biharmonic, $\boldsymbol{\psi}$ must be harmonic.

We also note that

$$\begin{aligned} \nabla^2(\boldsymbol{r}.\boldsymbol{\psi}) &= \nabla^2(x\psi_x + y\psi_y + z\psi_z) \\ &= 2\frac{\partial \psi_x}{\partial x} + 2\frac{\partial \psi_y}{\partial y} + 2\frac{\partial \psi_z}{\partial z} \\ &= 2 \text{ div } \boldsymbol{\psi} \ . \end{aligned} \tag{15.20}$$

Substituting for $\boldsymbol{\psi}$ from equation (15.19) into the right hand side of (15.20), we obtain

$$\begin{aligned} \nabla^2(\boldsymbol{r}.\boldsymbol{\psi}) &= -\text{div } \nabla^2 \boldsymbol{F} \\ &= -\nabla^2 \text{div } \boldsymbol{F} \ . \end{aligned} \tag{15.21}$$

It follows that

$$-\text{ div } \boldsymbol{F} = \boldsymbol{r}.\boldsymbol{\psi} + \phi \ , \tag{15.22}$$

where $\phi$ is an arbitrary harmonic function.

Finally, we substitute (15.19, 15.22) into the Galerkin vector representation (15.17) obtaining

$$2\mu \boldsymbol{u} = -4(1-\nu)\boldsymbol{\psi} + \boldsymbol{\nabla}(\boldsymbol{r}.\boldsymbol{\psi} + \phi) \ , \tag{15.23}$$

where $\boldsymbol{\psi}$ and $\phi$ are both harmonic functions. i.e.

$$\nabla^2 \boldsymbol{\psi} = 0; \quad \nabla^2 \phi = 0 \ . \tag{15.24}$$

This is known as the *Papkovich-Neuber solution* and it is widely used in modern treatments of three-dimensional elasticity problems We note that the function $\phi$ is the harmonic strain potential introduced in §15.1 above.

## 15.4 Completeness and uniqueness

It is a fairly easy matter to prove that the representations (15.17, 15.23) are complete — i.e. that they are capable of describing all possible elastic displacement fields in a three dimensional body[4]. The related problem of uniqueness presents greater difficulties. Both representations are redundant, in the sense that there are infinitely many combinations of functions which correspond to a given displacement field. In a non-rigorous sense, this is clear from the fact that a typical elasticity problem will be reduced to a boundary-value problem with three conditions at each point of the boundary (three tractions or three displacements or some combination of the two). We should normally expect such conditions to be sufficient to determine three potential functions in the interior, but (15.23) essentially contains four independent functions (three components of $\boldsymbol{\psi}$ and $\phi$).

Neuber gave a 'proof' that one of these four functions can always be set equal to zero, but it has since been found that his argument only applies under certain restrictions on the shape of the body. A related 'proof' for the Galerkin representation is given by Westergaard in §70. We shall give a summary of one such argument and show why it breaks down if the shape of the body does not meet certain conditions.

We suppose that a solution to a certain problem is known and that it involves all the three components $\psi_x, \psi_y, \psi_z$ of $\boldsymbol{\psi}$ and $\phi$. We wish to find another representation *of the same field* in which one of the components is everywhere zero. We could achieve this if we could construct a 'null' field — i.e. one which corresponds to zero displacement and stress at all points — for which the appropriate component is the same as in the original solution. The difference between the two solutions would then give the same displacements as in the original solution (since we have subtracted a null field only) and the appropriate component will have been reduced to zero.

Equation (15.23) shows that $\boldsymbol{\psi}$, $\phi$ will define a null field if

$$\nabla(\boldsymbol{r}.\boldsymbol{\psi} + \phi) - 4(1-\nu)\boldsymbol{\psi} = 0 \,. \tag{15.25}$$

Taking the curl of this equation, we obtain

$$\text{curl } \nabla(\boldsymbol{r}.\boldsymbol{\psi} + \phi) - 4(1-\nu)\text{curl } \boldsymbol{\psi} = 0 \tag{15.26}$$

and hence

$$\text{curl } \boldsymbol{\psi} = 0 \,, \tag{15.27}$$

since curl $\nabla$ of any function is identically zero.

We also note that by applying the operator div to equation (15.25), we obtain

$$\nabla^2(\boldsymbol{r}.\boldsymbol{\psi} + \phi) - 4(1-\nu)\text{div } \boldsymbol{\psi} = 0 \tag{15.28}$$

[4]See for example, Westergaard, *loc. cit.* §69 and R.D.Mindlin, Note on the Galerkin and Papcovich stress functions, Bull.Am.Math.Soc., Vol. 42 (1936), 373-376.

and hence, substituting from equation (15.20) for the first term,

$$2\text{div }\boldsymbol{\psi} - 4(1-\nu)\text{div }\boldsymbol{\psi} = -2(1-2\nu)\text{div }\boldsymbol{\psi} = 0\ . \tag{15.29}$$

Now equation (15.27) is precisely the condition which must be satisfied for the function $\boldsymbol{\psi}$ to be expressible as the gradient of a scalar function[5]. Hence, we conclude that there exists a function $H$ such that

$$\boldsymbol{\psi} = \boldsymbol{\nabla} H\ . \tag{15.30}$$

Furthermore, substituting (15.30) into (15.29), we see that $H$ must be a harmonic function.

In order to dispense with one of the components (say $\psi_z$) of $\boldsymbol{\psi}$ in the given solution, we need to be able to construct a harmonic function $H$ to satisfy the equation

$$\frac{\partial H}{\partial z} = \psi_z\ . \tag{15.31}$$

If we can do this, it is easily verified that the appropriate null field can be found, since on substituting into equation (15.25) we get

$$\boldsymbol{\nabla}(\boldsymbol{r}.\boldsymbol{\nabla} H + \phi - 4(1-\nu)H) = 0\ , \tag{15.32}$$

which can be satisfied by choosing $\phi$ so that

$$\phi = 4(1-\nu)H - \boldsymbol{r}.\boldsymbol{\nabla} H\ . \tag{15.33}$$

Noting that

$$\begin{aligned} \nabla^2(\boldsymbol{r}.\boldsymbol{\nabla} H) &= \nabla^2\left(x\frac{\partial H}{\partial x} + y\frac{\partial H}{\partial y} + z\frac{\partial H}{\partial z}\right) \\ &= 2\nabla^2 H\ , \end{aligned} \tag{15.34}$$

we see that $\phi$ will be harmonic, as required, since $H$ is harmonic.

### 15.4.1 Methods of partial integration

We therefore consider the problem of integrating equation (15.31) subject to the constraint that the integral be a harmonic function. We shall examine two methods of doing this which give some insight into the problem.

The given solution and hence the function $\psi_z$ is only defined in the finite region occupied by the elastic body which we denote by $V$. We also denote the boundary of $V$ (not necessarily connected) by $S$.

---

[5] See §7.1.1.

One candidate for the required function $H$ is obtained from the integral

$$H_1 = \int_{z_S}^{z} \psi_z dz , \tag{15.35}$$

where the integral is performed from a point on $S$ to a point in the body along a line in the $z$-direction. This function is readily constructed and it clearly satisfies equation (15.31), but it will not in general be harmonic. However,

$$\frac{\partial}{\partial z}\nabla^2 H_1 = \nabla^2 \frac{\partial H_1}{\partial z} = 0 \tag{15.36}$$

and hence

$$\nabla^2 H_1 = f(x,y) . \tag{15.37}$$

Now a harmonic function $H$ satisfying equation (15.31) can be constructed as

$$H = H_1 + H_2 , \tag{15.38}$$

where $H_2$ is any function of $x, y$ only that satisfies the equation

$$\nabla^2 H_2 = -f(x,y) . \tag{15.39}$$

A suitable choice for $H_2$ is the logarithmic potential due to the source distribution $f(x,y)$, which can be explicitly constructed as a convolution integral on the logarithmic (two-dimensional) point source solution. Notice that $H_2$ is not uniquely determined by (15.39) unless the domain of definition is infinite, but we are looking for a particular integral only, not a general solution. Indeed, this freedom of choice of solution could prove useful in extending the range of conditions under which the integration of (15.31) can be performed.

At first sight, the procedure described above seems to give a general method of constructing the required function $H$, but a difficulty is encountered with the definition of the lower limit $z_S$ in the integral (15.35). The path of integration is along a straight line in the $z$-direction from $S$ to the general point $(x,y,z)$. There may exist points for which such a line cuts the surface $S$ at more than two points — i.e. for which $z_S$ is multivalued. Such a case is illustrated as the point $P(x,y,z)$ in Figure 15.1. The integrand is undefined outside $V$ and hence we can only give a meaning to the definition (15.35) if we choose the point $A$ as the lower limit. However, by the same token, we must choose the point $C$ as the lower limit in the integral for the point $Q(x,y,z')$.

It follows that the function $f(x,y)$ is not truly a function of $x, y$ only, since it has different values at the points $P, Q$, which have the same coördinates $(x,y)$. The $f(x,y)$ appropriate to the region above the line $TT'$ will generally have a discontinuity across the line $TR$, where $T$ is the point of tangency of $S$ to lines in the $z$-direction (see Figure 15.1).

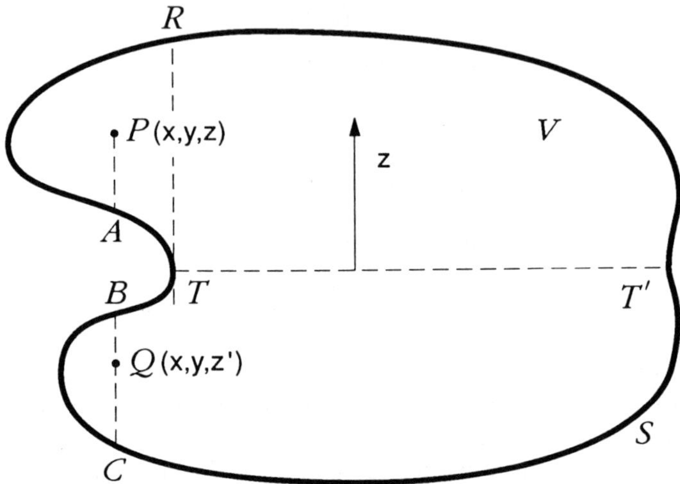


Figure 15.1: Path of integration for equation (15.35)

Notice, however that the freedom of choice of the integral of equation (15.39) may in certain circumstances permit us to eliminate this discontinuity.

Another way of developing the function $H_1$ is as follows:- We first find that distribution of sources which when placed on the surface $S$ in an infinite space would give the required potential $\psi_z$ in $V$. This distribution is unique and gives a form of continuation of $\psi_z$ into the rest of the infinite space which is everywhere harmonic *except* on the surface $S$. We then write $-\infty$ for the lower limit in equation (15.35).

The function defined by equation (15.35) will now be harmonic except when the path of integration passes through $S$, in which case $\nabla^2 H_1$ will contain a term proportional to the strength of the singularity at the appropriate point(s) of intersection. As these points will be the same for any given value of $(x, y)$, the resulting function will satisfy equation (15.36), but we note that again there will be different values for the function $f(x, y)$ if the appropriate path of integration cuts $S$ in more than two points. Furthermore, this method of construction shows that the difference between the two values is proportional to the strength of the singularities on the extra two intersections with $S$ (i.e. at points $B, C$ in Figure 15.1.). One of the consequences is that $f(x, y)$ in the upper region will then show a square root singular discontinuity to the left of the line $TR$, assuming that $S$, has no sharp corners.

This question is still a matter of active interest in elasticity[6].

## PROBLEM

1. A three-dimensional body is subjected to a body force $\boldsymbol{p}$ that can be described in terms of a potential function $V$ by the relation

$$\boldsymbol{p} = -\boldsymbol{\nabla} V \ .$$

Obtain a particular solution for the stress and displacement fields by representing the displacement $\boldsymbol{u}$ in the form

$$2\mu \boldsymbol{u} = \boldsymbol{\nabla} \psi \ .$$

Substitute this expression into the equilibrium equation (2.17) to determine the governing equation which must be satisfied by the scalar potential $\psi$. Notice that, since we are seeking a particular solution, any arbitrary constants may be discarded.

Determine the expressions for the stress and displacement components in Cartesian coördinates in terms of $\psi$.

[6] For further discussion of the question, see R.A.Eubanks and E.Sternberg, On the completeness of the Boussinesq-Papcovich stress functions, J.Rat.Mech.Anal., Vol. 5 (1956), 735-746., P.M.Naghdi and C.S.Hsu, On a representation of displacements in linear elasticity in terms of three stress functions, J.Math.Mech., Vol. 10 (1961), 233-246, T.Tran Cong and G.P.Steven, On the respresentation of elastic displacement fields in terms of three harmonic functions, J.Elasticity, Vol. 9 (1979), 325-333.

# Chapter 16

# THE BOUSSINESQ POTENTIALS

The Galerkin and Papkovich-Neuber solutions have the advantage of presenting a general solution to the problem of elasticity in a suitably compact notation, but they are not always the most convenient starting point for the solution of particular three-dimensional problems. If the problem has a plane of symmetry or particularly simple boundary conditions, it is often possible to develop a special solution of sufficient generality in one or two harmonic functions, which may or may not be components or linear combinations of components of the Papkovich-Neuber solution. For this reason, it is convenient to record detailed expressions for the displacement and stress components arising from the several terms separately and from certain related displacement potentials.

The first catalogue of solutions of this kind was compiled by Boussinesq[1] and is reproduced by Green and Zerna[2], whose terminology we use here. Boussinesq identified three categories of harmonic potential, one being the strain potential of §15.1, already introduced by Lamé and another comprising a set of three scalar functions equivalent to the three components of the Papkovich-Neuber vector, $\boldsymbol{\psi}$. The third category comprises three solutions particularly suited to torsional deformations about the three axes respectively. In view of the completeness of the Papkovich-Neuber solution, it is clear that these latter functions must be derivable from equation (15.23) and we shall show how this can be done in §16.3 below[3].

[1] J.Boussinesq, *Application des potentiels à l'étude de l'équilbre et du mouvement des solides élastiques*, Gauthier-Villars, Paris (1885).

[2] A.E.Green and W.Zerna, *Theoretical Elasticity*, 2nd.edn., Clarendon Press, Oxford, (1968), 165-176.

[3] Other relations between the Boussinesq potentials are demonstrated by J.P.Bentham, Note on the Boussinesq-Papkovich stress functions, J.Elasticity, Vol. 9 (1979), 201–206.

## 16.1 Solution A : The strain potential

Green and Zerna's Solution A is identical with the strain potential defined in §15.1 above (with the restriction that $\nabla^2\phi = 0$) and with the scalar potential $\phi$ of the Papkovich-Neuber solution (equation (15.23)).

We define

$$2\mu u_x = \frac{\partial \phi}{\partial x} \ ; \ 2\mu u_y = \frac{\partial \phi}{\partial y} \ ; \ 2\mu u_z = \frac{\partial \phi}{\partial z} \ , \tag{16.1}$$

and the dilatation

$$2\mu e = 2\mu \text{div } \boldsymbol{u} = 2\mu\nabla^2\phi = 0 \ . \tag{16.2}$$

Substituting in the stress-strain relations, we have

$$\sigma_{xx} = \lambda e + 2\mu e_{xx} = \frac{\partial^2 \phi}{\partial x^2} \tag{16.3}$$

and similarly

$$\sigma_{yy} = \frac{\partial^2 \phi}{\partial y^2} \ ; \ \sigma_{zz} = \frac{\partial^2 \phi}{\partial z^2} \ . \tag{16.4}$$

We also have

$$\sigma_{xy} = 2\mu e_{xy} = \frac{\partial^2 \phi}{\partial x \partial y} \ ; \tag{16.5}$$

$$\sigma_{yz} = \frac{\partial^2 \phi}{\partial y \partial z} \ ; \ \sigma_{zx} = \frac{\partial^2 \phi}{\partial z \partial x} \ . \tag{16.6}$$

## 16.2 Solution B

Green and Zerna's Solution B is obtained by setting to zero all the functions in the Papkovich-Neuber solution except the $z$-component of $\boldsymbol{\psi}$ , which we shall denote by $\omega$. In other words, in equation (15.23), we take

$$\boldsymbol{\psi} = \boldsymbol{k}\omega \ ; \quad \phi = 0 \ . \tag{16.7}$$

It follows that

$$2\mu u_x = z\frac{\partial \omega}{\partial x} \ ; \ 2\mu u_y = z\frac{\partial \omega}{\partial y} \ ; \tag{16.8}$$

$$2\mu u_z = z\frac{\partial \omega}{\partial z} - (3-4\nu)\omega \ ; \tag{16.9}$$

$$\begin{aligned} 2\mu e &= z\nabla^2\omega + \frac{\partial \omega}{\partial z} - (3-4\nu)\frac{\partial \omega}{\partial z} \\ &= -2(1-2\nu)\frac{\partial \omega}{\partial z} \ , \end{aligned} \tag{16.10}$$

since $\nabla^2\omega = 0$.

Substituting these expressions into the stress-strain relations and noting that

$$\lambda = \frac{2\mu\nu}{(1-2\nu)} , \tag{16.11}$$

we find

$$\sigma_{xx} = \lambda e + 2\mu e_{xx} = z\frac{\partial^2\omega}{\partial x^2} - 2\nu\frac{\partial\omega}{\partial z} ; \tag{16.12}$$

$$\sigma_{yy} = z\frac{\partial^2\omega}{\partial y^2} - 2\nu\frac{\partial\omega}{\partial z} ; \tag{16.13}$$

$$\sigma_{zz} = z\frac{\partial^2\omega}{\partial z^2} - 2(1-\nu)\frac{\partial\omega}{\partial z} ; \tag{16.14}$$

$$\sigma_{xy} = z\frac{\partial^2\omega}{\partial x\partial y} ; \tag{16.15}$$

$$\begin{aligned}\sigma_{yz} &= \frac{1}{2}\left(\frac{\partial\omega}{\partial y} + z\frac{\partial^2\omega}{\partial y\partial z} + z\frac{\partial^2\omega}{\partial y\partial z} - (3-4\nu)\frac{\partial\omega}{\partial y}\right) \\ &= z\frac{\partial^2\omega}{\partial y\partial z} - (1-2\nu)\frac{\partial\omega}{\partial y} ;\end{aligned} \tag{16.16}$$

$$\sigma_{zx} = z\frac{\partial^2\omega}{\partial z\partial x} - (1-2\nu)\frac{\partial\omega}{\partial x} . \tag{16.17}$$

This solution — if combined with solution A — gives a general solution for the torsionless axisymmetric deformation of a body of revolution. The earliest solution of this problem is due to Love[4] and leads to a single biharmonic function which is actually the component $F_z$ of the Galerkin vector (§15.2). Solution B is also particularly suitable for problems in which the plane surface $z=0$ is a boundary, since the expressions for stresses and displacements take simple forms on this surface.

It is clearly possible to write down similar solutions corresponding to Papkovich-Neuber vectors in the $x$- and $y$-directions by permuting suffices. Green and Zerna refer to these as Solutions C and D.

## 16.3 Solution E : Rotational deformation

Green and Zerna's Solution E can be derived (albeit somewhat indirectly) from the Papkovich-Neuber solution by taking the vector function $\boldsymbol{\psi}$ to be the curl of a vector field oriented in the $z$-direction. i.e.

$$\boldsymbol{\psi} = \text{curl } \boldsymbol{k}\chi . \tag{16.18}$$

[4] A.E.H.Love, *loc. cit.*

It follows that

$$\boldsymbol{r}.\boldsymbol{\psi} = x\frac{\partial \chi}{\partial y} - y\frac{\partial \chi}{\partial x} \tag{16.19}$$

and hence

$$\nabla^2(\boldsymbol{r}.\boldsymbol{\psi}) = 2\frac{\partial^2 \chi}{\partial x \partial y} - 2\frac{\partial^2 \chi}{\partial y \partial x} = 0 \; , \tag{16.20}$$

i.e. $\boldsymbol{r}.\boldsymbol{\psi}$ is harmonic. We can therefore choose $\phi$ such that

$$\boldsymbol{r}.\boldsymbol{\psi} + \phi \equiv 0 \tag{16.21}$$

and hence

$$2\mu\boldsymbol{u} = -4(1-\nu)\text{curl}\ \boldsymbol{k}\chi \; . \tag{16.22}$$

To avoid unnecessary multiplying constants, it is convenient to write

$$\Psi = -2(1-\nu)\chi \; , \tag{16.23}$$

in which case

$$2\mu u_x = 2\frac{\partial \Psi}{\partial y} \;\; ; \;\; 2\mu u_y = -2\frac{\partial \Psi}{\partial x} \;\; ; \;\; 2\mu u_z = 0 \; , \tag{16.24}$$

giving

$$2\mu e = 0 \; . \tag{16.25}$$

Substituting in the stress-strain relations, we find

$$\sigma_{xx} = 2\frac{\partial^2 \Psi}{\partial x \partial y} \;\; ; \;\; \sigma_{yy} = -2\frac{\partial^2 \Psi}{\partial x \partial y} \;\; ; \;\; \sigma_{zz} = 0 \; ; \tag{16.26}$$

$$\sigma_{xy} = \left(\frac{\partial^2 \Psi}{\partial y^2} - \frac{\partial^2 \Psi}{\partial x^2}\right) \;\; ; \;\; \sigma_{yz} = -\frac{\partial^2 \Psi}{\partial x \partial z} \;\; ; \;\; \sigma_{zx} = \frac{\partial^2 \Psi}{\partial y \partial z} \; . \tag{16.27}$$

As with Solution B, two additional solutions of this type can be constructed by permuting suffices.

## 16.4 Solutions obtained by superposition

The solutions obtained in the above sections can be superposed as required to provide a solution appropriate to the particular problem at hand. To assist with this, the stress and displacement equations for solutions A,B,E are tabulated in Table 16.1.

A case of particular interest is that in which the plane $z=0$ is one of the boundaries of the body — for example the semi-infinite solid $z>0$. We consider here two special cases.

Table 16.1: Green and Zerna's Solutions A, B and E

| | **Solution A** | **Solution B** | **Solution E** |
|---|---|---|---|
| $2\mu u_x$ | $\frac{\partial\phi}{\partial x}$ | $z\frac{\partial\omega}{\partial x}$ | $2\frac{\partial\Psi}{\partial y}$ |
| $2\mu u_y$ | $\frac{\partial\phi}{\partial y}$ | $z\frac{\partial\omega}{\partial y}$ | $-2\frac{\partial\Psi}{\partial x}$ |
| $2\mu u_z$ | $\frac{\partial\phi}{\partial z}$ | $z\frac{\partial\omega}{\partial z}-(3-4\nu)\omega$ | $0$ |
| $\sigma_{xx}$ | $\frac{\partial^2\phi}{\partial x^2}$ | $z\frac{\partial^2\omega}{\partial x^2}-2\nu\frac{\partial\omega}{\partial z}$ | $2\frac{\partial^2\Psi}{\partial x\partial y}$ |
| $\sigma_{xy}$ | $\frac{\partial^2\phi}{\partial x\partial y}$ | $z\frac{\partial^2\omega}{\partial x\partial y}$ | $\frac{\partial^2\Psi}{\partial y^2}-\frac{\partial^2\Psi}{\partial x^2}$ |
| $\sigma_{yy}$ | $\frac{\partial^2\phi}{\partial y^2}$ | $z\frac{\partial^2\omega}{\partial y^2}-2\nu\frac{\partial\omega}{\partial z}$ | $-2\frac{\partial^2\Psi}{\partial x\partial y}$ |
| $\sigma_{xz}$ | $\frac{\partial^2\phi}{\partial x\partial z}$ | $z\frac{\partial^2\omega}{\partial x\partial z}-(1-2\nu)\frac{\partial\omega}{\partial x}$ | $\frac{\partial^2\Psi}{\partial y\partial z}$ |
| $\sigma_{yz}$ | $\frac{\partial^2\phi}{\partial y\partial z}$ | $z\frac{\partial^2\omega}{\partial y\partial z}-(1-2\nu)\frac{\partial\omega}{\partial y}$ | $-\frac{\partial^2\Psi}{\partial x\partial z}$ |
| $\sigma_{zz}$ | $\frac{\partial^2\phi}{\partial z^2}$ | $z\frac{\partial^2\omega}{\partial z^2}-2(1-\nu)\frac{\partial\omega}{\partial z}$ | $0$ |
| $2\mu u_r$ | $\frac{\partial\phi}{\partial r}$ | $z\frac{\partial\omega}{\partial r}$ | $\frac{2}{r}\frac{\partial\Psi}{\partial\theta}$ |
| $2\mu u_\theta$ | $\frac{1}{r}\frac{\partial\phi}{\partial\theta}$ | $\frac{z}{r}\frac{\partial\omega}{\partial\theta}$ | $-2\frac{\partial\Psi}{\partial r}$ |
| $\sigma_{rr}$ | $\frac{\partial^2\phi}{\partial r^2}$ | $z\frac{\partial^2\omega}{\partial r^2}-2\nu\frac{\partial\omega}{\partial z}$ | $\frac{2}{r}\frac{\partial^2\Psi}{\partial r\partial\theta}-\frac{2}{r^2}\frac{\partial\Psi}{\partial\theta}$ |
| $\sigma_{r\theta}$ | $\frac{1}{r}\frac{\partial^2\phi}{\partial r\partial\theta}-\frac{1}{r^2}\frac{\partial\phi}{\partial\theta}$ | $\frac{z}{r}\frac{\partial^2\omega}{\partial r\partial\theta}-\frac{z}{r^2}\frac{\partial\omega}{\partial\theta}$ | $\frac{1}{r}\frac{\partial\Psi}{\partial r}-\frac{\partial^2\Psi}{\partial r^2}+\frac{1}{r^2}\frac{\partial^2\Psi}{\partial\theta^2}$ |
| $\sigma_{\theta\theta}$ | $\frac{1}{r}\frac{\partial\phi}{\partial r}+\frac{1}{r^2}\frac{\partial^2\phi}{\partial\theta^2}$ | $\frac{z}{r}\frac{\partial\omega}{\partial r}+\frac{z}{r^2}\frac{\partial^2\omega}{\partial\theta^2}-2\nu\frac{\partial\omega}{\partial z}$ | $-\frac{2}{r}\frac{\partial^2\Psi}{\partial r\partial\theta}+\frac{2}{r^2}\frac{\partial\Psi}{\partial\theta}$ |
| $\sigma_{rz}$ | $\frac{\partial^2\phi}{\partial r\partial z}$ | $z\frac{\partial^2\omega}{\partial r\partial z}-(1-2\nu)\frac{\partial\omega}{\partial r}$ | $\frac{1}{r}\frac{\partial^2\Psi}{\partial\theta\partial z}$ |
| $\sigma_{\theta z}$ | $\frac{1}{r}\frac{\partial^2\phi}{\partial\theta\partial z}$ | $\frac{z}{r}\frac{\partial^2\omega}{\partial\theta\partial z}-\frac{1-2\nu}{r}\frac{\partial\omega}{\partial\theta}$ | $-\frac{\partial^2\Psi}{\partial r\partial z}$ |

### 16.4.1 Solution F : Frictionless isothermal contact problems

If the half space is indented by a frictionless punch, so that the surface $z = 0$ is subjected to normal tractions only, a simple formulation can be obtained by combining solutions A and B and defining a relationship between $\phi$ and $\omega$ in order to satisfy identically the condition $\sigma_{zx} = \sigma_{zy} = 0$ on $z = 0$.

We write

$$\phi = (1-2\nu)\varphi; \quad \omega = \frac{\partial \varphi}{\partial z} \tag{16.28}$$

in solutions A, B respectively, obtaining

$$\sigma_{zx} = z\frac{\partial^3 \varphi}{\partial x \partial z^2} \ ; \ \sigma_{zy} = z\frac{\partial^3 \varphi}{\partial y \partial z^2} , \tag{16.29}$$

which vanish as required on the plane $z = 0$. The remaining stress and displacement components are listed in Table 16.2 and at this stage we merely note that the normal traction and normal displacement at the surface reduce to the simple expressions

$$\sigma_{zz} = -\frac{\partial^2 \varphi}{\partial z^2} \ ; \ 2\mu u_z = -2(1-\nu)\frac{\partial \varphi}{\partial z} . \tag{16.30}$$

It follows that indentation problems for the half-space can be reduced to classical boundary-value problems in the harmonic potential function $\varphi$. Various examples are considered by Green and Zerna[5], who also show that symmetric problems of the plane crack in an isotropic body can be treated in the same way. These problems are considered in Chapters 21–23 below.

### 16.4.2 Solution G: The surface free of normal traction

The counterpart of the preceding solution is that obtained by combining Solutions A, B and requiring that the *normal* tractions vanish on the surface $z = 0$. We shall find this solution useful in problems involving a plane interface between two dissimilar materials (Chapter 24) and also in certain axisymmetric crack and contact problems.

The normal traction on the plane $z = 0$ is

$$\sigma_{zz} = \frac{\partial^2 \phi}{\partial z^2} - 2(1-\nu)\frac{\partial \omega}{\partial z} , \tag{16.31}$$

from Table 16.1 and hence it can be made to vanish by writing

$$\phi = 2(1-\nu)\chi \ ; \ \omega = \frac{\partial \chi}{\partial z} , \tag{16.32}$$

where $\chi$ is a new harmonic potential function.

The resulting stress and displacement components are given as Solution G in Table 16.2.

[5]A.E.Green and W.Zerna, *loc. cit.* §§5.8-5.10.

Table 16.2: Solutions F and G

| | **Solution F**<br>Frictionless isothermal | **Solution G** |
|---|---|---|
| $2\mu u_x$ | $z\frac{\partial^2\varphi}{\partial x\partial z}+(1-2\nu)\frac{\partial\varphi}{\partial x}$ | $2(1-\nu)\frac{\partial\chi}{\partial x}+z\frac{\partial^2\chi}{\partial x\partial z}$ |
| $2\mu u_y$ | $z\frac{\partial^2\varphi}{\partial y\partial z}+(1-2\nu)\frac{\partial\varphi}{\partial y}$ | $2(1-\nu)\frac{\partial\chi}{\partial y}+z\frac{\partial^2\chi}{\partial y\partial z}$ |
| $2\mu u_z$ | $z\frac{\partial^2\varphi}{\partial z^2}-2(1-\nu)\frac{\partial\varphi}{\partial z}$ | $-(1-2\nu)\frac{\partial\chi}{\partial z}+z\frac{\partial^2\chi}{\partial z^2}$ |
| $\sigma_{xx}$ | $z\frac{\partial^3\varphi}{\partial x^2\partial z}+\frac{\partial^2\varphi}{\partial x^2}+2\nu\frac{\partial^2\varphi}{\partial y^2}$ | $2(1-\nu)\frac{\partial^2\chi}{\partial x^2}+z\frac{\partial^3\chi}{\partial x^2\partial z}-2\nu\frac{\partial^2\chi}{\partial z^2}$ |
| $\sigma_{xy}$ | $z\frac{\partial^3\varphi}{\partial x\partial y\partial z}+(1-2\nu)\frac{\partial^2\varphi}{\partial x\partial y}$ | $2(1-\nu)\frac{\partial^2\chi}{\partial x\partial y}+z\frac{\partial^3\chi}{\partial x\partial y\partial z}$ |
| $\sigma_{yy}$ | $z\frac{\partial^3\varphi}{\partial y^2\partial z}+\frac{\partial^2\varphi}{\partial y^2}+2\nu\frac{\partial^2\varphi}{\partial x^2}$ | $2(1-\nu)\frac{\partial^2\chi}{\partial y^2}+z\frac{\partial^3\chi}{\partial y^2\partial z}-2\nu\frac{\partial^2\chi}{\partial z^2}$ |
| $\sigma_{xz}$ | $z\frac{\partial^3\varphi}{\partial x\partial z^2}$ | $\frac{\partial^2\chi}{\partial x\partial z}+z\frac{\partial^3\chi}{\partial x\partial z^2}$ |
| $\sigma_{yz}$ | $z\frac{\partial^3\varphi}{\partial y\partial z^2}$ | $\frac{\partial^2\chi}{\partial y\partial z}+z\frac{\partial^3\chi}{\partial y\partial z^2}$ |
| $\sigma_{zz}$ | $z\frac{\partial^3\varphi}{\partial z^3}-\frac{\partial^2\varphi}{\partial z^2}$ | $z\frac{\partial^3\chi}{\partial z^3}$ |
| $2\mu u_r$ | $z\frac{\partial^2\varphi}{\partial r\partial z}+(1-2\nu)\frac{\partial\varphi}{\partial r}$ | $2(1-\nu)\frac{\partial\chi}{\partial r}+z\frac{\partial^2\chi}{\partial r\partial z}$ |
| $2\mu u_\theta$ | $\frac{z}{r}\frac{\partial^2\varphi}{\partial\theta\partial z}+\frac{(1-2\nu)}{r}\frac{\partial\varphi}{\partial\theta}$ | $\frac{2(1-\nu)}{r}\frac{\partial\chi}{\partial\theta}+\frac{z}{r}\frac{\partial^2\chi}{\partial\theta\partial z}$ |
| $\sigma_{rr}$ | $z\frac{\partial^3\varphi}{\partial r^2\partial z}+\frac{\partial^2\varphi}{\partial r^2}-2\nu\left(\frac{\partial^2\varphi}{\partial r^2}+\frac{\partial^2\varphi}{\partial z^2}\right)$ | $2(1-\nu)\frac{\partial^2\chi}{\partial r^2}+z\frac{\partial^3\chi}{\partial r^2\partial z}-2\nu\frac{\partial^2\chi}{\partial z^2}$ |
| $\sigma_{r\theta}$ | $\frac{z}{r}\frac{\partial^3\varphi}{\partial r\partial\theta\partial z}-\frac{z}{r^2}\frac{\partial^2\varphi}{\partial\theta\partial z}$ | $\frac{2(1-\nu)}{r}\left(\frac{\partial^2\chi}{\partial r\partial\theta}-\frac{1}{r}\frac{\partial\chi}{\partial\theta}\right)$ |
| | $+\frac{(1-2\nu)}{r}\left(\frac{\partial^2\varphi}{\partial r\partial\theta}-\frac{1}{r}\frac{\partial\varphi}{\partial\theta}\right)$ | $+\frac{z}{r}\frac{\partial^3\chi}{\partial z\partial r\partial\theta}-\frac{z}{r^2}\frac{\partial^2\chi}{\partial z\partial\theta}$ |
| $\sigma_{\theta\theta}$ | $-(1-2\nu)\frac{\partial^2\varphi}{\partial r^2}-\frac{\partial^2\varphi}{\partial z^2}$ | $-2(1-\nu)\frac{\partial^2\chi}{\partial r^2}-2\frac{\partial^2\chi}{\partial z^2}$ |
| | $-z\frac{\partial^3\varphi}{\partial r^2\partial z}-z\frac{\partial^3\varphi}{\partial z^3}$ | $+\frac{z}{r}\frac{\partial^2\chi}{\partial z\partial r}+\frac{z}{r^2}\frac{\partial^3\chi}{\partial z\partial\theta^2}$ |
| $\sigma_{rz}$ | $z\frac{\partial^3\varphi}{\partial r\partial z^2}$ | $\frac{\partial^2\chi}{\partial r\partial z}+z\frac{\partial^3\chi}{\partial r\partial z^2}$ |
| $\sigma_{\theta z}$ | $\frac{z}{r}\frac{\partial^3\varphi}{\partial\theta\partial z^2}$ | $\frac{1}{r}\frac{\partial^2\chi}{\partial\theta\partial r}+\frac{z}{r}\frac{\partial^3\chi}{\partial\theta\partial z^2}$ |

## 16.5 The plane strain solution in complex variables

Although this chapter is concerned principally with three-dimensional problems, it is clearly possible to use any of the above representations to solve problems in two-dimensions. For example, if we took solutions A and B and assumed both potential functions were independent of $x$, we could obtain a series of solutions to problems in plane strain.

A more powerful plane strain solution is that expressed in terms of complex variables and in the following section we shall show how this can be derived from the Papkovich-Neuber solution.

### 16.5.1 Preliminary mathematical results

We define the complex variable $\zeta$ and its conjugate $\bar{\zeta}$ by the equations

$$\zeta = x + iy \ ; \ \bar{\zeta} = x - iy \ , \tag{16.33}$$

with solution

$$x = \frac{1}{2}(\zeta + \bar{\zeta}) \ ; \ iy = \frac{1}{2}(\zeta - \bar{\zeta}) \tag{16.34}$$

and hence

$$\frac{\partial}{\partial \zeta} = \frac{\partial x}{\partial \zeta}\frac{\partial}{\partial x} + \frac{\partial y}{\partial \zeta}\frac{\partial}{\partial y} = \frac{1}{2}\left(\frac{\partial}{\partial x} - i\frac{\partial}{\partial y}\right) \ ; \tag{16.35}$$

$$\frac{\partial}{\partial \bar{\zeta}} = \frac{1}{2}\left(\frac{\partial}{\partial x} + i\frac{\partial}{\partial y}\right) \ ; \tag{16.36}$$

$$4\frac{\partial^2}{\partial \zeta \partial \bar{\zeta}} = \frac{\partial^2}{\partial x^2} + \frac{\partial^2}{\partial y^2} \ . \tag{16.37}$$

It follows that the Laplace equation ($\nabla^2 \phi = 0$) in two dimensions can be written

$$\frac{\partial^2 \phi}{\partial \zeta \partial \bar{\zeta}} = 0 \tag{16.38}$$

and has the general solution

$$\phi = f_1(\zeta) + f_2(\bar{\zeta}) \ , \tag{16.39}$$

where $f_1, f_2$ are arbitrary functions.

### 16.5.2 Representation of vectors

We represent vectors by equating the two components of the vector to the real and imaginary parts of a complex function. For example

$$\boldsymbol{u} \equiv \boldsymbol{i}u_x + \boldsymbol{j}u_y \tag{16.40}$$

is represented by the complex function

$$u = u_x + iu_y \ . \tag{16.41}$$

It follows that the vector operator

$$\nabla \equiv \boldsymbol{i}\frac{\partial}{\partial x} + \boldsymbol{j}\frac{\partial}{\partial y} = \frac{\partial}{\partial x} + i\frac{\partial}{\partial y} = 2\frac{\partial}{\partial \bar{\zeta}} \ , \tag{16.42}$$

from equation (16.36). Also

$$\text{div } \boldsymbol{u} \equiv \frac{\partial u_x}{\partial x} + \frac{\partial u_y}{\partial y} = \frac{\partial u}{\partial \zeta} + \frac{\partial \bar{u}}{\partial \bar{\zeta}} \ , \tag{16.43}$$

as can be verified by expansion, using equations (16.35, 16.36, 16.41).

Notice that in the above equation, we use the notation $\bar{u}$ for the complex conjugate of $u$ — i.e. for the complex quantity obtained by writing $-i$ for $i$ in $u$, wherever it occurs.

### 16.5.3 Representation of displacement

We can now write the Papkovich-Neuber solution in complex variable form, assuming plane strain conditions ($u_z = 0$). We write

$$\psi = \psi_x + i\psi_y \tag{16.44}$$

and satisfy the requirement that it be harmonic by expanding it in the form[6]

$$\psi = \chi(\zeta) + \bar{\omega}(\bar{\zeta}) \ . \tag{16.45}$$

Now

$$\begin{aligned} \boldsymbol{r}.\boldsymbol{\psi} &= x\psi_x + y\psi_y \\ &= x\chi_x + y\chi_y + x\omega_x - y\omega_y \ , \end{aligned} \tag{16.46}$$

where we write $\chi_x, \chi_y$ respectively for the real and imaginary parts of $\chi$, in view of the analogy between complex functions and vectors. Note also that the imaginary part of $\bar{\omega}(\bar{\zeta})$ is $-\omega_y$.

[6]Notice that we denote the function of the complex conjugate $\bar{\omega}(\bar{\zeta})$ with a bar over the function as well as the argument. This is a redundant notation, but it permits us to drop the argument for brevity in long derivations.

Also, we have

$$\begin{aligned}\bar{\zeta}\chi(\zeta) &= (x-iy)(\chi_x+i\chi_y)\\ &= x\chi_x+y\chi_y+i(x\chi_y-y\chi_x)\end{aligned} \tag{16.47}$$

and hence

$$\frac{1}{2}(\bar{\zeta}\chi(\zeta)+\zeta\bar{\chi}(\bar{\zeta})) = x\chi_x+y\chi_y\,. \tag{16.48}$$

Similarly

$$\begin{aligned}\bar{\zeta}\bar{\omega}(\bar{\zeta}) &= (x-iy)(\omega_x-i\omega_y)\\ &= x\omega_x-y\omega_y-i(x\omega_y+y\omega_x)\end{aligned} \tag{16.49}$$

and hence

$$\frac{1}{2}(\bar{\zeta}\bar{\omega}(\bar{\zeta})+\zeta\omega(\zeta)) = x\omega_x-y\omega_y\,. \tag{16.50}$$

It follows that

$$\boldsymbol{r}.\boldsymbol{\psi} = \frac{1}{2}(\bar{\zeta}\chi+\zeta\bar{\chi}+\bar{\zeta}\bar{\omega}+\zeta\omega) \tag{16.51}$$

and that the Papkovich-Neuber solution (15.23) can be expressed in the form

$$\begin{aligned}2\mu u &= 2\frac{\partial}{\partial\bar{\zeta}}\left[\frac{1}{2}(\bar{\zeta}\chi+\zeta\bar{\chi}+\bar{\zeta}\bar{\omega}+\zeta\omega)+\phi_1+\bar{\phi}_2\right]-4(1-\nu)(\chi+\bar{\omega})\\ &= \chi+\zeta\bar{\chi}'+\bar{\zeta}\bar{\omega}+\bar{\omega}+\bar{\phi}_2'-4(1-\nu)(\chi+\bar{\omega})\,.\end{aligned} \tag{16.52}$$

We can simplify this representation by grouping all the functions of $\bar{\zeta}$ in the definition of a new function

$$\bar{\theta} \equiv -4(1-\nu)\bar{\omega}+\bar{\zeta}\bar{\omega}+\bar{\omega}+\bar{\phi}_2'\,, \tag{16.53}$$

giving

$$2\mu u = -(3-4\nu)\chi+\zeta\bar{\chi}'+\bar{\theta}\,. \tag{16.54}$$

This constitutes a general representation of a plane strain displacement field in complex variable form.

### 16.5.4 Expressions for stresses

The stress components are conveniently expressed in terms of the groups

$$\Theta \equiv \sigma_{xx}+\sigma_{yy}\,; \tag{16.55}$$

$$\Phi \equiv \sigma_{xx}+2i\sigma_{xy}-\sigma_{yy}\,, \tag{16.56}$$

since $\Theta$ is invariant with respect to coördinate transformation and $\Phi$ transforms according to the rule

$$\Phi_\alpha = e^{2i\alpha}\Phi\,, \tag{16.57}$$

where $\Phi_\alpha$ is defined in a coördinate system rotated anticlockwise through an angle $\alpha$ with respect to $x, y$.

Substituting (16.54) into (16.43), we find

$$2\mu \text{div } \boldsymbol{u} = -2(1-2\nu)(\chi' + \bar{\chi}') . \tag{16.58}$$

We can now substitute in the stress-strain relations, obtaining

$$\begin{aligned} \Theta &= 2(\lambda+\mu)\text{div } \boldsymbol{u} \\ &= -2(\chi' + \bar{\chi}') , \end{aligned} \tag{16.59}$$

$$\begin{aligned} \Phi &= 2\mu \left[ \frac{\partial u_x}{\partial x} - \frac{\partial u_y}{\partial y} + i \left( \frac{\partial u_x}{\partial y} + \frac{\partial u_y}{\partial x} \right) \right] \\ &= 4\mu \frac{\partial u}{\partial \bar{\zeta}} = 2(\zeta \bar{\chi}'' + \bar{\theta}') . \end{aligned} \tag{16.60}$$

The solution given above can be used in much the same way as the Airy stress function to solve problems in which the stresses vary as polynomials in $x, y$ or as Fourier series in polar coördinates. However, the major advantage of the complex variable approach is seen when it is combined with the technique of conformal transformation and the Cauchy integral representation of an analytic function in terms of its boundary values[7].

## PROBLEMS

1. Use solution E to solve the elementary torsion problem for a cylindrical bar of radius $b$ transmitting a torque $T$, the surface $r = b$ being traction-free. You will need to find a suitable harmonic potential function — it will be an axisymmetric polynomial.

2. Find a way of taking a combination of solutions A,B,E so as to satisfy identically the global conditions

$$\sigma_{zy} = \sigma_{zz} = 0 \ ; \quad \text{all } \ x, y \ ; \quad z = 0.$$

Thus, $\sigma_{zx}$ should be the only non-zero traction component on the surface.

---

[7]The complex variable method for plane strain problems in elasticity was developed simultaneously by Muskhelishvili in the Soviet Union and Stevenson and Green in England during the 1940s. The classical texts on this early work are N.I.Muskhelishvili, *Some Basic Problems of the Mathematical Theory of Elasticity*, (English translation by J.R.M.Radok, Noordhoff, (1953) and A.E.Green and W.Zerna, *Theoretical Elasticity*, 2nd.edn., Clarendon Press, Oxford, (1968), Chapter 8. More approachable treatments for the engineering reader are given by A.H.England, *Complex Variable Methods in Elasticity*, John Wiley, London, (1971), S.P.Timoshenko and J.N.Goodier, *loc. cit.* Chapter 6 and D.S.Dugdale and C.Ruiz, *Elasticity for Engineers*, McGraw-Hill, London (1971), Chapter 2.

3. An important class of frictional contact problems involves the steady sliding of an indenter in the $x$-direction across the surface of the half-space $z > 0$. Assuming Coulomb's friction law to apply, with coefficient of friction, $f$, the appropriate boundary conditions are

$$\sigma_{xz} = -f\sigma_{zz} \ ; \ \sigma_{zy} = 0 \ , \tag{16.61}$$

in the contact area, the rest of the surface $z=0$ being traction-free.

Show that (16.61) are actually ***global*** conditions for this problem and find a way of combining Solutions A,B,E so as to satisfy them identically.

# Chapter 17

# THERMOELASTIC DISPLACEMENT POTENTIALS

As in the two-dimensional case (Chapter 14), three-dimensional problems of thermoelasticity are conveniently treated by finding a *particular solution* — i.e. a solution which satisfies the field equations without regard to boundary conditions — and completing the general solution by superposition of an appropriate representation for the general isothermal problem, such as the Papkovich-Neuber solution.

In this section, we shall show that a particular solution can always be obtained in the form of a strain potential — i.e. by writing

$$2\mu \boldsymbol{u} = \boldsymbol{\nabla}\phi \, . \tag{17.1}$$

The thermoelastic stress-strain relations (14.3, 14.4) can be solved to give

$$\begin{aligned} \sigma_{xx} &= \lambda e + 2\mu e_{xx} - (3\lambda + 2\mu)\alpha T \, ; && (17.2) \\ \sigma_{xy} &= 2\mu e_{xy} \, , && (17.3) \end{aligned}$$

etc.

Using the strain-displacement relations and substituting for $\boldsymbol{u}$ from (17.1), we then find

$$\begin{aligned} 2\mu e &= \nabla^2\phi \, ; && (17.4) \\ \sigma_{xx} &= \frac{\lambda}{2\mu}\nabla^2\phi + \frac{\partial^2\phi}{\partial x^2} - (3\lambda + 2\mu)\alpha T \, ; && (17.5) \\ \sigma_{xy} &= \frac{\partial^2\phi}{\partial x \partial y} \, , && (17.6) \end{aligned}$$

etc. and hence the equilibrium equation (2.2) requires that

$$\frac{\partial}{\partial x}\left(\frac{\lambda}{2\mu}\nabla^2\phi + \nabla^2\phi - (3\lambda + 2\mu)\alpha T\right) = 0 \, . \tag{17.7}$$

Two similar equations are obtained from (2.3, 2.4) and, since we are only looking for a particular solution, we can satisfy them by choosing

$$\nabla^2\phi = \frac{2\mu(3\lambda+2\mu)\alpha T}{(\lambda+2\mu)} = \frac{2\mu(1+\nu)\alpha T}{(1-\nu)} . \tag{17.8}$$

Equation (17.8) defines the potential $\phi$ due to a source distribution proportional to the given temperature, $T$. Such a function can be found for all $T$ in any geometric domain[1] and hence a particular solution of the thermoelastic problem in the form (17.1) can always be found.

Substituting for $T$ from equation (17.8) back into the stress equation (17.5), we obtain

$$\begin{aligned} \sigma_{xx} &= \left(\frac{\lambda}{2\mu} - \frac{(\lambda+2\mu)}{2\mu}\right)\nabla^2\phi + \frac{\partial^2\phi}{\partial x^2} \\ &= -\frac{\partial^2\phi}{\partial y^2} - \frac{\partial^2\phi}{\partial z^2} . \end{aligned} \tag{17.9}$$

The particular solution can therefore be summarized in the equations

$$\nabla^2\phi = \frac{2\mu(1+\nu)\alpha T}{(1-\nu)} ; \tag{17.10}$$

$$2\mu\boldsymbol{u} = \boldsymbol{\nabla}\phi ; \tag{17.11}$$

$$\sigma_{xx} = -\frac{\partial^2\phi}{\partial y^2} - \frac{\partial^2\phi}{\partial z^2} ; \quad \sigma_{yy} = -\frac{\partial^2\phi}{\partial z^2} - \frac{\partial^2\phi}{\partial x^2} ; \quad \sigma_{zz} = -\frac{\partial^2\phi}{\partial x^2} - \frac{\partial^2\phi}{\partial y^2} ; \tag{17.12}$$

$$\sigma_{xy} = \frac{\partial^2\phi}{\partial x\partial y} ; \quad \sigma_{yz} = \frac{\partial^2\phi}{\partial y\partial z} ; \quad \sigma_{zx} = \frac{\partial^2\phi}{\partial z\partial x} . \tag{17.13}$$

## 17.1 Plane problems

If $\phi$ is independent of $z$, the non-zero stress components reduce to

$$\sigma_{xx} = -\frac{\partial^2\phi}{\partial y^2} ; \quad \sigma_{xy} = \frac{\partial^2\phi}{\partial x\partial y} ; \quad \sigma_{yy} = -\frac{\partial^2\phi}{\partial x^2} ; \tag{17.14}$$

$$\sigma_{zz} = -\frac{\partial^2\phi}{\partial x^2} - \frac{\partial^2\phi}{\partial y^2} \tag{17.15}$$

and the solution corresponds to a state of plane strain. In fact, the relations (17.14) are identical to the Airy function definitions, apart from a difference of sign, and similarly, (17.10) reduces to a particular integral of (14.7). However, an important difference is that the present solution also gives explicit relations for the displacements and can therefore be used for problems with displacement boundary conditions, including those arising for multiply-connected bodies.

[1] For example, as a convolution integral on the point source solution $1/R$.

### 17.1.1 Axisymmetric problems for the cylinder

We consider the cylinder $b < r < a$ in a state of plane strain with a prescribed temperature distribution. The example of §14.1.1 and the treatment of isothermal problems in Chapters 8, 9 show that this problem could be solved for a fairly general temperature distribution, but a case of particular importance is that in which the temperature is an axisymmetric function, $T(r)$.

If $\phi$ is also taken to depend upon $r$ only, we can write (17.10) in the form

$$\frac{1}{r}\frac{d}{dr}r\frac{d\phi}{dr} = \frac{2\mu(1+\nu)\alpha T(r)}{(1-\nu)} , \tag{17.16}$$

which has the particular integral

$$\frac{d\phi}{dr} = \frac{2\mu(1+\nu)\alpha}{(1-\nu)r}\int_b^r rT(r)dr , \tag{17.17}$$

where we have selected the inner radius, $b$, as the lower limit of integration, since the required generality will be introduced later through the superposed isothermal solution.

The corresponding displacement and stress components are then obtained as

$$u_r = \frac{1}{2\mu}\frac{d\phi}{dr} = \frac{(1+\nu)\alpha}{(1-\nu)r}\int_b^r rT(r)dr \; ; \tag{17.18}$$

$$\sigma_{rr} = -\frac{1}{r}\frac{d\phi}{dr} = -\frac{2\mu(1+\nu)\alpha}{(1-\nu)r^2}\int_b^r rT(r)dr \; ; \tag{17.19}$$

$$\sigma_{\theta\theta} = -\frac{d^2\phi}{dr^2} = \frac{2\mu(1+\nu)\alpha}{(1-\nu)}\left(\frac{1}{r^2}\int_b^r rT(r)dr - T(r)\right) \; ; \tag{17.20}$$

$$\sigma_{r\theta} = 0 \; ; \; u_\theta = 0 \, . \tag{17.21}$$

The general solution is now obtained by superposing those axisymmetric terms from Tables 8.1, 9.1 that give single-valued displacements, with the result

$$u_r = \frac{(1+\nu)\alpha}{(1-\nu)r}\int_b^r rT(r)dr + A(1-2\nu)r - \frac{B}{r} \; ; \tag{17.22}$$

$$\sigma_{rr} = -\frac{2\mu(1+\nu)\alpha}{(1-\nu)r^2}\int_b^r rT(r)dr + 2\mu A + \frac{2\mu B}{r^2} \; ; \tag{17.23}$$

$$\sigma_{\theta\theta} = \frac{2\mu(1+\nu)\alpha}{(1-\nu)}\left(\frac{1}{r^2}\int_b^r rT(r)dr - T(r)\right) + 2\mu A - \frac{2\mu B}{r^2} \; ; \tag{17.24}$$

$$\sigma_{r\theta} = 0 \; ; \; u_\theta = 0 \, . \tag{17.25}$$

The constants $A, B$ permit arbitrary axisymmetric boundary conditions to be satisfied at $r = a, b$.

The above results are appropriate for plane strain. The corresponding plane stress results (for the thin disk) can be obtained by omitting the factor $(1-\nu)$ in each of (17.22–17.24) and replacing $(1-2\nu)$ in the second term of (17.22) by $(1-\nu)/(1+\nu)$.

The same equations can also be used for the solid cylinder or disk ($b = 0$), provided the constant $B$ associated with the singular term is set to zero.

A similar method can be used for the hollow or solid sphere with spherically symmetric boundary conditions and whose temperature is a function of radius only. The analysis is given by Timoshenko and Goodier, *loc. cit.*, §152.

## 17.2 Solution T : Steady-state temperature

If the temperature distribution is in a steady state, the temperature $T$ is harmonic and hence the potential function $\phi$ is biharmonic from equation (17.10). It is convenient to express this function in terms of a harmonic function $\chi$ through the relation

$$\phi = z\chi \tag{17.26}$$

in which case,

$$\nabla^2\phi = 2\frac{\partial\chi}{\partial z} \tag{17.27}$$

and hence

$$T = \frac{(1-\nu)}{\mu\alpha(1+\nu)}\frac{\partial\chi}{\partial z} \tag{17.28}$$

from equation (17.10). Expressions for the displacements and stresses are easily obtained by substituting (17.26) into equations (17.11–17.13) and are tabulated as Solution T in Table 17.1

We also note that the heat flux

$$q_z = -K\frac{\partial T}{\partial z} = -\frac{(1-\nu)K}{\mu\alpha(1+\nu)}\frac{\partial^2\chi}{\partial z^2} \tag{17.29}$$

where $K$ is the thermal conductivity of the material.

The solution here described was first obtained by Williams[2] and was used by him for the solution of some thermoelastic crack problems.

### 17.2.1 Thermoelastic plane stress

An important thermoelastic solution is that appropriate to the steady-state thermoelastic deformation of a traction-free half-space, which can be obtained by superposing solutions A,B,T and taking

$$\phi = 2(1-\nu)\psi; \quad \omega = -\chi = \frac{\partial\psi}{\partial z}\,. \tag{17.30}$$

[2]W. E. Williams, A solution of the steady-state thermoelastic equations, Z. angew. Math. Phyz. (ZAMP), Vol. 12 (1961), 452–455.

Table 17.1: Solutions T and P

| | **Solution T**<br>Williams' solution | **Solution P**<br>Thermoelastic Plane Stress |
|---|---|---|
| $2\mu u_x$ | $z\frac{\partial\chi}{\partial x}$ | $2(1-\nu)\frac{\partial\psi}{\partial x}$ |
| $2\mu u_y$ | $z\frac{\partial\chi}{\partial y}$ | $2(1-\nu)\frac{\partial\psi}{\partial y}$ |
| $2\mu u_z$ | $\chi+z\frac{\partial\chi}{\partial z}$ | $-2(1-\nu)\frac{\partial\psi}{\partial z}$ |
| $\sigma_{xx}$ | $z\frac{\partial^2\chi}{\partial x^2}-2\frac{\partial\chi}{\partial z}$ | $-2(1-\nu)\frac{\partial^2\psi}{\partial y^2}$ |
| $\sigma_{xy}$ | $z\frac{\partial^2\chi}{\partial x\partial y}$ | $2(1-\nu)\frac{\partial^2\psi}{\partial x\partial y}$ |
| $\sigma_{yy}$ | $z\frac{\partial^2\chi}{\partial y^2}-2\frac{\partial\chi}{\partial z}$ | $-2(1-\nu)\frac{\partial^2\psi}{\partial x^2}$ |
| $\sigma_{xz}$ | $\frac{\partial\chi}{\partial x}+z\frac{\partial^2\chi}{\partial x\partial z}$ | 0 |
| $\sigma_{yz}$ | $\frac{\partial\chi}{\partial y}+z\frac{\partial^2\chi}{\partial y\partial z}$ | 0 |
| $\sigma_{zz}$ | $z\frac{\partial^2\chi}{\partial z^2}$ | 0 |
| $2\mu u_r$ | $z\frac{\partial\chi}{\partial r}$ | $2(1-\nu)\frac{\partial\psi}{\partial r}$ |
| $2\mu u_\theta$ | $\frac{z}{r}\frac{\partial\chi}{\partial\theta}$ | $\frac{2(1-\nu)}{r}\frac{\partial\psi}{\partial\theta}$ |
| $\sigma_{rr}$ | $z\frac{\partial^2\chi}{\partial r^2}-2\frac{\partial\chi}{\partial z}$ | $2(1-\nu)\left(\frac{1}{r}\frac{\partial\psi}{\partial r}+\frac{1}{r^2}\frac{\partial^2\psi}{\partial\theta^2}\right)$ |
| $\sigma_{r\theta}$ | $\frac{z}{r}\frac{\partial^2\chi}{\partial r\partial\theta}-\frac{z}{r^2}\frac{\partial\chi}{\partial\theta}$ | $2(1-\nu)\left(\frac{1}{r}\frac{\partial^2\psi}{\partial r\partial\theta}-\frac{1}{r^2}\frac{\partial\psi}{\partial\theta}\right)$ |
| $\sigma_{\theta\theta}$ | $\frac{z}{r}\frac{\partial\chi}{\partial r}-\frac{z}{r^2}\frac{\partial^2\chi}{\partial\theta^2}-2\frac{\partial\chi}{\partial z}$ | $-2(1-\nu)\frac{\partial^2\psi}{\partial r^2}$ |
| $\sigma_{rz}$ | $\frac{\partial\chi}{\partial r}+z\frac{\partial^2\chi}{\partial r\partial z}$ | 0 |
| $\sigma_{\theta z}$ | $\frac{z}{r}\frac{\partial^2\chi}{\partial\theta\partial z}+\frac{1}{r}\frac{\partial\chi}{\partial\theta}$ | 0 |
| $T$ | $\frac{1-\nu}{\mu\alpha(1+\nu)}\frac{\partial\chi}{\partial z}$ | $-\frac{(1-\nu)}{\mu\alpha(1+\nu)}\frac{\partial^2\psi}{\partial z^2}$ |
| $q_z$ | $-\frac{(1-\nu)K}{\mu\alpha(1+\nu)}\frac{\partial^2\chi}{\partial z^2}$ | $\frac{(1-\nu)K}{\mu\alpha(1+\nu)}\frac{\partial^3\psi}{\partial z^3}$ |

The resulting expressions are given in Table 16.1 as Solution P. A striking result is that the three stress components $\sigma_{zx}, \sigma_{zy}, \sigma_{zz}$ are zero, not merely on the plane $z=0$, where we have forced them to zero by the relations (17.30), but throughout the half-space. For this reason, the solution is also appropriate to the problem of a thick plate $a<z<b$ with traction-free faces. Because of this feature, the solution is henceforth referred to as *thermoelastic plane stress.*

Notice also the similarity of the expressions for the non-zero stresses $\sigma_{xx}, \sigma_{xy}, \sigma_{yy}$ to the corresponding expressions obtained from the Airy stress function. However, the present solution is not two-dimensional — i.e. the temperature and hence all the non-zero stress and displacement components can be functions of all three coördinate directions — and it is exact, whereas the two-dimensional plane stress solution is generally approximate.

A second result of interest is

$$\frac{\partial^2 u_z}{\partial x^2}+\frac{\partial^2 u_z}{\partial y^2}=-\frac{(1-\nu)}{\mu}\left(\frac{\partial^3\psi}{\partial x^2\partial z}+\frac{\partial^3\psi}{\partial y^2\partial z}\right)=\frac{(1-\nu)}{\mu}\frac{\partial^3\psi}{\partial z^3}$$

(because $\psi$ and hence $\partial\psi/\partial z$ is harmonic)

$$=\frac{\alpha(1+\nu)q_z}{K}=\delta q_z \tag{17.31}$$

where $\delta$ is the *thermal distortivity* (see §14.3.1).

In other words, the sum of the principal curvatures of any distorted $z$-plane is proportional to the local heat flux across that plane[3]. The corresponding two-dimensional result was proved in §14.3.1.

## PROBLEM

1. During a test on an automotive disk brake, it was found that the temperature $T$ of the disk could be approximated by the expression

$$T=\frac{5}{4}(1-e^{-\tau})-\frac{r^2}{2a^2}(1-e^{-\tau})+\frac{r^4}{8a^4}\tau e^{-\tau}$$

where $\tau=t/t_0$ is a dimensionless time, $a$ is the radius of the disk, $r$ is the distance from the axis, $t$ is time and $t_0$ is a constant.

The disk is continuous at $r=0$ and all its surfaces can be assumed to be traction-free.

Derive expressions for the radial and circumferential thermal stresses as functions of radius and time, and hence show that the maximum value of the stress difference $(\sigma_{\theta\theta}-\sigma_{rr})$ at any given time always occurs at the outer edge of the disk.

[3] Further consequences of this result are discussed by J.R.Barber, Some implications of Dundurs' theorem for thermoelastic contact and crack problems, J.Strain Anal., Vol. 22 (1980), 229-232.

# Chapter 18

# SINGULAR SOLUTIONS

Singular solutions have a special place in the historical development of potential theory in general and of Elasticity in particular. Many of the important early solutions were obtained by appropriate superposition of singular potentials, noting that any form of singularity is permitted provided that the singular point is not a point of the body[1].

The generalization of this technique to allow continuous distributions of singularities in space (either at the boundary of the body or in a region of space not occupied by it) is still one of the most widely used methods of treating three-dimensional problems.

In this chapter, we shall consider some elementary forms of singular solution and examine the problems that they solve when used in the various displacement function representations of Chapter 16. The solutions will be developed in a rather *ad hoc* way, starting with those that are mathematically most straightforward and progressing to more complex forms. However, once appropriate forms have been introduced, a systematic way of developing them from first principles will be discussed in the next chapter.

## 18.1 The source solution

A convenient starting point is the most elementary singular harmonic function

$$\phi = \frac{1}{R} = \frac{1}{\sqrt{x^2+y^2+z^2}} , \tag{18.1}$$

which is easily demonstrated to be harmonic (except at the origin) by substitution into Laplace's equation. We shall refer to this solution as the 'source' solution, since it describes the steady-state temperature in an infinite body with a point heat source at

[1]If it *is* a point of the body, certain restrictions must be imposed on the strength of the singularity, as discussed in §11.2.1.

the origin or alternatively the potential in an infinite space with a point electric charge at the origin. It is clearly spherically symmetric — i.e. the potential depends only on the distance $R$ from the origin — and it can be developed from first principles by casting Laplace's equation in spherical polar coördinates $(R, \theta, \alpha)$ and then imposing the condition that $\phi$ be a function of $R$ only. The governing equation then reduces to

$$\frac{d^2\phi}{dR^2} + \frac{2}{R}\frac{d\phi}{dR} = 0 \qquad (18.2)$$

of which (18.1) is clearly the only singular solution

We shall examine the stress fields corresponding to the use of the source solution in Solutions A and B of Table 16.1.

### 18.1.1 The centre of dilatation

The strain potential solution (Solution A) is *isotropic* — i.e., the stress and displacement definitions preserve the same form in any Cartesian coördinate transformation — and hence the spherically symmetric function $1/R$ must correspond to a spherically symmetric state of stress and displacement.

Expressing Solution A in spherical polar coördinates and dropping the non-symmetric terms, we obtain

$$2\mu u_R = \frac{d\phi}{dR} = -\frac{1}{R^2} \; ; \qquad (18.3)$$

$$2\mu u_\theta = 2\mu u_\alpha = 0 \; ; \qquad (18.4)$$

$$\sigma_{RR} = \frac{d^2\phi}{dR^2} = \frac{2}{R^3} \; ; \qquad (18.5)$$

$$\sigma_{\theta\theta} = \sigma_{\alpha\alpha} = \frac{1}{R}\frac{d\phi}{dR} = -\frac{1}{R^3} \; ; \qquad (18.6)$$

$$\sigma_{\theta\alpha} = \sigma_{\alpha R} = \sigma_{R\theta} = 0 \; , \qquad (18.7)$$

where we have used ordinary derivatives to emphasise that these expressions relate only to the one-dimensional case of spherical symmetry.

This is the singularity known as the *centre of dilatation* or centre of compression[2]. Combined with a state of uniform hydrostatic stress, it can be used to describe the stresses and displacements in a thick-walled spherical pressure vessel loaded by arbitrary uniform pressure at the inside and outside surfaces. This solution was first obtained by Lamé.

The centre of dilatation is not admissable at an interior point of a continuous body — it would correspond to an infinitesimal hole containing a fluid at infinite pressure. However, a *distribution* of centres of dilatation can be used to describe dilatational effects such as those arising from thermal expansion. In fact, the thermoelastic potential of §15.5 could be regarded as being derived from Solution A by distributing

[2] S.P.Timoshenko and J.N.Goodier, *loc. cit.*, §136.

sources of strength proportional to $\alpha T$ throughout the space occupied by the body, to take account of the thermal dilatation. A similar technique can be used for dilatation arising from other sources, such as phase transformation.

### 18.1.2 The Kelvin solution

In Solution B, the stress components contain terms that are first derivatives of the potential $\omega$ or second derivatives multiplied by $z$. Hence, if we substitute $\omega=1/R$, we anticipate that the stresses will vary inversely with the square of the distance from the origin and hence that the force resultant over a spherical surface of radius $a$, with centre at the origin will be independent of $a$. In fact, this potential corresponds to the three-dimensional Kelvin problem, in which a concentrated force in the $z$-direction is applied at the origin in an infinite elastic body.

We shall demonstrate this by evaluating the stress components $\sigma_{zz}, \sigma_{rz}$ and considering the equilibrium of the region $-h<z<h$, which includes the origin.

Noting that

$$\frac{\partial}{\partial z}\left(\frac{1}{R}\right) = -\frac{z}{R^3} \; ; \; \frac{\partial}{\partial r}\left(\frac{1}{R}\right) = -\frac{r}{R^3} \, , \tag{18.8}$$

we obtain

$$\begin{aligned} \sigma_{zz} &= z\frac{\partial^2 \omega}{\partial z^2} - 2(1-\nu)\frac{\partial \omega}{\partial z} \\ &= z\left(\frac{3z^2}{R^5} - \frac{1}{R^3}\right) + 2(1-\nu)\frac{z}{R^3} \\ &= (1-2\nu)\frac{z}{R^3} + \frac{3z^3}{R^5} \; ; \end{aligned} \tag{18.9}$$

$$\begin{aligned} \sigma_{rz} &= z\frac{\partial^2 \omega}{\partial r \partial z} - (1-2\nu)\frac{\partial \omega}{\partial r} \\ &= (1-2\nu)\frac{r}{R^3} + \frac{3rz^2}{R^5} \, . \end{aligned} \tag{18.10}$$

We now consider the equilibrium of the cylinder $r<a$, $-h<z<h$, which is shown in Figure 18.1. The curved surfaces $r=a$ of this cylinder experience a force in the $z$-direction because of the stress $\sigma_{rz}$, but this force decays to zero as $a\to 0$, since, although the area increases with $a$ (being equal to $2\pi ah$), the stress component decreases with $a^2$. We can therefore restrict attention to the surface $z=\pm h$ if $a$ is sufficiently large.

The stress component $\sigma_{zz}$ is odd in $z$ and hence the forces transmitted across the two surfaces are equal and have the total value

$$\begin{aligned} F_z &= 2\int_0^\infty 2\pi r\left[\frac{(1-2\nu)h}{R^3} + \frac{3h^3}{R^5}\right]dr \\ &= 4\pi\left[-\frac{(1-2\nu)h}{R} - \frac{h^3}{R^3}\right]_{r=0}^{r=\infty} = 8\pi(1-\nu) \, , \end{aligned} \tag{18.11}$$

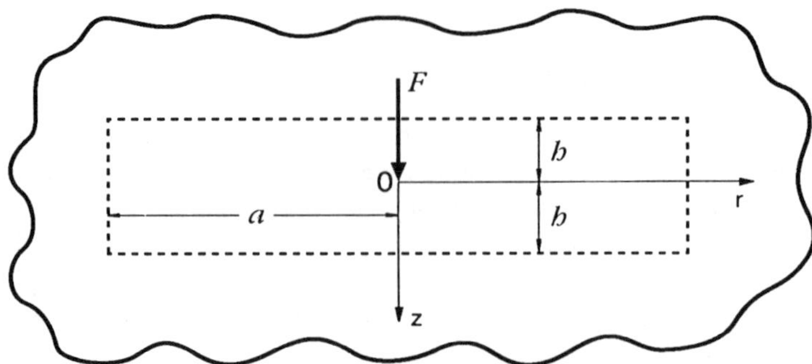


Figure 18.1: The Kelvin problem

where we note that $R = |z|$ when $r = 0$.

It follows that there must be an equal and opposite force at the origin. Hence, the function

$$\omega = -\frac{F}{8\pi(1-\nu)R} \tag{18.12}$$

in Solution B corresponds to the problem of a force $F$ acting in the $z$-direction at the origin in the infinite elastic body. The complete stress field is easily obtained by substituting (18.12) into the appropriate expressions from Table 16.1.

## 18.2 Dimensional considerations

In the last section, we developed the Kelvin solution 'accidentally' by examining the stress field due to the source potential in Solution B. However, a more deductive development of the solution can be made from equilibrium and dimensional considerations.

We first note that the Kelvin problem is self-similar (see §12.1), since there is no inherent length scale. It follows that the form of the variation of the solution with $\theta, \alpha$ must be independent of the distance from the origin $R$ and hence that the solution can be expressed in separated variable form

$$f(R, \theta, \alpha) = g(R)h(\alpha, \theta) \; , \tag{18.13}$$

where $f$ is any field quantity such as a stress or displacement component.

Given this result, we can deduce from equilibrium considerations that the function $g(R)$ appropriate to the stress components must be $R^{-2}$, since the total resultant

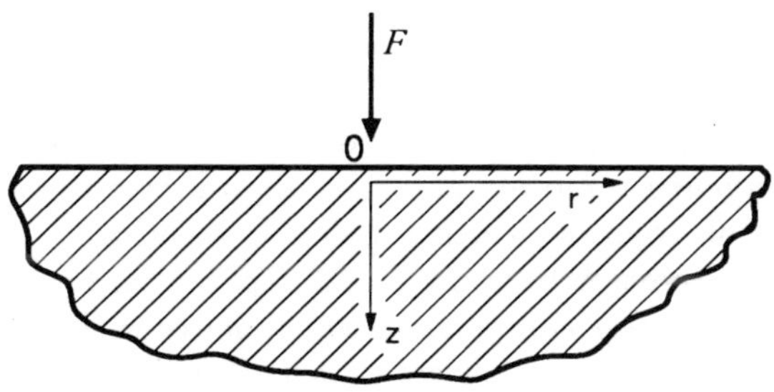


Figure 18.2: The Boussinesq problem

force across any one of a class of self-similar surfaces (i.e. surfaces of the same shape but different size) must be the same and the surface area of such surfaces will be proportional to the square of their linear dimensions. Thus, any stress component must decay with distance from the origin according to $R^{-2}$.

We now examine solutions A and B to see what type of singular function would satisfy this condition. On dimensional grounds, we find that the potential would have to be of order $R^0$ in Solution A and of order $R^{-1}$ in Solution B. We shall see in the next section that there *are* singular potentials of order $R^0$, but that they involve singularities not merely at the origin, but along at least half of the $z$-axis. Thus, they cannot be used for a problem involving the whole infinite body. We are therefore left with the function $1/R$ — which we know to be harmonic — in Solution B, which can then be developed as in §18.1.2.

### 18.2.1 The Boussinesq solution

We shall now apply similar arguments to solve the Boussinesq problem, in which a point force, $F$, in the $z$-direction is applied at the origin $R=0$ on the surface $z=0$ of the semi-infinite body (or *half-space*) $z>0$, as shown in Figure 18.2. In particular, it is clear that this problem is also self-similar and equilibrium arguments again demand that the stress components decay with distance from the origin with $R^{-2}$.

This problem is treated by Timoshenko and Goodier[3] by superposing further singular solutions on the Kelvin solution so as to make the surface $z=0$ free of tractions except at the origin. A more direct (and modern) approach is to seek a special potential function solution which identically satisfies two of the three boundary

[3] *loc. cit.*, §138.

conditions at the surface.

We note that the force is normal to the surface, so that there is no *tangential* traction at any point on the surface[4]. i.e.

$$\sigma_{zx} = \sigma_{zy} = 0; \quad \text{all } x, y; \; z = 0 \tag{18.14}$$

Since this condition applies at all points of the surface $z = 0$ we shall refer to it as a *global boundary condition* and it is appropriate to satisfy it by constructing a potential function solution which satisfies it identically, this being Solution F of Table 16.2. It then remains to find a suitable potential function which when used in Solution F will yield stress components which decay with $R^{-2}$.

Examination of Solution F shows that $1/R$ will not be suitable in the present instance, since the stresses are obtained by two differentiations of the stress function and hence we require $\varphi$ to vary with $R^0$. We therefore seek a suitable partial integral of $1/R$ — to be dimensionless in $R$ and singular at the origin, but otherwise to be continuous and harmonic in $z>0$.

It is easily verified that the function

$$\varphi = \int_{-\infty}^{0} \frac{d\zeta}{\sqrt{x^2 + y^2 + (z-\zeta)^2}} = \log(R + z) \tag{18.15}$$

satisfies these requirements. Notice that this function is singular on the negative $z$-axis, where $R = -z$, but not on the positive $z$-axis, where $R = z$. In fact, it can be regarded as the potential due to a uniform distribution of sources on the negative $z$-axis ($r=0$, $z<0$).

Substituting into Solution F, we find

$$\sigma_{zz} = z\frac{\partial^3\varphi}{\partial z^3} - \frac{\partial^2\varphi}{\partial z^2} = \frac{3z^3}{R^5} \tag{18.16}$$

and hence the surface $z=0$ is free of traction except at the origin as required[5].

The force applied at the origin is

$$\begin{aligned} F &= -2\pi\int_0^\infty r\sigma_{zz}(r,h)dr \\ &= -6\pi h^3 \int_0^\infty \frac{rdr}{R^5} = -2\pi \end{aligned} \tag{18.17}$$

and hence the stress field due to a force $F$ in the $z$-direction applied at the origin is obtained from the potential

$$\varphi = -\frac{F}{2\pi}\log(R+z)\,. \tag{18.18}$$

---

[4]The normal traction is zero everywhere except at the origin, where there is a delta function loading, i.e. $\sigma_{zz} = -F\delta(x)\delta(y)$.

[5]The condition $\sigma_{zr} = 0$ on the surface is guaranteed by the use of Solution F.

The displacements at the surface $z=0$ are of particular interest, in view of the application to contact problems[6]. We have

$$2\mu u_r = (1-2\nu)\frac{\partial\varphi}{\partial r} = -\frac{F(1-2\nu)}{2\pi r} \quad ; \quad z=0\, ; \qquad (18.19)$$

$$2\mu u_z = -2(1-\nu)\frac{\partial\varphi}{\partial z} = \frac{F(1-\nu)}{\pi r} \quad ; \quad z=0\, . \qquad (18.20)$$

Thus, both displacements vary inversely with distance from the point of application of the force $F$, as indeed we could have deduced from dimensional considerations. The singularity in displacement at the origin is not serious, since in any practical application, the force will be distributed over a finite area. If the solution for such a distributed force is found by superposition using the above result, the singularity will be integrated out.

We note that the displacements are bounded at infinity, in contrast to the two-dimensional case (see Chapter 12). It is therefore possible to regard the point at infinity as a reference for rigid-body displacements, if appropriate.

The radial surface displacement $u_r$ is negative, indicating that a force $F$ directed into the half-space — i.e. a compressive force — causes the surrounding surface to move towards the origin[7]. We can define a *surface dilatation* $e_s$ on the surface $z=0$, such that

$$e_s \equiv \frac{\partial u_x}{\partial x} + \frac{\partial u_y}{\partial y} = \frac{\partial u_r}{\partial r} + \frac{u_r}{r} + \frac{1}{r}\frac{\partial u_\theta}{\partial \theta}\, . \qquad (18.21)$$

Substituting for the displacement from equation (18.19), we obtain

$$e_s = -\frac{F(1-2\nu)}{4\pi\mu}\left[-\frac{1}{r^2} + \frac{1}{r^2}\right] = 0\, . \qquad (18.22)$$

In other words, the normal force $F$ does not cause any dilatation of the surface $z=0$, except at the origin. However, if we draw a circle of radius $a$, centre the origin, on the surface, it is clear that the radius of this circle gets smaller by the amount of the inward radial displacement and hence the area of the circle $A$ increases by

$$\delta A = 2\pi a u_r = -\frac{F(1-2\nu)}{2\mu}\, . \qquad (18.23)$$

This result is independent of $a$ — i.e. all circles reduce in area by the same amount[8] and all of this reduction must therefore be concentrated at the origin, where in a sense

[6]cf. Chapter 12.

[7]Most engineers would say that this is what they would intuitively expect, but the validity of this intuition is suspect, since the corresponding result for a tensile force — for which the radial displacement is directed away from the origin — is counter-intuitive. Our expectation here is probably determined by the thought that forces of either direction will stretch the surface and hence draw points towards the origin, but this is a second order (non-linear) effect and cannot be admitted in the linear theory.

[8]as could be argued from the fact that the surface dilatation is zero.

a small amount of the surface is lost. We can generalize this result by superposition to state that, if there is an arbitrary distribution of purely normal traction on the surface $z=0$ of the half-space, the area $A$ enclosed by any closed curve will change by $\delta A$ of equation (18.23), where $F$ is now to be interpreted as the resultant of the tractions acting *within* the area $A$.

## 18.3 Other singular solutions

We have already shown how the singular solution $\log(R+z)$ can be obtained from $1/R$ by partial integration, which of course is a form of superposition. A whole sequence of axially symmetric solutions can be obtained in the same way. Defining

$$\phi_0 = \frac{1}{R}, \tag{18.24}$$

we can develop the sequence of solutions

$$\begin{aligned} \phi_{-1} &= \log(R+z)\ ;\ \phi_{-2} = z\log(R+z) - R\ ; \\ \phi_{-3} &= (3z^2 - R^2)\log(R+z) - 3Rz + 2z^2;\ \ldots \end{aligned} \tag{18.25}$$

by the operation

$$\phi_{n-1}(r,z) = \int_{-\infty}^{0} \phi_n(r,(z-\zeta))d\zeta\ . \tag{18.26}$$

The inverse operation is one of differentiation, so that

$$\phi_n = \frac{\partial \phi_{n-1}}{\partial z}\ . \tag{18.27}$$

We can therefore extend the sequence to functions with stronger singularities such as

$$\phi_1 = -\frac{z}{R^3}\ ;\ \phi_2 = \frac{3z^2}{R^5} - \frac{1}{R^3}\ ;\ \ldots \tag{18.28}$$

All of these functions are harmonic except at the origin and on the negative $z$-axis in the case of those involving the term $\log(R+z)$.

We can also generate non-axisymmetric potential functions by differentiation with respect to $x$ or $y$. For example, the function

$$\frac{\partial}{\partial x}\left(-\frac{z}{R^3}\right) = \frac{3xz}{R^5} = \frac{3\sin 2\alpha\cos\theta}{2R^3} \tag{18.29}$$

in spherical coordinates, is a non-axisymmetric harmonic function. However, these solutions can be obtained more systematically in terms of Legendre Polynomials, as will be shown below in Chapter 19.

The integrated solutions solutions defined by equation (18.26) can be generalized to permit an arbitrary distribution of sources along a line or surface. For example, a fairly general axisymmetric harmonic potential can be written in the form

$$\phi(r,z) = \int_a^b \frac{g(t)dt}{\sqrt{r^2+(z+t)^2}} , \tag{18.30}$$

which defines the potential due to a distribution of sources of strength $g(z)$ along the negative $z$-axis in the range $-b<z<-a$. Such a solution defines bounded stresses and displacements in any body which does not include this line segment. When the boundary conditions of the problem are expressed in terms of a representation such as (18.30), the problem is essentially reduced to the solution of an integral equation or a set of integral equations. This technique has a long history in the solution of axisymmetric flow problems around smooth bodies and it can be used for the related elasticity problem in which a uniform tensile stress is perturbed by an axisymmetric cavity. We shall also use an adaptation of the same method for solving crack and contact problems for the half-space.

## PROBLEMS

1. Use dimensional arguments to determine a suitable *axisymmetric* harmonic singular potential in the solution of Problem 2, Chapter 16 to solve the problem of a concentrated *tangential* force $F$ in the $x$-direction, applied to the surface of the half-space at the origin.

2. Find the stresses and displacements corresponding to the use of $\Psi=1/R$ in Solution E. Show that the surfaces of the cone, $r=z\tan\alpha$ are free of traction and hence use the solution[9] to determine the stresses in a conical shaft of semi-angle $\alpha$, transmitting a torque, $T$.

Find the maximum shear stress at the section $z=c$, and compare it with the maximum shear stress in a cylindrical bar of radius $a=c\tan\alpha$. **Note** that (i) the *maximum* shear stress will not generally be either $\sigma_{z\theta}$ or $\sigma_{r\theta}$, but must be found by appropriate coördinate transformation, and (ii) it does not *necessarily* occur at the outer radius of the cone, $r=c\tan\alpha$.

3. An otherwise uniform cylindrical bar of radius $b$ has a small spherical hole of radius $a$ on the axis.

Combine the function $\Psi=Az/R^3$ in solution E with the elementary torsion solution for the bar without a hole (see, for example, Chapter 16, Problem 1) and show that, with a suitable choice of the arbitrary constant $A$, the surface of the hole, $R=a$ can be made traction-free. Hence deduce the stress field near the hole for this problem.

---

[9]This solution is known as the *centre of rotation*.

Assume that $b \gg a$, so that the perturbation due to the hole has a negligible influence on the stresses near the outer surface, $r = b$.

4. The area $A$ on the surface of an elastic half-space is subjected to a purely normal pressure $p(x, y)$ corresponding to the total force, $F$. The loaded region is now expanded in a self-similar manner by multiplying all its linear dimensions by the same ratio, $\lambda$. Each point in the new contact area is subjected to a self-similar loading such that the pressure at the point $(\lambda x, \lambda y)$ is $Cp(x, y)$, where the constant $C$ is chosen to ensure that the total force $F$ is independent of $\lambda$.

Show that the deflection at corresponding points $(\lambda x, \lambda y, \lambda z)$ will then be proportional to $F/\lambda$.

# Chapter 19

# SPHERICAL HARMONICS

The singular solutions introduced in the last chapter are all particular cases of a class of functions known as spherical harmonics. In this chapter, we shall develop these solutions in a more formal way and show how they constitute an orthogonal set of functions appropriate to the problem of the sphere with given surface tractions or to the infinite body with a spherical hole. These problems are of course the three-dimensional counterparts of those considered in Chapter 8.

## 19.1 Fourier series solution

The Laplace equation in cylindrical polar coördinates takes the form

$$\nabla^2\phi \equiv \frac{\partial^2\phi}{\partial r^2} + \frac{1}{r}\frac{\partial\phi}{\partial r} + \frac{1}{r^2}\frac{\partial^2\phi}{\partial\theta^2} + \frac{\partial^2\phi}{\partial z^2} = 0 \; . \tag{19.1}$$

We first perform a Fourier decomposition with respect to the variable $\theta$, giving a series of terms involving $\sin m\theta, \cos m\theta,\;\; m = 0, 1, 2, \ldots, \infty$. For the sake of brevity, we restrict attention to the cosine terms, since the sine terms will be of the same form and can be reintroduced at the end of the analysis. We therefore assume that the function $\phi$ can be written

$$\phi = \sum_{m=0}^{\infty} f_m(r, z) \cos m\theta \; . \tag{19.2}$$

Substituting this series into (19.1), we find that the functions $f_m$ must satisfy the equation

$$\frac{\partial^2 f}{\partial r^2} + \frac{1}{r}\frac{\partial f}{\partial r} - \frac{m^2 f}{r^2} + \frac{\partial^2 f}{\partial z^2} = 0 \; . \tag{19.3}$$

## 19.2 Reduction to Legendre's equation

We now make the change of variable

$$R = \sqrt{r^2 + z^2} \ ; \ x = \frac{z}{R} , \tag{19.4}$$

with solution

$$r = R\sqrt{1 - x^2} \ ; \ z = Rx . \tag{19.5}$$

We wish to cast equation (19.3) in terms of the new variables, for which we need the relations

$$\begin{aligned} \frac{\partial}{\partial r} &= \frac{\partial}{\partial R}\frac{\partial R}{\partial r} + \frac{\partial}{\partial x}\frac{\partial x}{\partial r} \\ &= \sqrt{1-x^2}\frac{\partial}{\partial R} - \frac{x\sqrt{1-x^2}}{R}\frac{\partial}{\partial x} ; \qquad (19.6) \\ \frac{\partial}{\partial z} &= x\frac{\partial}{\partial R} + \frac{1-x^2}{R}\frac{\partial}{\partial x} . \qquad (19.7) \end{aligned}$$

After routine but lengthy substitutions, these results enable us to to express equation (19.3) in the form

$$\frac{\partial^2 f}{\partial R^2} + \frac{2}{R}\frac{\partial f}{\partial R} - \frac{m^2 f}{R^2(1-x^2)} - \frac{2x}{R^2}\frac{\partial f}{\partial x} + \frac{(1-x^2)}{R^2}\frac{\partial^2 f}{\partial x^2} = 0 . \tag{19.8}$$

Finally, noting that (19.8) is homogeneous in powers of $R$, we expand $f_m(R, x)$ as a power series — i.e.

$$f_m(R, x) = \sum_{n=-\infty}^{\infty} g_{mn}(x)R^n , \tag{19.9}$$

which on substitution into equation (19.8) requires that the functions $g_{mn}(x)$ satisfy the ordinary differential equation

$$(1-x^2)\frac{d^2 g}{dx^2} - 2x\frac{dg}{dx} + \left[n(n+1) - \frac{m^2}{(1-x^2)}\right] g = 0 . \tag{19.10}$$

This is the standard form of *Legendre's equation*[1]. It is a second order ordinary differential equation, so it has two linearly independent solutions. If $m$ is an even integer, one of these solutions is a finite polynomial which can readily be determined by writing $g$ in polynomial form, substituting into equation (19.10) and equating coefficients to obtain a set of recurrence relations.

[1] See for example I.S.Gradshteyn and I.M.Ryzhik, *Tables of Integrals, Series and Products*, Academic Press, New York (1980), §8.70.

## 19.3 Legendre functions

The two linearly independent solutions of equation (19.10) are denoted by $P_n^m(x)$, $Q_n^m(x)$ and are known as *Legendre functions.* In view of the above derivations, it follows that the functions

$$\begin{array}{lcl} P_n^m(z/R)R^n \cos m\theta & ; & Q_n^m(z/R)R^n \cos m\theta \; ; \\ P_n^m(z/R)R^n \sin m\theta & ; & Q_n^m(z/R)R^n \sin m\theta \end{array} \tag{19.11}$$

are harmonic for all $m, n$. These functions are known as *spherical harmonics.*

### 19.3.1 Legendre polynomials

In the special case where $m = 0$, it is generally omitted — i.e.

$$P_n^0(x) \equiv P_n(x); \quad Q_n^0(x) \equiv Q_n(x) \; . \tag{19.12}$$

The finite polynomial $P_n(x)$ is of degree $n$ in $x$ and is known as a *Legendre polynomial.* The first few Legendre polynomials are

$$\begin{aligned} P_0(x) &= 1 \; ; \\ P_1(x) &= x = \cos\alpha \; ; \\ P_2(x) &= \frac{1}{2}(3x^2-1) = \frac{1}{4}(3\cos 2\alpha + 1) \; ; \\ P_3(x) &= \frac{1}{2}(5x^3-3x) = \frac{1}{8}(5\cos 3\alpha + 3\cos\alpha) \; ; \\ P_4(x) &= \frac{1}{8}(35x^4 - 30x^2 + 3) = \frac{1}{64}(35\cos 4\alpha + 20\cos 2\alpha + 9) \; ; \\ P_5(x) &= \frac{1}{8}(63x^5 - 70x^3 + 15x) = \frac{1}{128}(63\cos 5\alpha + 35\cos 3\alpha + 30\cos\alpha) \; , \end{aligned} \tag{19.13}$$

where

$$\cos\alpha = x = \frac{z}{R} \tag{19.14}$$

and $\alpha$ is the polar angle in spherical polar coördinates.

The sequence can be extended to higher values of $n$ by using the recurrence relation[2]

$$(n+1)P_{n+1}(x) - (2n+1)xP_n(x) + nP_{n-1}(x) = 0 \tag{19.15}$$

or the differential definition

$$P_n(x) = \frac{1}{2^n n!}\frac{d^n}{dx^n}(x^2-1)^n \; . \tag{19.16}$$

[2]See I.S.Gradshteyn and I.M.Ryzhik, *loc. cit.* §§8.81, 8.82, 8.91 for this and more results concerning Legendre functions and polynomials.

### 19.3.2 Singular spherical harmonics

Legendre functions are generally defined only for $n \geq 0$, in which case the corresponding spherical harmonics will be bounded at the origin, $R=0$. However, if we set

$$n = -p - 1 \,, \tag{19.17}$$

where $p \geq 0$, we find that

$$n(n+1) = (-p-1)(-p) = p(p+1) \tag{19.18}$$

and hence equation (19.10) becomes

$$(1-x^2)\frac{d^2 g}{dx^2} - 2x\frac{dg}{dx} + \left[p(p+1) - \frac{m^2}{(1-x^2)}\right] g = 0 \,. \tag{19.19}$$

In other words, the equation is of the same form, with $p$ replacing $n$. It follows that

$$\begin{array}{lcl} P_n^m(z/R)R^{-n-1}\cos m\theta & ; & Q_n^m(z/R)R^{-n-1}\cos m\theta \,; \\ P_n^m(z/R)R^{-n-1}\sin m\theta & ; & Q_n^m(z/R)R^{-n-1}\sin m\theta \,, \end{array} \tag{19.20}$$

where $n \geq 0$, are harmonic functions, which are singular at $R=0$. If $n$ is allowed to take all non-negative values, the two sets of functions (19.11, 19.20) define two sine and two cosine harmonics for all integer powers of $R$, as indicated formally by the series (19.9)[3].

## 19.4 Axisymmetric potentials

Axisymmetric potentials are obtained by setting $m=0$ in the spherical harmonics with cosine multipliers. In particular, if we set $m=n=0$ in the first of (19.20), we obtain

$$\phi = P_0(z/R)R^{-1} = \frac{1}{R} \,, \tag{19.21}$$

which we recognize as the source solution of the previous chapter.

Furthermore, if we next set $n=1$ in the same term, we obtain

$$\phi = P_1(z/R)R^{-2} = \frac{z}{R^3} \,, \tag{19.22}$$

which is proportional to the function $\phi_1$ of (18.28). Similar relations exist between the sequence $\phi_0, \phi_1, \phi_2, \ldots$, and the higher order singular spherical harmonics.

[3]Those harmonics with the multiplier $\sin m\theta$ degenerate to zero in the axisymmetric case $m=0$. As in Chapter 8, this degeneracy is not important if the body is continuous in $\theta$, but if it is bounded by planes such as $\theta = \pm\beta$, we may need to repair the deficit by developing some special solutions by the technique of §§10.3, 10.4

**The $Q$ series**

When $m = 0$, the functions $P_n(x)$ are finite polynomials, but the functions $Q_n(x)$ involve logarithms. The first few functions are

$$
\begin{aligned}
Q_0(x) &= \frac{1}{2}\log\left(\frac{1+x}{1-x}\right) ; \\
Q_1(x) &= \frac{x}{2}\log\left(\frac{1+x}{1-x}\right) - 1 ; \\
Q_2(x) &= \frac{1}{4}(3x^2-1)\log\left(\frac{1+x}{1-x}\right) - \frac{3x}{2} ; \\
Q_3(x) &= \frac{1}{4}(5x^3-3x)\log\left(\frac{1+x}{1-x}\right) - \frac{5x^2}{2} + \frac{2}{3} ; \\
Q_4(x) &= \frac{1}{16}(35x^4-30x^2+3)\log\left(\frac{1+x}{1-x}\right) - \frac{35x^3}{8} + \frac{55x}{24} ; \\
Q_5(x) &= \frac{1}{16}(63x^5-70x^3+15x)\log\left(\frac{1+x}{1-x}\right) - \frac{63x^4}{8} + \frac{49x^2}{8} - \frac{8}{15} .
\end{aligned}
\tag{19.23}
$$

The recurrence relation

$$(n+1)Q_{n+1}(x) - (2n+1)xQ_n(x) + nQ_{n-1}(x) = 0 \tag{19.24}$$

can be used to extend this sequence to higher values of $n$.

The attentive reader will recognize a connection between the logarithmic terms in this series and the function $\log(R+z)$ obtained from the source solution by partial integration in §18.3. However, the two solutions are not identical. If we construct the axisymmetric spherical harmonic

$$\phi = R^0 Q_0(z/R) = \frac{1}{2}\log\left(\frac{R+z}{R-z}\right) , \tag{19.25}$$

we find that it differs from $\log(R+z)$ by

$$
\begin{aligned}
\phi^* &= \log(R+z) - \frac{1}{2}\log\left(\frac{R+z}{R-z}\right) \\
&= \frac{1}{2}\log(R^2-z^2) = \log r ,
\end{aligned}
\tag{19.26}
$$

which we have already encountered as a two-dimensional harmonic potential.

More generally, we find that all the axisymmetric potentials derived form the functions $Q_n(x)$ include the term $\log[(R+z)(R-z)]$ and hence are singular everywhere on the $z$-axis.

We may wish to develop a related function which is singular only on one half of the $z$-axis, as in §18.3, where we used such a function to define the stress field in a body which did not include any of the singular points. We can always construct such a function.

We first note that the logarithmic term in $Q_n(x)$ is always multiplied by a Legendre polynomial. Indeed, we can write

$$Q_n(x) = \frac{1}{2}P_n(x)\log\left(\frac{1+x}{1-x}\right) - W_{n-1}(x) , \tag{19.27}$$

where

$$W_{n-1}(x) = \sum_{k=1}^{n} \frac{1}{k} P_{k-1}(x)P_{n-k}(x) \tag{19.28}$$

is a finite polynomial of degree of degree $n-1$.

It follows that the corresponding spherical harmonic is the sum of the bounded ($P$ series) spherical harmonic of the same degree multiplied by $\log[(R+z)/(R-z)]$ and a finite polynomial in $R, z$, i.e.

$$\phi = \frac{R^n}{2}P_n(z/R)\log\left(\frac{R+z}{R-z}\right) - R^n W_{n-1}(z/R) . \tag{19.29}$$

In order to construct a related function which is singular only on one half of the $z$-axis, we need to superpose a generalized form of equation (19.26), which we can write as

$$\phi^* = \frac{R^n}{2}P_n(z/R)\log r + F(r,z) , \tag{19.30}$$

where $F(r,z)$ is a finite polyomial to be determined from the condition that (19.30) be harmonic.

We substitute (19.30) into the Laplace equation, obtaining

$$\nabla^2\phi^* = \nabla^2\left(\frac{R^n}{2}P_n(z/R)\right)\log r + \frac{1}{r}\frac{\partial}{\partial r}[(R^n P_n(z/R)] + \nabla^2 F . \tag{19.31}$$

The first term in this equation is clearly zero, since $R^n P_n(z/R)$ is harmonic. Thus, for $\phi^*$ to be harmonic, we require

$$\nabla^2 F = -\frac{1}{r}\frac{\partial}{\partial r}[(R^n P_n(z/R)] \tag{19.32}$$

$$= \frac{1}{R}\left(\frac{\partial}{\partial R} - \frac{x}{R}\frac{\partial}{\partial x}\right)R^n \tag{19.33}$$

$$= R^{n-2}\left(nP_n(x) - x\frac{d}{dx}P_n(x)\right) . \tag{19.34}$$

Alternatively, if we write $F(R,z) = R^n S_n(x)$, the function $S$ must satisfy the equation

$$(1-x^2)\frac{d^2 S_n}{dx^2} - 2x\frac{dS_n}{dx} + n(n+1)S_n = \left(nP_n(x) - x\frac{d}{dx}P_n(x)\right) . \tag{19.35}$$

Now if we take $S_n$ to be a finite polynomial of degree $n$, both sides of this equation will be polynomials of degree $n$ and hence we can determine $S_n$ by equating coefficients.

In view of the results of this section, we conclude that it is always possible to construct a spherical harmonic related to the $Q$ series harmonic, but which is singular on only half of the $z$-axis and, further, that the resulting function will have the same multiplier on the logarithmic term, but will have a different finite polynomial in place of $W_{n-1}$ in equation (19.27). In practice, it is usually quicker to assume such a form from the beginning, substitute into the Laplace equation and hence develop a set of recurrence relations for the residual finite polynomial. The first few potentials of this form are given in equation (18.25) above.

## 19.5 Non-axisymmetric harmonics

When $m \neq 0$, equations (19.11, 19.20) will define harmonic functions that are not axisymmetric. The corresponding Legendre functions can be developed from the axisymmetric forms by the relations

$$P_n^m(x) = (-1)^m(1-x^2)^{m/2}\frac{d^m}{dx^m}P_n(x) ; \tag{19.36}$$

$$Q_n^m(x) = (-1)^m(1-x^2)^{m/2}\frac{d^m}{dx^m}Q_n(x) , \tag{19.37}$$

for $m \leq n$.

For $m > n$, equation (19.35) can still be used to determine the function $Q_n^m(x)$, but (19.34) would degenerate, since a finite polynomial of degree $n$ can only be differentiated $n$ times. For this case, an alternative representation for $P_n^m(x)$ is

$$P_n^m(x) = (1-x^2)^{-m/2}\int_x^1 \cdots \int_x^1 P_n(x)(dx)^m . \tag{19.38}$$

The first few non-axisymmetric functions in the $P$ series are

$$\begin{aligned}
P_1^1(x) &= -(1-x^2)^{\frac{1}{2}} = -\sin\alpha ; \\
P_2^1(x) &= -3x(1-x^2)^{\frac{1}{2}} = -\frac{3}{2}\sin 2\alpha ; \\
P_2^2(x) &= 3(1-x^2) = \frac{3}{2}(1-\cos 2\alpha) ; \\
P_3^1(x) &= -\frac{3}{2}(5x^2-1)(1-x^2)^{\frac{1}{2}} = -\frac{3}{8}(\sin\alpha + 5\sin 3\alpha) ; \\
P_3^2(x) &= 15x(1-x^2) = \frac{15}{4}(\cos\alpha - \cos 3\alpha) ; \\
P_3^3(x) &= -15(1-x^2)^{\frac{3}{2}} = -\frac{15}{4}(3\sin\alpha - \sin 3\alpha) .
\end{aligned} \tag{19.39}$$

A different but related way of developing non-axisymmetric harmonic functions is to differentiate axisymmetric functions in $(r,\theta,z)$ with respect to $x$ or $y$, as in equation (18.29) above. In general, each differentiation with respect to $x$ will introduce a multiple of $\cos\theta$ and each differentiation with respect to $y$, a multiple of $\sin\theta$. Thus, any required Fourier term in $\theta$ can be obtained by combining appropriate differentiations, using the expansions of $\cos(m\theta), \sin(m\theta)$ in powers of $\cos\theta, \sin\theta$. The leading terms in these expansions are of order $m$ and hence to generate a function with the multiplier $\cos(m\theta)$, it will be necessary to differentiate an axisymmetric polynomial $m$ times[4]. If we want to end up with a function of order $R^n$, we must therefore start with the axisymmetric function $R^{(m+n)}P_{(m+n)}(z/R)$.

The spherical harmonic solutions are most useful in problems involving an axisymmetric body, though the boundary tractions need not necessarily be axisymmetric. All the problems discussed in Chapters 5-14 have three-dimensional axisymmetric counterparts, in which the coördinates $x, y$ are replaced by $r, z$. Thus, we can use techniques similar to those of Chapter 5 to develop solutions for polynomial loading on the curved faces of the hollow or solid cylinder and the problems of §§8.3.2, 8.4.1 have their counterparts in the stress concentration due to a small spherical hole.

[4]This process can be used to generate non-axisymmetric from axisymmetric solutions in other problems. See, for example, J.R.Barber, Some ploynomial solutions for the non-axisymmetric Boussinesq problem, J.Elasticity, Vol. 14 (1984), 217–221.

# Chapter 20

# AXISYMMETRIC PROBLEMS

Axisymmetric problems tend to be algebraically more complicated than their two-dimensional counterparts. One reason for this is that the stress field is no longer independent of Poisson's ratio and hence terms with similar polynomial variation cannot always be combined into simpler expressions. We shall therefore restrict attention to the simplest examples in the present chapter.

## 20.1 The solid cylinder

We consider the case of the solid cylinder ($0 \leq r < a, -L < z < L$) subjected to the traction distribution

$$\sigma_{rr} = -p_0(L^2 - z^2) \;\; ; \quad \sigma_{zr} = 0; \;\; r = a \; ; \tag{20.1}$$

$$\sigma_{zr} = \sigma_{zz} = 0 \;\; ; \quad z = \pm L \; . \tag{20.2}$$

As in the problems of Chapter 5, we cannot satisfy the boundary conditions in the strong sense on all the boundaries. We therefore satisfy the boundary conditions on the ends ($z = \pm L$) only in the weak, force resultant, sense. The resulting solution will be in error in the vicinity of the ends, but there will be a central region in which it is quite accurate, providing the cylinder is long[1] (i.e. $L \gg a$). End effect solutions can also be obtained as in Chapter 6 to remove the residual error due to this approximation.

The problem is even in $z$ and axisymmetric, so the normal stresses will be even polynomial functions of $r$ and $z$. Also, since the loading produces no bending or axial force, we deduce from the quadratic variation of the tractions that the highest polynomial term in the boundary conditions will be of order 2. Referring to Table

[1]The other limiting case of the *short* cylinder ($L \ll a$) can be treated by satisfying the conditions on the curved boundary $r = a$ in the weak sense, but in the present problem, this would lead to a trivial result in which the cylinder was in a state of biaxially hydrostatic tension, since the ends on which the strong conditions would be applied are traction-free.

16.1, we note that quadratic stress components will be derived from polynomials of order 4 in Solution A and order 3 in Solution B. We therefore propose the functions

$$\phi = A_4R^4P_4(z/R) + A_2R^2P_2(z/R) \; ; \tag{20.3}$$
$$\omega = B_3R^3P_3(z/R) + B_1RP_1(z/R) \; , \tag{20.4}$$

for use in Solutions A,B respectively[2].

Using the definitions of Legendre polynomials (equation 19.13), these expressions can be expanded in cylindrical coördinates in the form

$$\phi = A_4\left(z^4 - 3r^2z^2 + \frac{3}{8}r^4\right) + A_2\left(z^2 - \frac{1}{2}r^2\right) \; ; \tag{20.5}$$
$$\omega = B_3\left(z^3 - \frac{3}{2}zr^2\right) + B_1z \; , \tag{20.6}$$

and substitution into the expressions for the stress components from Table 16.1 gives

$$\sigma_{rr} = A_4\left(-6z^2 + \frac{9}{2}r^2\right) - A_2 + B_3(-3(1+2\nu)z^2 + 3\nu r^2) - 2\nu B_1 \; ; \tag{20.7}$$
$$\sigma_{rz} = -12A_4rz - 6B_3\nu rz \; ; \tag{20.8}$$
$$\sigma_{zz} = A_4(12z^2 - 6r^2) + 2A_2 + B_3(6\nu z^2 + 3(1-\nu)r^2) - 2B_1(1-\nu) \; . \tag{20.9}$$

The two strong boundary conditions (20.1) require that

$$12A_4 + 6B_3\nu = 0 \; ; \tag{20.10}$$
$$-6A_4 - 3B_3(1+2\nu) = p_0 \; ; \tag{20.11}$$
$$\frac{9}{2}A_4a^2 - A_2 + 3B_3\nu a^2 - 2B_1\nu = -p_0L^2 \; , \tag{20.12}$$

with solution

$$B_3 = -\frac{p_0}{3(1+\nu)} \; ; \quad A_4 = \frac{p_0\nu}{6(1+\nu)} \; . \tag{20.13}$$

To complete the solution, we impose the weak boundary conditions on the ends $z=\pm L$, which in this case amounts to the statement that there be no axial force, i.e.

$$\int_0^a r\sigma_{zz}dr = 0 \; . \tag{20.14}$$

From the boundary conditions, it is clear that if this condition is satisfied at $z=L$, it will also be satisfied[3] at any other value of $z$, so we can simplify the algebra by

[2]Notice that the admissible function $A_3R^0P_0(z/R)$ in Solution A leads to a null stress field and is therefore trivial.

[3]since the equilibrium of any part of the body is guaranteed by the harmonic potential function representation.

imposing it at $z=0$ with the result

$$-\frac{3}{2}A_4a^2 + A_2 + \frac{3}{4}B_3(1-\nu)a^2 - B_1(1-\nu) = 0 \; . \tag{20.15}$$

It is now a routine matter to solve for the remaining constants $A_2, B_1$ and substitute into the stress equations (20.7–20.9), with the result

$$\sigma_{rr} = \frac{p_0\nu(a^2-r^2)}{4(1+\nu)} + p_0(z^2-L^2) \; ; \tag{20.16}$$

$$\sigma_{rz} = 0 \; ; \tag{20.17}$$

$$\sigma_{zz} = \frac{p_0}{2(1+\nu)}(2a^2-r^2) \; . \tag{20.18}$$

The same methods can be applied to a wide range of problems for the solid cylinder and also for the hollow cylinder defined by $a<r<b$, in which case we also require the $Q$-series functions, which are singular on the axis $z=0$. Equilibrium considerations dictate that no bending moment be generated by an axisymmetric system of tractions, but an axial force can be developed in response to uniform shear tractions, as in Problem 1 at the end of this chapter. Problems involving bending are of course non-axisymmetric and can be treated using the same technique except that (i) the corresponding polynomials involving $\cos\theta, \sin\theta$ must be used (i.e. $m=1$) and (ii) we must also include Solution E.

We also note here that Solution E can be used alone with axisymmetric polynomial solutions to treat problems involving torsion of the solid or hollow cylinder due to shear tractions $\sigma_{r\theta}, \sigma_{z\theta}$ on the curved boundaries and the ends respectively.

## 20.2 The spherical hole

Another important class of problems concerns the effect of a small spherical hole of radius $a$ in a large body. Suppose for example that the far boundaries of the body are subjected to a uniform tensile stress, in which case the boundary conditions are most simply stated in the form

$$\sigma_{zz} = S \; ; \; \sigma_{rz} = \sigma_{rr} = 0 \; ; \; R \to \infty \; ; \tag{20.19}$$

$$\sigma_{RR} = \sigma_{R\alpha} = 0 \; ; \; R = a \tag{20.20}$$

As in the corresponding two-dimensional problem, it is easiest to cast the far field conditions in cylindrical polar coördinates and the conditions at the boundary of the hole in spherical polars. The problem is of course axisymmetric and hence all the stress components with a $\theta$-suffix except $\sigma_{\theta\theta}$ will be identically zero throughout the body and need not be considered.

By analogy with the two dimensional problem of §8.4.1, we anticipate that the resulting field will involve a spherically symmetric component and a component varying with $\cos(2\alpha)$. The full solution will consist of a trivial *unperturbed* uniaxial tensile field ($\sigma_{zz}=S$) and a *corrective* solution which decays with $R$ and which will therefore involve singular Legendre polynomials. Substitution of such functions into Solutions A, B shows that an appropriate formulation of the corrective solution can be obtained in terms of the singular potentials

$$\phi = \frac{A_1}{R} + A_2\left(\frac{3z^2}{R^5} - \frac{1}{R^3}\right) ; \tag{20.21}$$

$$\omega = -\frac{B_1 z}{R^3} . \tag{20.22}$$

The non-trivial stress components can now be obtained from Table 16.1 in the form

$$\sigma_{zz} = A_1\left(-\frac{1}{R^3} + \frac{3z^2}{R^5}\right) + A_2\left(\frac{9}{R^5} - \frac{90z^2}{R^7} + \frac{105z^4}{R^9}\right) + B_1\left(\frac{9z^2}{R^5} - \frac{15z^4}{R^7} - 2(1-\nu)\left[-\frac{1}{R^3} + \frac{3z^2}{R^5}\right]\right) + S ; \tag{20.23}$$

$$\sigma_{rz} = \frac{3A_1rz}{R^5} + A_2\left(-\frac{45zr}{R^7} + \frac{105rz^3}{R^9}\right) + B_1\left(\frac{3rz}{R^5} - \frac{15rz^3}{R^7} - \frac{3(1-2\nu)rz}{R^5}\right) ; \tag{20.24}$$

$$\sigma_{rr} = A_1\left(-\frac{1}{R^3} + \frac{3r^2}{R^5}\right) + A_2\left(-\frac{12}{R^5} + \frac{105z^2r^2}{R^9}\right) + B_1\left(\frac{3z^2}{R^5} - \frac{15z^2r^2}{R^7} - 2\nu\left[-\frac{1}{R^3} + \frac{3z^2}{R^5}\right]\right) , \tag{20.25}$$

where the expression for $\sigma_{zz}$ also includes the unperturbed uniaxial tensile field, $S$.

It remains to determine the unknown constants $A_1, A_2, B_1$ from the boundary conditions (20.20) at $R=a$ which can be cast in cylindrical polar coördinates in the form

$$\sigma_{zz}\cos\alpha + \sigma_{rz}\sin\alpha = 0 \; ; \; R=a ; \tag{20.26}$$

$$\sigma_{zr}\cos\alpha + \sigma_{rr}\sin\alpha = 0 \; ; \; R=a , \tag{20.27}$$

— i.e.

$$z\sigma_{zz} + r\sigma_{rz} = 0 \; ; \; R=a ; \tag{20.28}$$

$$z\sigma_{zr} + r\sigma_{rr} = 0 \; ; \; R=a . \tag{20.29}$$

Substituting for the stress components and using the result $r^2+z^2=R^2$, we find that these conditions will be satisfied everywhere on $R=a$, provided that

$$2A_1 - \frac{36A_2}{a^2} + 2B_1(1+2\nu) = -Sa^3 ; \tag{20.30}$$

$$\frac{5A_2}{a^2} - B_1 = 0 ; \tag{20.31}$$

$$2A_1 - \frac{12A_2}{a^2} + 2\nu B_1 = 0 , \tag{20.32}$$

which have the solution

$$A_1 = \frac{Sa^3(6-5\nu)}{2(7-5\nu)} \; ; \; A_2 = \frac{Sa^5}{2(7-5\nu)} \; ; \; B_1 = \frac{5Sa^3}{2(7-5\nu)} . \tag{20.33}$$

The complete stress field can now be recovered by substituting back into equations (20.23–20.25). In particular, we note that the maximum tensile stress[4] occurs at the equator ($z=0$, $r=a$) and is

$$\sigma_{zz} = \frac{S(27-15\nu)}{2(7-5\nu)} . \tag{20.34}$$

## PROBLEMS

1. The solid cylinder $0 \leq r < a$, $0 < z < L$ is supported at the end $z=L$ and subjected to a uniform shear traction $\sigma_{rz}=T$ on the curved boundary $r=a$. The end $z=0$ is traction-free.

   Find an approximate solution for the stress field, based on the use of the strong boundary conditions on the curved boundary and weak conditions at the ends.

2. The *hollow* cylinder $a<r<b$, $0<z<L$ is supported at the end $z=L$ and subjected to a uniform shear traction $\sigma_{rz} = T$ on the outer boundary $r=b$, the inner boundary $r=a$ and the end $z=0$ being traction-free.

   Find an approximate solution for the stress field, based on the use of the strong boundary conditions on the curved boundaries and weak conditions at the ends.

3. Investigate the steady-state thermal stress field due to the disturbance of an otherwise uniform heat flux, $q_z=Q$, by an insulated spherical hole of radius $a$.

   You must first determine the perturbed temperature field, noting (i) that the temperature $T$ is harmonic and hence can be represented in terms of spherical harmonics and (ii) that the insulated boundary condition implies that $\frac{\partial T}{\partial R}=0$ at $R=a$.

   The thermal stress field can now be obtained using Solution T supplemented by appropriate functions in Solutions A,B, chosen so as to satisfy the traction-free condition at the hole.

4. A state of uniaxial tension, $\sigma_{zz}=S$ in a large body is perturbed by the presence of a rigid spherical inclusion in the region $0 \leq R < a$. The inclusion is perfectly bonded

[4]More detail of the resulting stress field is given by S.P.Timoshenko and J.N.Goodier, *loc. cit.*, §137.

to the surrounding material, so that the boundary condition at the interface in a suitable frame of reference is one of zero displacement ($u_r = u_z = 0$; $R = a$).

Develop a solution for the stress field and find the stress concentration factor, which is the ratio between the maximum tensile stress ($\sigma_{zz}$ at $r = a, z = 0$) and $S$.

5. A solid sphere of density $\rho$ and radius $a$ rotates at constant angular velocity $\Omega$ about the polar axis, which we take to coincide with the $z$-axis in cylindrical polar coördinates $r, \theta, z$. The surface $R = a$ is traction-free[5].

Use the results of Problem 15.1 to obtain a particular solution for the body forces[6] and then superpose appropriate non-singular spherical harmonics in Solutions A,B of Table 16.1 to satisfy the traction-free boundary condition.

Determine the location and magnitude of (i) the maximum tensile stress and (ii) the maximum shear stress in the rotating sphere. Determine also the change in the diameter of the sphere across the poles ($r = 0$) and across the equator ($z = 0$).

**Note**: Three-dimensional problems tend to become algebraically complicated because of the dependence on Poisson's ratio. If you have access to a symbolic processor such as Mathematica or Reduce, this presents no serious difficulty, but if not, problem 5 might be simplified by restricting attention to the case $\nu = 0$. Alternatively, a numerical method could be used to solve the resulting simultaneous equations for the coefficients on the spherical harmonics and compute the stresses at selected points for various values of $\nu$.

[5]This problem is of some interest in connection with the stresses and displacements in the earth's crust due to diurnal rotation.

[6]If in doubt about the appropriate differential or vector operators in polar coördinates, refer to §8.1. A suitable value of $V$ is that given in equation (7.23).

# Chapter 21

# FRICTIONLESS CONTACT

As we noted in §16.4.1, Green and Zerna's Solution F is ideally suited to the solution of frictionless contact problems for the half-space, since it identically satisfies the condition that the shear tractions be zero at the surface $z=0$. In fact the *surface* tractions for this solution take the form

$$\sigma_{zz} = -\frac{\partial^2 \varphi}{\partial z^2} \; ; \; \sigma_{zx} = \sigma_{zy} = 0 \; ; \; z = 0 \; , \qquad (21.1)$$

whilst the surface displacements are

$$u_x = \frac{(1-2\nu)}{2\mu}\frac{\partial \varphi}{\partial x} \; ; \; u_y = \frac{(1-2\nu)}{2\mu}\frac{\partial \varphi}{\partial y} \; ; \; u_z = -\frac{(1-\nu)}{\mu}\frac{\partial \varphi}{\partial z} \; ; \; z = 0 \; , \qquad (21.2)$$

from Table 16.2.

## 21.1 Boundary conditions

Following §12.5, we can now formulate the general problem of indentation of the half-space by a frictionless rigid punch whose profile is denoted by a function $u_0(x,y)$ (see Figure 12.5). Notice that since the problem is three-dimensional, the punch profile is now a function of two variables $x, y$ and the contact area will be an extended region of the plane $z=0$, which we denote by $A$.

Within the contact area, the normal displacement of the half-space must equal the indentation of the punch and hence

$$u_z = u_0(x,y) + C_0 + C_1 x + C_2 y \quad \text{in } A \; , \qquad (21.3)$$

whereas, outside the contact area, there must be no normal tractions. i.e.

$$\sigma_{zz} = 0 \quad \text{in } \bar{A} \; , \qquad (21.4)$$

where we denote the complement of $A$ — i.e. that part of the surface $z=0$ which is *not* in contact — by $\bar{A}$.

We now use equations (21.1, 21.2) to write these conditions in terms of the potential function $\varphi$ with the result

$$\frac{\partial\varphi}{\partial z} = -\frac{\mu}{(1-\nu)}u_0(x,y) \; ; \; \text{in } A \; ; \tag{21.5}$$

$$\frac{\partial^2\varphi}{\partial z^2} = 0 \; ; \; \text{in } \bar{A} \, , \tag{21.6}$$

where for brevity we have omitted the rigid-body displacement terms in equation (21.3), since these can easily be reintroduced as required or subsumed under the function $u_0(x,y)$.

### 21.1.1 Mixed boundary-value problems

Equations (21.5, 21.6) define a *mixed boundary-value problem* for the potential function $\varphi$ in the region $z > 0$. The word 'mixed' here refers to the fact that different derivatives of the function are specified at different parts of the boundary. More specifically, it is a *two-part* mixed boundary-value problem[1], since the boundary conditions are specified over two complementary regions of the boundary, $A$, $\bar{A}$.

Similar mixed boundary-value problems arise in many fields of engineering mechanics, such as heat conduction, electrostatics and fluid mechanics. For example, in heat conduction, a mathematically similar problem arises if the region $A$ is raised to a prescribed temperature whilst $\bar{A}$ is insulated. They can be reduced to integral equations using a Green's function formulation. Thus, considering the normal traction $\sigma_{zz}(x,y,0) \equiv p_z(x,y)$ in $A$ as composed as a set of point forces $F = p_z(x,y)dxdy$ and using equation (18.20) for the normal surface displacement due to a concentrated normal force at the surface, we can write

$$u_z(x,y,0) = \frac{(1-\nu)}{2\pi\mu}\int\int_A \frac{p_z(\xi,\eta)d\xi d\eta}{r[(x-\xi),(y-\eta)]} \, , \tag{21.7}$$

where

$$r[(x-\xi),(y-\eta)] = \sqrt{(x-\xi)^2+(y-\eta)^2} \tag{21.8}$$

is the distance between the points $(\xi,\eta,0)$ and $(x,y,0)$. Substitution in (21.3) now yields a double integral equation for the unknown contact pressure $p_z(x,y)$. This is of course a generalization of the method used in Chapter 12. However, in three-dimensional problems, the resulting double integral equation is generally of rather intractable form. We shall find that other methods of solution are more efficient (see Chapter 22 below).

[1]Strictly this terminology is restricted to cases where the two regions $A$, $\bar{A}$ are connected. A special case where this condition is not satisfied is the indentation by an annular punch for which $\bar{A}$ has two unconnected regions — one inside the annulus and one outside. This would be referred to as a three-part problem.

The fact that Solution F is described in terms of a single harmonic function enables us to use results from potential theory to prove some interesting results. For example, we note that the derivative $\partial\varphi/\partial z$ must also be harmonic in the half-space $z > 0$. Suppose we wish to locate the point where $\partial\varphi/\partial z$ is a maximum. It cannot be inside the region $z > 0$, since at a such a maximum we would need

$$\nabla^2 \frac{\partial \varphi}{\partial z} < 0 \tag{21.9}$$

and the left hand side of this expression is zero everywhere. It follows that the maximum must occur somewhere on the boundary $z = 0$ or else at infinity. Furthermore, if it occurs on the boundary, all immediately adjacent points must have lower values and in particular, we must have

$$\frac{\partial^2 \varphi}{\partial z^2} < 0 \tag{21.10}$$

at the maximum point.

Recalling equations (21.1, 21.2), we conclude that the maximum value of the surface displacement $u_z$ must occur either at infinity or at a point where the contact traction $\sigma_{zz}$ is compressive. If we choose a frame of reference such that the displacement at infinity is zero, it follows that the maximum can only occur there if the surface displacement is negative throughout the finite domain, which can only arise if the net force on the half-space is tensile. Thus, when the half-space is loaded by a net compressive force, the maximum surface displacement must occur in a region where the surface traction is compressive, and must be positive.

## 21.2 Determining the contact area

In many contact problems, the contact area is not known *a priori*, but has to be determined as part of the solution[2]. As in Chapter 12, the contact area has to be chosen so as to satisfy the two inequalities

$$\sigma_{zz} \;<\; 0 \;;\; \text{in } A \;; \tag{21.11}$$

$$u_z \;>\; u_0(x,y) + C_0 + C_1 x + C_2 y \;;\; \text{in } \bar{A} \,, \tag{21.12}$$

which state respectively that the contact traction should never be tensile and that the gap between the indenter and the half-space should always be positive. It is worth noting that physical considerations demand that these conditions be satisfied in *all* problems — not only those in which the contact area is initially unknown. However, if the indenter is rigid and has sharp corners, the contact area will usually be identical with the plan-form of the indenter and conditions (21.11, 21.12) need not be explicitly imposed. Notice however that if there is also distortion due to thermoelastic

[2]See for example §12.5.3.

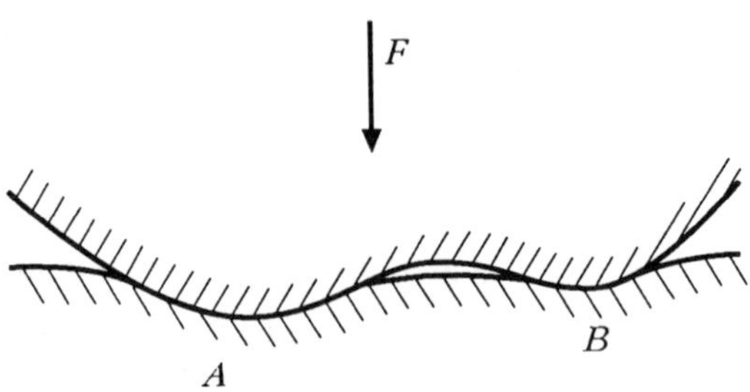


Figure 21.1: Contact problem which will have two unconnected multiply-connected contact areas in an intermediate load range

effects or if there is a sufficiently large rigid body rotation, the contact area may not be coextensive with the plan-form of the indenter. In general, conditions (21.11, 21.12) define a *solution space* in which the mathematical solution will be physically meaningful. If any further restrictions are imposed in the interests of simplicity, it is always wise to check the solution at the end to ensure that the inequalities are satisfied.

In treating the Hertzian problem in §12.5.3, we were able to satisfy the inequalities by choosing the contact area in such a way that the contact traction tended to zero at the edge. A similar technique can be applied in three-dimensional problems provided that the profile of the indenter is smooth, but now we have to determine the equation of a boundary line in the plane $z=0$ rather than two points defining the ends of a strip. This is generally a formidable problem and closed form solutions are known only for the case where the contact area is either an ellipse or a circle.

Boundedness of the contact traction at the edge of the contact area is a necessary but not sufficient condition for the inequalities to be satisfied. Suppose for example the indenter has concave regions as shown in Figure 21.1. For some values of the load, there will be contact in two discrete regions near to P and Q. However, we could construct a mathematical solution assuming contact near P only, in which the traction was bounded throughout the edge of the contact area and in which there would be interpenetration (i.e. violation of inequality (21.12)) near Q. In general, the boundedness condition is unable to determine a unique solution if the correct contact area is not simply-connected.

In a numerical solution, this difficulty can be avoided by using the inequalities directly to determine the contact area by iteration. A simple approach is to guess the extent of the contact area, $A$; solve the resulting contact problem numerically and then examine the solution to determine whether the inequalities (21.11, 21.12) are violated at any point. Any points in $A$ at which the contact traction is tensile are then excluded from $A$ in the next iteration, whereas any points in $\bar{A}$ at which the gap is negative are included in $A$.

This method is found to converge rapidly, but it involves the calculation of tractions or displacements at all surface nodes. An alternative method is to formulate a function which is minimized when the solution satisfies the inequalities and then treat the iterative process as an optimization problem. Kalker[3] and Kikuchi and Oden[4] have explored this technique at length and used it to develop finite element solutions to the contact problem.

A simpler method is available for the special problem of the indentation of an elastic half-space by a rigid indenter. Suppose we select a value at random for the contact area, $A$, solve the resulting indentation problem, and then determine the corresponding total indentation force

$$P(A) = -\int\int_A \sigma_{zz}(x, y, 0)dA \; , \tag{21.13}$$

where $\sigma_{zz}$ is the contact traction associated with the 'wrong' contact area, $A$. It can be shown the maximum value of $P(A)$ occurs when $A$ is chosen such that the inequalities (21.11, 21.12) are satisfied — i.e. when $A$ takes its correct value. Thus, we can formulate the iteration process in terms of an optimization problem for $P(A)$. The proof of the theorem is as follows:-

### Proof

Consider the effect of increasing $A$ by a small element $\delta A$.

If $P(A)$ is thereby increased, the corresponding *differential* contact traction distribution — i.e. the difference between the final and the initial traction — amounts to a net compressive force on the half-space. Now we proved in §21.1.1 that in such cases, the maximum surface displacement must occur in a region where the traction is compressive and be positive in sign. Thus, the maximum differential surface displacement $\delta u_z$ must occur somewhere in $A$ or $\delta A$ and be positive in sign. However, $\delta u_z$ is zero throughout $A$, since the process of including a further region in the contact area does not affect the displacement boundary condition in $A$. It therefore follows that the differential displacement in $\delta A$ is positive and that the contact traction there is compressive.

---

[3]J.J.Kalker, Variational principles of contact elastostatics, J.Inst.Math.Appl., Vol. 20 (1977), 199–219.

[4]N.Kikuchi and J.T.Oden, *Contact Problems in Elasticity: A Study of Variational Inequalities and Finite Element Methods*, SIAM, Philadelphia, (1988).

We know that the final value of $u_z$ in $\delta A$ is $f(x,y)+C_0+C_1x+C_2y$. The effect of the differential contact traction distribution has therefore been shown to be an *increase* of $u_z$ at $\delta A$ to the value $f(x,y)+C_0+C_1x+C_2y$ and it follows that $\delta A$ must be an area for which initially

$$u_z < f(x,y)+C_0+C_1x+C_2y \tag{21.14}$$

— i.e. for which the gap is negative.

Hence, an increase in $P(A)$ can only be achieved by increasing $A$ if there are some areas $\delta A$ outside $A$ satisfying (21.14). By a similar argument, we can show that an increase in $P(A)$ can follow from a *reduction* in $A$ only if there exist areas $\delta A$ *inside* $A$ for which

$$\sigma_{zz} > 0 \tag{21.15}$$

— i.e. for which the contact traction is tensile.

It follows that the maximum value of $P(A)$ must occur when $A$ is chosen so that there are no regions satisfying (21.14, 21.15) — i.e. when the original inequalities (21.11, 21.12) are satisfied everywhere[5].

The above method has the formal advantage that it replaces the intractable inequality conditions by a variational statement, but in order to use it we need to have a way of determining the total load $P(A)$ on the indenter for an arbitrary contact area[6] $A$. We shall show in Chapter 25 how the reciprocal theorem can be used to simplify this problem[7].

## PROBLEM

1. Using the result of Problem 18.4 or otherwise, show that if the punch profile $u_0(x,y)$ has the form

$$u_0 = r^p f(\theta)$$

where $p>0$ and $f(\theta)$ is any function of $\theta$, the resulting frictionless contact problems at different indentation forces will be self-similar[8].

Show also that if $l$ is a representative dimension of the contact area, the indentation force, $F$, and the rigid-body indentation, $d$, will vary according to

$$F \sim l^{p+1} \;\; ; \;\; d \sim l^p$$

---

[5]For more details of this argument see J.R.Barber, Determining the contact area in elastic contact problems, J.Strain Analysis, Vol. 9 (1974), 230-232.

[6]Even then, the variational problem is far from trivial.

[7]R.T.Shield, Load-displacement relations for elastic bodies, Z.angew.Math.Phyz. (ZAMP), Vol. 18 (1967), pp. 682–693.

[8]D.A.Spence, An eigenvalue problem for elastic contact with finite friction, Proc. Camb. Phil. Soc., Vol. 73 (1973), 249–268, has shown that this argument also extends to problems with Coulomb friction at the interface, in which case, the zones of stick and slip also remain self-similar with monotonically increasing indentation force.

and hence that the indentation has a stiffening load-displacement relation

$$F \sim d^{1+\frac{1}{p}}$$

Verify that the Hertzian contact relations (§22.2.5 below) agree with this result.

# Chapter 22

# THE BOUNDARY-VALUE PROBLEM

The simplest frictionless contact problem of the class defined by equations (21.3–21.6) is that in which the contact area $A$ is the circle $0 < r < a$ and the indenter is axisymmetric, in which case we have to determine a harmonic function $\varphi(r, z)$ to satisfy the mixed boundary conditions

$$\frac{\partial\varphi}{\partial z} = -\frac{\mu}{(1-\nu)}u_0(r) \ ; \ 0 \leq r < a \ , \ z = 0 \ ; \tag{22.1}$$

$$\frac{\partial^2\varphi}{\partial z^2} = 0 \ ; \ r > a \ , \ z = 0 \ . \tag{22.2}$$

## 22.1 Hankel transform methods

This is a classical problem and many solution methods have been proposed. The most popular is the Hankel transform method developed by Sneddon[1] and discussed in the application to contact problems by Gladwell[2]. The function

$$f(r, z) = \exp(-\xi z)J_0(\xi r) \ , \tag{22.3}$$

where $\xi$ is a constant, is harmonic and hence a more general function can be written in the form

$$f(r, z) = \int_0^\infty A(\xi)\exp(-\xi z)J_0(\xi r)d\xi \ , \tag{22.4}$$

where $A(\xi)$ is an unknown function to be determined from the boundary conditions.

[1]I.N.Sneddon, Note on a boundary value problem of Reissner and Sagoci, J.Appl.Phys., Vol. 18 (1947), 130–132.

[2]G.M.L.Gladwell, *Contact Problems in the Classical Theory of Elasticity*, Sijthoff and Noordhoff, Alphen aan den Rijn, (1980).

Using this representation for $\varphi$ and substituting into equations (22.1, 22.2), we find that

$$-\int_0^\infty \xi A(\xi)J_0(\xi r)d\xi = -\frac{\mu}{(1-\nu)}u_0(r) \;;\; 0 \le r < a\;; \tag{22.5}$$

$$\int_0^\infty \xi^2 A(\xi)J(\xi r)d\xi = 0 \;;\; r > a\,. \tag{22.6}$$

These constitute a pair of *dual integral equations* for the function $A(\xi)$. Sneddon used the method of Titchmarsh[3] and Busbridge[4] to reduce equations of this type to a single equation, but a more recent solution by Sneddon[5] and formalized by Gladwell[6] effects this reduction more efficiently for the classes of equation considered here.

## 22.2 Collins' Method

A related method, which has the advantage of yielding a single integral equation in elementary functions directly, was introduced by Green and Zerna[7] and applied to a wide range of axisymmetric boundary-value problems by Collins[8].

### 22.2.1 Indentation by a flat punch

To introduce Collins' method, we first examine the simpler problem in which the punch is flat and hence $u_0(r)$ is a constant. This was first solved by Boussinesq in the 1880s.

A particularly elegant solution was developed by Love[9], using a series of complex harmonic potential functions generated from the real Legendre polynomial solutions of §19.4 by substituting $(z+ia)$ for $z$. This is tantamount to putting the origin at the 'imaginary' point $(0,-ia)$. The real and imaginary parts of the resulting functions are separately harmonic and have discontinuities at $r=a$ on the plane $z=0$.

For example, if we start with the harmonic function $\log(R+z)$ from equation (18.30) and replace $z$ by $z+ia$, we can define the new harmonic function

$$\phi = \phi_1 + i\phi_2 = \log(R^* + z + ia)\,, \tag{22.7}$$

[3] E.C.Titchmarsh, *Introduction to the theory of Fourier integrals*, Clarendon Press, Oxford, (1937).

[4] I.W.Busbridge, Dual integral equations, Proc.London Math. Soc., Ser.2 Vol. 44 (1938), 462

[5] I.N.Sneddon, The elementary solution of dual integral equations, Proc. Glasgow Math. Assoc., Vol. 4 (1960), 108–110.

[6] G.M.L.Gladwell, *loc. cit.*, Chapters 5,10.

[7] A.E.Green and W.Zerna, *loc. cit.*.

[8] W.D.Collins, On the solution of some axisymmetric boundary-value problems by means of integral equations, II: Further problems for a circular disc and a spherical cap, Mathematika, Vol. 6 (1959), 120–133.

[9] A.E.H.Love, Boussinesq's problem for a rigid cone, Q.J.Math., Vol. 10 (1939), 161–175.

where

$$R^{*2} = r^2 + (z + ia)^2 \,. \tag{22.8}$$

We also record the first derivative of $\phi$ which is

$$\frac{\partial \phi}{\partial z} = \frac{1}{R^*} \,. \tag{22.9}$$

On the plane $z = 0$, $R^* \rightarrow \sqrt{r^2 - a^2}$ and hence

$$\phi(r,0) = \log(\sqrt{r^2 - a^2} + ia) \,; \tag{22.10}$$

$$\frac{\partial \phi}{\partial z}(r,0) = \frac{1}{\sqrt{r^2 - a^2}} \,. \tag{22.11}$$

Both the real and imaginary parts of these functions have discontinuities at $r = a$ on the plane $z = 0$. For example

$$\phi_2(r,0) \equiv \Im\phi(r,0) = \frac{\pi}{2} \;;\; 0 \le r < a \,; \tag{22.12}$$

$$= \sin^{-1}\frac{a}{r} \;;\; r > a \,, \tag{22.13}$$

whilst

$$\frac{\partial \phi_2}{\partial z}(r,0) = -\frac{1}{\sqrt{a^2 - r^2}} \;;\; 0 \le r < a \,; \tag{22.14}$$

$$= 0 \;;\; r > a \,. \tag{22.15}$$

Comparing these results with the boundary conditions (22.1, 22.2), we see that the function

$$\frac{\partial \varphi}{\partial z} = -\frac{2\mu u_0}{\pi(1-\nu)}\Im[\log(R^* + z + ia)] \tag{22.16}$$

identically satisfies both boundary conditions for the case where the function $u_0(r)$ is a constant — i.e. if the rigid indenter is flat and is pressed a distance $u_0$ into the half-space, as shown in Figure 22.1.

The contact pressure distribution is then immediately obtained from equations (21.1, 22.14, 22.16) and is

$$p(r) = -\sigma_{zz}(r,0) = \frac{2\mu u_0}{\pi(1-\nu)\sqrt{a^2 - r^2}} \;;\; 0 \le r < a \,. \tag{22.17}$$

We also use (22.13) to record the surface displacement of the half-space *outside* the contact area, which is

$$u_z(r,0) = \frac{2u_0}{\pi}\sin^{-1}\frac{a}{r} \;;\; r > a \tag{22.18}$$

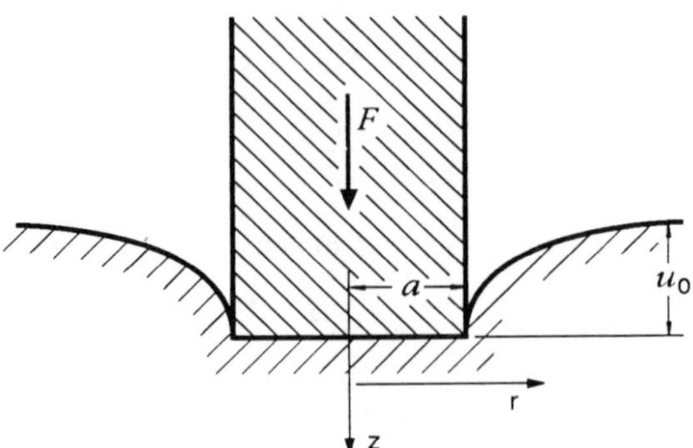


Figure 22.1: Indentation by a flat-ended cylindrical rigid punch

and the total indenting force

$$P = 2\pi \int_0^a rp(r)dr = \frac{4\mu u_0}{(1-\nu)} \int_0^a \frac{rdr}{\sqrt{a^2-r^2}} = \frac{4\mu u_0 a}{(1-\nu)} , \tag{22.19}$$

The remaining stress and displacement components can be obtained by substituting (22.16) into the corresponding expressions for Solution F from Table 16.2.

This is Love's solution of the flat punch problem. To gain an appreciation of its elegance, it is really necessary to compare it with the original Hankel transform solution[10] which is extremely complicated.

### 22.2.2 Integral representation

Green and Zerna[11] extended the method of the last section to give a general solution of the boundary-value problem of equations (22.1, 22.2), using the integral representation

$$\frac{\partial \varphi}{\partial z} = \Re \int_0^a \frac{g(t)dt}{\sqrt{r^2+(z+it)^2}} = \frac{1}{2} \int_{-a}^a \frac{g(t)dt}{\sqrt{r^2+(z+it)^2}} \tag{22.20}$$

where $g(t)$ is an even function of $t$. It can be shown that equation (22.20) satisfies (22.2) identically and the remaining boundary condition (22.1) will yield an integral equation for the unknown function $g(t)$.

In effect, equation (22.20) is a superposition of solutions of the form $\Re(r^2+(z+it)^2)^{-1/2}$ derived from the source solution $1/R$. We could therefore describe $\varphi$ as the

[10]I.N.Sneddon, Boussinesq's problem for a flat-ended cylinder, Proc.Cambridge Phil.Soc., Vol. 42 (1946), 29–39.

[11]A.E.Green and W.Zerna, *loc.cit.*, §§5.8–5.10.

potential due to an arbitrary distribution $g(t)$ of point sources along the *imaginary* $z$-axis between (0,0) and $(0, ia)$. It is therefore a logical development of the classical method of obtaining axisymmetric potential functions by distributing singularities along the axis of symmetry[12]. A closely related solution is given by Segedin[13], who develops it as a convolution integral of an arbitrary kernel function with the Boussinesq solution of §22.2.1. He uses this method to obtain the solution for a power law punch ($u_0(r) \sim r^n$) and treats more general problems by superposition after expanding the punch profile as a power series in $r$. It should be noted that Segedin's solution is restricted to indentation by a punch of continuous profile[14] in which case the contact pressure tends to zero at $r=a$. The idea of representing a general solution by superposition of Boussinesq-type solutions for different values of $a$ has also been used as a direct numerical method by Maw et al.[15].

Green's method was extensively developed by Collins, who used it to treat many interesting problems including the indentation problem for an annular punch[16] and a problem involving 'radiation' boundary conditions[17].

### 22.2.3 Basic forms and surface values

To represent harmonic potential functions we shall use suitable combinations of the four basic forms

$$\begin{aligned} \phi_1 &= \Re \int_0^a g_1(t) F(r,z,t) dt \, ; \\ \phi_2 &= \Re \int_a^\infty g_2(t) F(r,z,t) dt \, ; \\ \phi_3 &= \Im \int_0^a g_3(t) F(r,z,t) dt \, ; \\ \phi_4 &= \Im \int_a^\infty g_4(t) F(r,z,t) dt \, , \end{aligned} \tag{22.21}$$

where

$$F(r,z,t) = \log\left(\sqrt{r^2+(z+it)^2} + z + it\right) \, . \tag{22.22}$$

[12] see §18.3 and particularly equation (18.30).

[13] C.M.Segedin, The relation between load and penetration for a spherical punch, Mathematika, Vol. 4 (1957), 156–161.

[14] see §21.2.

[15] N.Maw, J.R.Barber and J.N.Fawcett, The oblique impact of elastic spheres, Wear, Vol. 38 (1976), 101–114.

[16] W.D.Collins, On the solution of some axisymmetric boundary-value problems by means of integral equations, IV: Potential problems for a circular annulus, Proc. Edinburgh Math. Soc., Vol. 13 (1963), 235–246.

[17] W.D.Collins, On the solution of some axisymmetric boundary value problems by means of integral equations, II: Further problems for a circular disc and a spherical cap, Mathematika, Vol. 6 (1959), 120–133.

The square root in (22.22) is interpreted as

$$\sqrt{r^2+(z+it)^2} = \rho e^{iv/2} , \tag{22.23}$$

where

$$\rho = \sqrt[4]{(r^2+z^2-t^2)^2+4z^2t^2} \ ; \ \ v = \tan^{-1}\left(\frac{2zt}{r^2+z^2-t^2}\right) \tag{22.24}$$

and $\rho \geq 0$, $0 \leq v < \pi$.

Equations (22.21) can be written in two alternative forms which for $\phi_1$ are

$$\phi_1 = \frac{1}{2}\int_0^a g_1(t)[F(r,z,t)+F(r,z,-t)]dt \tag{22.25}$$

and

$$\phi_1 = \frac{1}{2}\int_{-a}^a g_1(t)F(r,z,t)dt \, . \tag{22.26}$$

We note that (22.25) is exactly equivalent to (22.26) if and only if $g_1$ is an even function of $t$. If the boundary values of $\phi, \partial\phi/\partial z$ etc. specified at $z=0$ are even in $r$, it will be found that $g_1, g_2$ are even and $g_3, g_4$ odd functions of $t$ and forms like (22.26) can be used. The majority of problems fall into this category, but there are important exceptions such as the conical punch (where $u_z$ is proportional to $r$) and problems with Coulomb friction for which $\sigma_{zr}$ is proportional to $\sigma_{zz}$ in some region. In these problems, forms like (22.26) can only be used if $g_i(t)$ is extended into $t<0$ by a definition with the required symmetry.

Expressions for the important derivatives of the functions $\phi_i$ at the surface $z=0$ are given in Table 22.1.

In certain cases, higher derivatives are required — notably in thermoelastic problems where heat flux is proportional to $\partial^3\phi/\partial z^3$ (see Solution P, Table 17.1). Higher derivatives are most easily obtained by differentiating *within* the plane $z=0$, making use of the fact that for an axisymmetric harmonic function $f$,

$$\frac{\partial^2 f}{\partial z^2} = -\frac{1}{r}\frac{\partial}{\partial r}r\frac{\partial f}{\partial r} \, . \tag{22.27}$$

The reader can verify that this result permits the expressions for $\partial^2\phi_i/\partial z^2$ in Table 22.1 to be obtained from those for $\partial\phi_i/\partial r$.

### 22.2.4 Reduction to an Abel equation

Table 22.1 shows that we can satisfy the boundary condition (22.2) identically if we represent $\varphi$ in the form of $\phi_1$ of equations (22.21). Substitution in the remaining boundary condition (22.1) then gives the integral equation

$$\int_0^r \frac{g_1(t)dt}{\sqrt{r^2-t^2}} = -\frac{\mu}{(1-\nu)}u_0(r) \ ; \ \ 0 \leq r < a \, , \tag{22.28}$$

Table 22.1: Surface values of the derivatives of the functions $\phi_i$ (Equations (22.21))

| | $0 < r < a$ | $r > a$ |
|---|---|---|
| $\frac{\partial\phi_1}{\partial r}$ | $\frac{1}{r}\int_0^a g_1(t)dt - \frac{1}{r}\int_r^a \frac{tg_1(t)dt}{\sqrt{t^2-r^2}}$ | $\frac{1}{r}\int_0^a g_1(t)dt$ |
| $\frac{\partial\phi_1}{\partial z}$ | $\int_0^r \frac{g_1(t)dt}{\sqrt{r^2-t^2}}$ | $\int_0^a \frac{g_1(t)dt}{\sqrt{r^2-t^2}}$ |
| $\frac{\partial^2\phi_1}{\partial z^2}$ | $\frac{1}{r}\frac{d}{dr}\int_r^a \frac{tg_1(t)dt}{\sqrt{t^2-r^2}}$ | $0$ |
| $\frac{\partial\phi_2}{\partial r}$ | $\frac{1}{r}\int_a^\infty g_2(t)dt - \frac{1}{r}\int_a^\infty \frac{tg_2(t)dt}{\sqrt{t^2-r^2}}$ | $\frac{1}{r}\int_a^\infty g_2(t)dt - \frac{1}{r}\int_r^\infty \frac{tg_2(t)dt}{\sqrt{t^2-r^2}}$ |
| $\frac{\partial\phi_2}{\partial z}$ | $0$ | $\int_a^r \frac{g_2(t)dt}{\sqrt{r^2-t^2}}$ |
| $\frac{\partial^2\phi_2}{\partial z^2}$ | $\frac{1}{r}\frac{d}{dr}\int_a^\infty \frac{tg_2(t)dt}{\sqrt{t^2-r^2}}$ | $\frac{1}{r}\frac{d}{dr}\int_r^\infty \frac{tg_2(t)dt}{\sqrt{t^2-r^2}}$ |
| $\frac{\partial\phi_3}{\partial r}$ | $-\frac{1}{r}\int_0^r \frac{tg_3(t)dt}{\sqrt{r^2-t^2}}$ | $-\frac{1}{r}\int_0^a \frac{tg_3(t)dt}{\sqrt{r^2-t^2}}$ |
| $\frac{\partial\phi_3}{\partial z}$ | $-\int_r^a \frac{g_3(t)dt}{\sqrt{t^2-r^2}}$ | $0$ |
| $\frac{\partial^2\phi_3}{\partial z^2}$ | $\frac{1}{r}\frac{d}{dr}\int_0^r \frac{tg_3(t)dt}{\sqrt{r^2-t^2}}$ | $\frac{1}{r}\frac{d}{dr}\int_0^a \frac{tg_3(t)dt}{\sqrt{r^2-t^2}}$ |
| $\frac{\partial\phi_4}{\partial r}$ | $0$ | $-\frac{1}{r}\int_a^r \frac{tg_4(t)dt}{\sqrt{r^2-t^2}}$ |
| $\frac{\partial\phi_4}{\partial z}$ | $-\int_a^\infty \frac{g_4(t)dt}{\sqrt{t^2-r^2}}$ | $-\int_r^\infty \frac{g_4(t)dt}{\sqrt{t^2-r^2}}$ |
| $\frac{\partial^2\phi_4}{\partial z^2}$ | $0$ | $\frac{1}{r}\frac{d}{dr}\int_a^r \frac{tg_4(t)dt}{\sqrt{r^2-t^2}}$ |

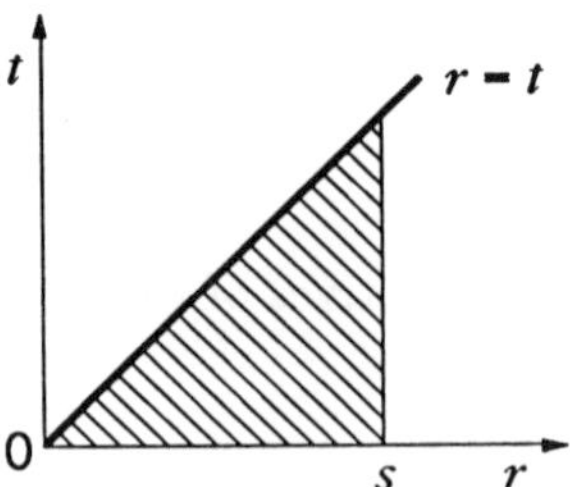


Figure 22.2: Domain of integration for equation (22.29)

for the unknown function $g_1(t)$. Notice that there is a one-to-one correspondence between the points $0 \leq t < a$ in which $g_1(t)$ is defined and $0 \leq r < a$ in which the integral equation is to be satisfied. This is a necessary condition for the equation to be solvable[18].

Equation (22.28) is an *Abel integral equation.* Abel equations can be inverted explicitly. We operate on both sides of the equation with

$$\int_0^s \frac{rdr}{\sqrt{s^2 - r^2}},$$

obtaining the equation

$$\int_0^s \int_0^r \frac{rg_1(t)dtdr}{\sqrt{(s^2 - r^2)(r^2 - t^2)}} = -\frac{\mu}{(1-\nu)} \int_0^s \frac{ru_0(r)dr}{\sqrt{s^2 - r^2}} \quad ; \quad 0 \leq r < a \, . \tag{22.29}$$

Since the unknown function, $g_1(t)$ does not depend upon $r$, it is advantageous to reverse the order of integration on the left-hand-side of (22.29) and perform the inner integration. To obtain the limits on the new double integral, we consider $r, t$ as coördinates defining the two-dimensional region of Figure 22.2, in which the domain of integration is shown shaded. It follows from this Figure that when we reverse the order of integration the new limits are

$$\int_0^s \int_0^r \frac{rg_1(t)dtdr}{\sqrt{(s^2 - r^2)(r^2 - t^2)}} = \int_0^s g_1(t)dt \int_t^s \frac{rdr}{\sqrt{(s^2 - r^2)(r^2 - t^2)}} \, . \tag{22.30}$$

[18]It is akin to the condition that the number of unknowns and the number of equations in a system of linear equations be equal.

Table 22.2: Inversions of Abel integral equations

| | | |
|---|---|---|
| If $f(x) =$ | $\int_0^x \frac{g(t)dt}{\sqrt{x^2-t^2}}$ | $\int_x^a \frac{g(t)dt}{\sqrt{t^2-x^2}}$ |
| then $g(x) =$ | $\frac{2}{\pi}\frac{d}{dx}\int_0^x \frac{tf(t)dt}{\sqrt{x^2-t^2}}$ | $-\frac{2}{\pi}\frac{d}{dx}\int_x^a \frac{tf(t)dt}{\sqrt{t^2-x^2}}$ |
| If $f(x) =$ | $\int_a^x \frac{g(t)dt}{\sqrt{x^2-t^2}}$ | $\int_x^\infty \frac{g(t)dt}{\sqrt{t^2-x^2}}$ |
| then $g(x) =$ | $\frac{2}{\pi}\frac{d}{dx}\int_a^x \frac{tf(t)dt}{\sqrt{x^2-t^2}}$ | $-\frac{2}{\pi}\frac{d}{dx}\int_x^\infty \frac{tf(t)dt}{\sqrt{t^2-x^2}}$ |

The inner integral can now be evaluated as

$$\int_t^s \frac{rdr}{\sqrt{(s^2-r^2)(r^2-t^2)}} = \frac{\pi}{2} \tag{22.31}$$

and hence from equations (22.30, 22.31) we have

$$\int_0^s g_1(t)dt = -\frac{2\mu}{\pi(1-\nu)}\int_0^s \frac{ru_0(r)dr}{\sqrt{s^2-r^2}} \tag{22.32}$$

and on differentiation,

$$g_1(s) = -\frac{2\mu}{\pi(1-\nu)}\frac{d}{ds}\int_0^s \frac{ru_0(r)dr}{\sqrt{s^2-r^2}}\,, \tag{22.33}$$

which is the explicit solution of the Abel equation (22.29).

In general, we shall find that the use of the standard forms of equations (22.21) in two-part boundary-value problems, leads to the four types of Abel equations whose solutions — obtained by a process similar to that given above — are given in Table 22.2.

Once $g_1$ is known, the complete stress field can be written down using equation (22.21) and Solution F of Table 16.2. A particular result of some interest is that for the total indenting force

$$\begin{aligned} P &= -2\pi\int_0^a r\sigma_{zz}(r)dr = 2\pi\int_0^a \frac{\partial^2\varphi}{\partial z^2}rdr \\ &= 2\pi\int_0^a\left(\frac{1}{r}\frac{d}{dr}\int_r^a \frac{tg_1(t)dt}{\sqrt{t^2-r^2}}\right)rdr = 2\pi\left[\int_r^a \frac{tg_1(t)dt}{\sqrt{t^2-r^2}}\right]_{r=0}^{r=a} \\ &= -2\pi\int_0^a g_1(t)dt\,. \end{aligned} \tag{22.34}$$

### 22.2.5 Example – The Hertz problem

As an example, we consider the problem in which a half-space is indented by a rigid sphere of radius $R$, the function $u_0$ now being defined by

$$u_0 = d - \frac{r^2}{2R} \; . \tag{22.35}$$

Substituting in equation (22.33) and performing the integration and differentiation, we obtain

$$g_1(s) = -\frac{2\mu}{\pi(1-\nu)}\left(d - \frac{s^2}{R}\right) \; . \tag{22.36}$$

The contact pressure distribution is then given by

$$p(r) = -\sigma_{zz} = \frac{\partial^2 \varphi}{\partial z^2} = \frac{1}{r}\frac{d}{dr}\int_r^a \frac{t g_1(t)dt}{\sqrt{t^2 - r^2}} \; , \tag{22.37}$$

from Table 22.1

$$= -\frac{2\mu}{\pi(1-\nu)R}\left(\frac{Rd + a^2 - 2r^2}{\sqrt{a^2 - r^2}}\right) \; . \tag{22.38}$$

In this problem, the contact area $a$ is unknown and can be determined either by the argument of §21.2 or by forcing the contact pressure to zero at $r = a$. From equation (22.38) it is clear that the latter condition will be satisfied if and only if

$$a^2 = Rd \; , \tag{22.39}$$

in which case the contact pressure distribution simplifies to

$$p(r) = \frac{4\mu\sqrt{a^2 - r^2}}{\pi(1-\nu)R} \; . \tag{22.40}$$

Alternatively, applying the condition $dP/dA = 0$ to equation (22.34), we see that in a smooth contact problem, we must have $g_1(a) = 0$ at the edge of the contact area, which with (22.36) immediately gives $a^2 = Rd$.

## 22.3 Choice of form

Each of the functions $\phi_i$ in Table 22.1 has a zero in $\partial\phi/\partial z$ or $\partial^2\phi/\partial z^2$ either in $0 < r < a$ or in $r > a$. Thus, in the above examples it would have been possible to satisfy (22.2) by choosing $\partial\varphi/\partial z = \phi_3$ instead of $\varphi = \phi_1$.

The choice is best made by examining the requirements of continuity imposed at $r = a$, $z = 0$ by the physical problem. It can be shown that, if $g_i(a)$ is bounded, the expressions in Table 22.1 will define continuous values of $\partial\phi_i/\partial z$, but $\partial^2\phi_1/\partial z^2$ will be discontinuous unless $g_i(a) = 0$. Thus if, as in the present case, the function $\partial\varphi/\partial z$

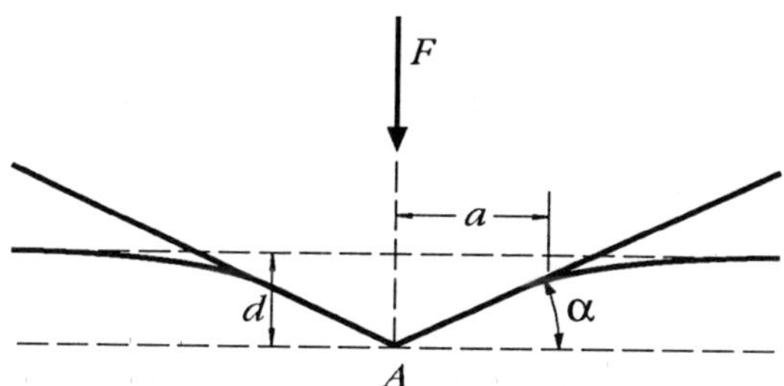


Figure 22.3: Indentation by a conical punch

represents a physical quantity like displacement or temperature that is required to be continuous, it is appropriate to choose $\partial\varphi/\partial z = \partial\phi_1/\partial z$. The alternative choice of $\partial\varphi/\partial z = \phi_3$ imposes too strong a continuity condition at $r = a$, since it precludes discontinuities in second derivatives and hence in the stress components.

Of course, in problems involving contact between continuous surfaces — such as the Hertzian problem of §22.2.5 — the normal traction $\sigma_{zz}$ must also be continuous at the edge of the contact region. The most straightforward treatment is that given above in which the continuity condition furnishes an extra condition to determine the radius of the contact region, which is not known *a priori*. However, it would also be possible to force the required continuity through the formulation by using $\partial\varphi/\partial z = \phi_3$ in which case the contact radius is prescribed and the rigid-body indentation $d$ of the punch must be allowed to float. This is essentially the technique used by Segedin[19].

## PROBLEMS

1. (i) An elastic half-space is indented by a conical frictionless rigid punch as shown in Figure 22.3, the penetration of the apex being $d$. Find the contact pressure distribution, the total load $F$ and the contact radius $a$. The semi-angle of the cone is $\frac{\pi}{2} - \alpha$, where $\alpha \ll 1$.

   (ii) What is the nature of the singularity in pressure at the vertex of the cone (point $A$).

   (iii) Now suppose that the cone is truncated as shown in Figure 22.4. *Without finding the new contact pressure distribution*, find the new relationship between $F$ and $a$. *Note*: Take the radius of the truncated end of the cone as $b$ and refer the profile to the point $A$ where the vertex would have been before truncation.

[19]C.M.Segedin *loc. cit.*.

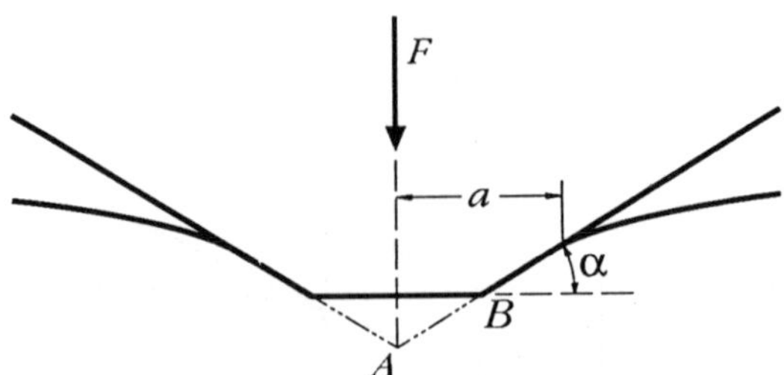


Figure 22.4: The truncated conical punch

(iv) Would you expect a singularity in contact pressure at $B$ and if so of what form. (The process of determining $p(r)$ is algebraically tedious. Try to find a simpler way to answer the question. A clue may be obtained from the asymptotic arguments of §11.2.)

2. A harmonic potential function $\omega(r,z)$ in the region $z>0$ is defined by the surface values $\omega(r,0)=A(a^2-r^2)$; $0\leq r<a$ ; $\omega(r,0)=0$; $r>a$. Select a suitable form for the function using Table 22.1 and then use the boundary conditions to determine the corresponding unknown function $g_i(t)$.

3. Use equation (22.27) to derive the expressions for $\partial^2\phi_1/\partial z^2$ in Table 22.1 from those for $\partial\phi_1/\partial r$.

# Chapter 23

# THE PENNY-SHAPED CRACK

As in the two-dimensional case, we shall find considerable similarities in the formulation and solution of contact and crack problems. In particular, we shall find that problems for the plane crack can be reduced to boundary-value problems which in the case of axisymmetry can be solved using the method of Green and Collins developed in §22.2.

## 23.1 The penny-shaped crack in tension

The simplest axisymmetric crack problem is that in which a state of uniform tension $\sigma_{zz} = S$ in an infinite isotrotropic homogeneous solid is perturbed by a plane crack occupying the region $0 \leq r < a,\ z = 0$. Thus, the crack has the shape of a circular disk. This geometry has come to be known as the *penny-shaped* crack[1].

As in Chapter 13, we seek the solution in terms of an unperturbed uniform stress field $\sigma_{zz} = S$ (constant) and a corrective solution which tends to zero at infinity. The boundary conditions on the corrective solution are therefore

$$\sigma_{zz} = -S;\ \sigma_{zr} = 0\ ;\ 0 \leq r < a\ ,\ z = 0\ ; \tag{23.1}$$

$$\sigma_{zz}, \sigma_{rz}, \sigma_{rr}, \sigma_{\theta\theta} \rightarrow 0\ ;\ R \rightarrow \infty\ . \tag{23.2}$$

The corrective solution corresponds to the problem in which the crack is opened by an internal pressure $-S$ and the body is not loaded at infinity.

The problem is symmetrical about the plane $z = 0$ and it follows that on that plane there can be no shear stress $\sigma_{zx}, \sigma_{zy}$ and no normal displacement $u_z$. We can therefore reduce the problem to a boundary-value problem for the half-space $z > 0$ defined by the boundary conditions

$$\sigma_{zx} = \sigma_{zy} = 0\ ;\ \text{all } r, z = 0\ ; \tag{23.3}$$

[1] I.N.Sneddon, The distribution of stress in the neighbourhood of a crack in an elastic solid, Proc.Roy.Soc. (London), Vol. A187 (1946), 226–260.

$$\begin{aligned} \sigma_{zz} &= -S \ ; \ 0 \leq r < a, z = 0 \ ; && (23.4) \\ u_z &= 0 \ ; \ r > a, \ z = 0 \ . && (23.5) \end{aligned}$$

Equation (23.3) is a global condition and can be satisfied identically[2] by using Solution F of Table 16.2. The remaining conditions (23.4, 23.5) then define the mixed boundary-value problem

$$\begin{aligned} \frac{\partial^2 \varphi}{\partial z^2} &= S \ ; \ 0 \leq r < a, z = 0 \ ; && (23.6) \\ \frac{\partial \varphi}{\partial z} &= 0 \ ; \ r > a, z = 0 \ . && (23.7) \end{aligned}$$

This problem can be solved by the method of §22.2. We satisfy condition (23.7) identically by representing $\varphi$ in the form of $\phi_3$ of (22.21). i.e.

$$\varphi = \Im \int_0^a g(t) F(r, z, t) dt \tag{23.8}$$

and the remaining boundary condition (23.6) then reduces to

$$\frac{1}{r} \frac{d}{dr} \int_0^r \frac{t g(t) dt}{\sqrt{r^2 - t^2}} = S \ ; \ 0 \leq r < a \ . \tag{23.9}$$

We now multiply both sides of the equation by $r$ and integrate, obtaining

$$\int_0^r \frac{t g(t) dt}{\sqrt{r^2 - t^2}} = \frac{S r^2}{2} + C \ ; \ 0 \leq r < a \ , \tag{23.10}$$

where $C$ is an arbitrary constant.

Equation (23.10) is an Abel integral equation similar to (22.28) and can be inverted in the same way (see Table 22.2), with the result

$$x g(x) = \frac{2}{\pi} \frac{d}{dx} \int_0^x \left( \frac{S r^2}{2} + C \right) \frac{r dr}{\sqrt{x^2 - r^2}} \tag{23.11}$$

and hence

$$g(x) = \frac{2 S x}{\pi} + \frac{2C}{\pi x} \ . \tag{23.12}$$

The second term in this expression is singular at $x \to 0$ and it is readily verified from Table 22.1 that $\partial \varphi / \partial z$ and hence $u_z$ would be unbounded at the origin if $C \neq 0$. We therefore set $C = 0$ to retain continuity of displacement at the origin.

[2]More generally, any problem involving a crack of arbitrary cross-section $A$ on the plane $z = 0$ and loaded by a uniform tensile stress $\sigma_{zz}$ at $z = \pm\infty$ has the same symmetry and can also be formulated using Solution F. In the general case, we obtain a two-part boundary-value problem for the half-space $z > 0$, in which conditions (23.6) and (23.7) are to be satisfied over $A$ and $\bar{A}$ respectively, where $\bar{A}$ is the complement of $A$ in the plane $z = 0$.

We can then recover the expression for the stress $\sigma_{zz}$ in the region $r > a, z = 0$ which is

$$\begin{aligned}\sigma_{zz} &= -\frac{\partial^2\varphi}{\partial z^2} = -\frac{2S}{\pi r}\frac{d}{dr}\int_0^a \frac{x^2dx}{\sqrt{r^2-x^2}} \\ &= -\frac{2S}{\pi}\left(\sin^{-1}\frac{a}{r} - \frac{a}{\sqrt{r^2-a^2}}\right) . \qquad (23.13)\end{aligned}$$

This of course is the corrective solution. To find the stress in the original crack problem, we must superpose the uniform tensile stress $\sigma_{zz} = S$, with the result

$$\sigma_{zz} = \frac{2S}{\pi}\left(\cos^{-1}\frac{a}{r} + \frac{a}{\sqrt{r^2-a^2}}\right) . \qquad (23.14)$$

The stress intensity factor is defined as

$$K_I = \lim_{r\to a^+} \sigma_{zz}\sqrt{r-a} = \frac{\sqrt{2a}S}{\pi} . \qquad (23.15)$$

We could have obtained this result directly from $g(x)$, without computing the complete stress distribution. We have

$$\begin{aligned}\sigma_{zz} &= -\frac{1}{r}\frac{d}{dr}\int_0^a \frac{xg(x)dx}{\sqrt{r^2-x^2}} \qquad (23.16)\\ &= \frac{g(a)}{\sqrt{r^2-a^2}} - \frac{g(0)}{r} - \int_0^a \frac{g'(x)dx}{\sqrt{r^2-x^2}} . \qquad (23.17)\end{aligned}$$

Now, unless $g'(x)$ is singular at $x = a$, the only singular term in (23.17) will be the first and the stress intensity factor is therefore

$$K_I = \lim_{r\to a^+} \sigma_{zz}\sqrt{r-a} = \frac{g(a)}{\sqrt{2a}} . \qquad (23.18)$$

In the present example, $g(a) = 2Sa/\pi$, leading to (23.15) as before.

We also compute the crack opening displacement $u_z$ at $0 \le r < a$, $z = 0$, which is

$$\begin{aligned}u_z &= -\frac{(1-\nu)}{\mu}\frac{\partial\varphi}{\partial z} = \frac{(1-\nu)}{\mu}\int_r^a \frac{g(x)dx}{\sqrt{x^2-r^2}} \\ &= \frac{2(1-\nu)S\sqrt{a^2-r^2}}{\pi\mu} . \qquad (23.19)\end{aligned}$$

## 23.2 Thermoelastic problems

In addition to acting as stress concentrations in an otherwise uniform stress field, a crack will obstruct the flow of heat and generate a perturbed temperature field in components of thermal machines. The simplest investigation of this effect assumes that the crack acts as a perfect insulator, so that the heat is forced to flow around it. For the penny-shaped crack[3], the appropriate thermal boundary conditions are then

$$q_z = -K\frac{\partial T}{\partial z} = 0 \; ; \; 0 \leq r < a, \; z = 0 \; ; \tag{23.20}$$
$$\rightarrow Q \; ; \; R \rightarrow \infty \; . \tag{23.21}$$

As in isothermal problems, we construct the temperature field as the sum of a uniform heat flux and a corrective solution, the boundary conditions on which become

$$q_z = -K\frac{\partial T}{\partial z} = -Q \; ; \; 0 \leq r < a, \; z = 0 \; ; \tag{23.22}$$
$$\rightarrow 0 \; ; \; R \rightarrow \infty \; . \tag{23.23}$$

This problem is antisymmetric about the plane $z=0$ and hence the temperature must be zero in the region $r>a$, $z=0$. We can therefore convert the heat conduction problem into a boundary-value problem for the half-space $z>0$ with boundary conditions

$$q_z = -Q \; ; \; 0 \leq r < a \; ; \tag{23.24}$$
$$T = 0 \; ; \; r > a \; . \tag{23.25}$$

The temperature field and a particular thermoelastic solution can be constructed using Solution P of Table 17.1, in terms of which (23.4, 23.5) define the boundary-value problem

$$\frac{\partial^3 \psi}{\partial z^3} = -\frac{\mu\alpha(1+\nu)Q}{(1-\nu)K} \; ; \; 0 \leq r < a \; ; \tag{23.26}$$
$$\frac{\partial^2 \psi}{\partial z^2} = 0 \; ; \; r > a \; . \tag{23.27}$$

We require that the temperature and hence $\partial^2\psi/\partial z^2$ be continuous at $r=a$, $z=0$ and hence in view of the arguments of §22.3, we choose to satisfy (23.27) using the function $\phi_3$ — i.e.

$$\frac{\partial \psi}{\partial z} = \Im \int_0^a g(t)F(r,z,t)dt \; . \tag{23.28}$$

Condition (23.26) then defines the Abel equation

$$\frac{1}{r}\frac{d}{dr}\int_0^r \frac{tg(t)dt}{\sqrt{r^2-t^2}} = -\frac{\mu\alpha(1+\nu)Q}{(1-\nu)K} \; ; \; 0 \leq r < a \; , \tag{23.29}$$

[3]This problem was first solved by A.L.Florence and J.N.Goodier, The linear thermoelastic problem of uniform heat flow disturbed by a penny-shaped insulated crack, Int.J.Engng.Sci., Vol. 1 (1963), 533–540.

which has the solution

$$g(t) = -\frac{2\mu\alpha(1+\nu)Qt}{\pi(1-\nu)K} . \tag{23.30}$$

This defines a particular thermoelastic solution. We recall from Chapter 14 that the homogeneous solution corresponding to thermoelasticity is the general solution of the isothermal problem. Thus we now must superpose a suitable isothermal stress field to satisfy the mechanical boundary conditions of the problem.

The mechanical boundary conditions are that (i) the surfaces of the crack be traction-free and (ii) the stress field should decay to zero at infinity. However, the antisymmetry of the problem once again permits us to define boundary conditions on the half-space $z>0$ which are

$$\sigma_{zz} = \sigma_{zr} = 0 \ ; \ 0 \leq r < a, z = 0 \ ; \tag{23.31}$$

$$\sigma_{zz} = 0 \ ; \ u_r = 0 \ ; \ r > a, z = 0 \ . \tag{23.32}$$

Equations (23.31) state that the crack surfaces are traction-free and (23.32) are symmetry conditions.

We notice that taken together, these conditions imply that $\sigma_{zz}=0$ throughout the plane $z=0$. This is therefore a global condition and can be satisfied by the choice of form. It is already satisfied by the particular solution (see Table 17.1) and hence it must also be satisfied by the additional isothermal solution, for which we therefore use Solution G of Table 16.2.

The complete solution is therefore obtained by superposing Solutions P and G, the surface values of $\sigma_{zr}$ and $u_r$ being

$$\sigma_{zr} = \frac{\partial^2\chi}{\partial r\partial z} \ ; \tag{23.33}$$

$$2\mu u_r = 2(1-\nu)\left(\frac{\partial\psi}{\partial r} + \frac{\partial\chi}{\partial r}\right) . \tag{23.34}$$

The mixed conditions in (23.31, 23.32) then require

$$\frac{\partial^2\chi}{\partial r\partial z} = 0 \ ; \ 0 \leq r < a, \ z = 0 \ ; \tag{23.35}$$

$$\frac{\partial\psi}{\partial r} + \frac{\partial\chi}{\partial r} = 0 \ ; \ r > a, \ z = 0 \ . \tag{23.36}$$

We can satisfy (23.36) by defining $\chi$ such that

$$\chi = -\psi + \Re\int_0^a g_1(t)F(r,z,t)dt \ , \tag{23.37}$$

with the auxiliary condition

$$\int_0^a g_1(t)dt = 0 \ , \tag{23.38}$$

(see Table 22.1).

The other boundary condition can then be integrated to give

$$\frac{\partial \chi}{\partial z} = C \ ; \ 0 \le r < a, z = 0 \ , \tag{23.39}$$

where $C$ is an arbitrary constant of integration, and hence

$$\int_0^r \frac{g_1(t)dt}{\sqrt{r^2 - t^2}} = \frac{\partial \psi}{\partial z} + C \ ; \ 0 \le r < a \ . \tag{23.40}$$

To solve for $g_1(t)$, we use the result

$$\frac{1}{r}\frac{d}{dr}r\frac{d}{dr}\left(\frac{\partial \psi}{\partial z}\right) = -\frac{\partial^3 \psi}{\partial z^3} = \frac{\mu\alpha(1+\nu)Q}{(1-\nu)K} \tag{23.41}$$

(see equation (22.27)), which can be integrated to give

$$\frac{\partial \psi}{\partial z} = \frac{\mu\alpha(1+\nu)Qr^2}{4(1-\nu)K} + B \ , \tag{23.42}$$

where $B$ is an arbitrary constant[4].

Substituting (23.42) into (23.40) and solving the resulting Abel integral equation using Table 22.2, we obtain

$$g_1(x) = \frac{2}{\pi}\left(B + C + \frac{\mu\alpha(1+\nu)Qx^2}{2(1-\nu)K}\right) \ . \tag{23.43}$$

Since $B$ and $C$ are two arbitrary constants and occur as a sum, there is no loss in generality in setting $B$ to zero. The auxiliary condition (23.38) can then be used to determine $C$ giving

$$C = -\frac{\mu\alpha(1+\nu)Qa^2}{6(1-\nu)K} \ . \tag{23.44}$$

Finally, we can recover the expression for the stress $\sigma_{zr}$ on $r > a$, $z = 0$, which is

$$\sigma_{zr} = \frac{\partial^2 \chi}{\partial z \partial r} = \frac{d}{dr}\int_0^a \frac{g_1(t)dt}{\sqrt{r^2 - t^2}} + \frac{1}{r}\int_0^a \frac{tg(t)dt}{\sqrt{r^2 - t^2}} \tag{23.45}$$

$$= -\frac{2\mu\alpha(1+\nu)Qa^3}{3\pi(1-\nu)Kr\sqrt{r^2 - a^2}} \ , \tag{23.46}$$

from Table 22.1 and equations (23.30, 23.43, 23.44).

Notice that the antisymmetry of the problem ensures that the crack tip is loaded in shear only — we argued earlier that $\sigma_{zz}$ would be zero throughout the plane $z = 0$.

[4]The other constant of integration leads to a term which is singular at the origin and has therefore been set to zero.

It also follows from a similar argument that the crack remains closed — there is a displacement $u_z$ at the crack plane, but it is the same on both sides of the crack.

The stress intensity factor is of mode II (shear) form and is

$$K_{II} = \lim_{r\to a^+} \sigma_{zr}\sqrt{r-a} = -\frac{\sqrt{2}\mu\alpha(1+\nu)Qa^{3/2}}{3\pi(1-\nu)K} . \tag{23.47}$$

## PROBLEMS

1. An infinite homogeneous body contains an axisymmetric *external* crack — i.e. the crack extends over the region $r > a$, $z = 0$. Alternatively, the external crack can be considered as two half-spaces bonded together over the region $0 \le r < a$, $z = 0$.

The body is loaded in tension, such that the total tensile force transmitted is $F$. Find an expression for the stress field and in particular determine the mode I stress intensity factor $K_I$.

2. An infinite homogeneous body containing a penny-shaped crack is subjected to torsional loading $\sigma_{z\theta} = Cr$, $R \to \infty$, where $C$ is a constant[5]. Use Solution E (Table 16.1) to formulate the problem and solve the resulting boundary-value problem using the methods of Chapter 22.

[5]This is known as the Reissner-Sagoci problem.

# Chapter 24

# THE INTERFACE CRACK

The problem of a crack at the interface between dissimilar elastic media is of considerable contemporary importance in Elasticity, because of its relevance to the problem of debonding of composite materials and structures. Figure 24.1 shows the case where such a crack occurs at the plane interface between two elastic half-spaces. We shall use the suffices 1,2 to distinguish the stress functions and mechanical properties for the lower and upper half-spaces, $z>0$, $z<0$, respectively.

The difference in material properties destroys the symmetry that we exploited in the previous chapter and we therefore anticipate shear stresses as well as normal stresses at the interface, even when the far-field loading is a state of uniform tension. However, we shall show that we can still use the same methods to reduce the problem to a mixed boundary-value problem for the half-space.

## 24.1 The uncracked interface

The presence of the interface causes the problem to be non-trivial, even if there is no crack and the two bodies are perfectly bonded.

Consider the tensile loading of the composite bar of Figure 24.2($a$). If we assume a state of uniform uniaxial tension $\sigma_{zz}=S$ exists throughout the bar, the Poisson's ratio strains, $e_{xx}=e_{yy}=-\nu_i S/E_i$ will imply an inadmissible discontinuity in displacements $u_x, u_y$ at the interface, unless $\nu_1/E_1=\nu_2/E_2$. In all other cases, a locally non-uniform stress field will be developed involving shear stresses on the interface and the bar will deform as shown in Figure 24.2($b$). Indeed, for many material combinations, there will be a stress singularity at the edges $A, B$, which can be thought of locally as two bonded orthogonal wedges[1] and analyzed by the method of §11.2.2.

We shall not pursue the perfectly-bonded problem here — a solution for the corresponding two-dimensional problem is given by Bogy[2] — but we note that as long

[1]D.B.Bogy, Edge-bonded dissimilar orthogonal elastic wedges under normal and shear loading, ASME J.Appl.Mech., Vol. 35 (1968), 460–466.

[2]D.B.Bogy, The plane solution for joined dissimilar elastic semistrips under tension, ASME

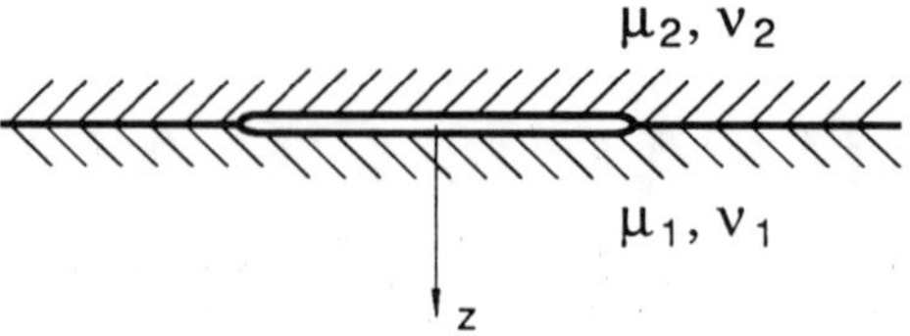


Figure 24.1: The plane interface crack

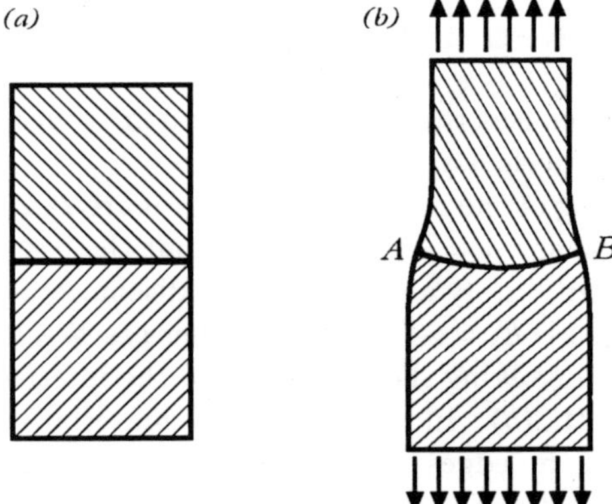


Figure 24.2: The composite bar in tension

as the solution for the uncracked interface is known, the corresponding solution when a crack is present can be obtained as in §§13.3.2, 23.1, by superposing a corrective solution in which tractions are imposed on the crack face, equal and opposite to those transmitted in the uncracked state, and the distant boundaries of the body are traction-free. Furthermore, if the crack is small compared with the other linear dimensions of the body — notably the distance from the crack to the free boundary — it can be conceived as occurring at the interface between two bonded half-spaces, as in Figure 24.1.

## 24.2 The corrective solution

The corrective solution is therefore defined by the boundary conditions

$$\sigma_{xz1}(x,y,0^+) = \sigma_{xz2}(x,y,0^-) = -S_1(x,y)\ ; \qquad (24.1)$$

$$\sigma_{yz1}(x,y,0^+) = \sigma_{yz2}(x,y,0^-) = -S_2(x,y)\ ; \qquad (24.2)$$

$$\sigma_{zz1}(x,y,0^+) = \sigma_{zz2}(x,y,0^-) = -S_3(x,y)\ , \qquad (24.3)$$

for $(x,y) \in A$ and

$$\sigma_{xxi}, \sigma_{yyi}, \sigma_{zzi}, \sigma_{xyi}, \sigma_{yzi}, \sigma_{zxi} \to 0\ ;\ \ R \to \infty\ , \qquad (24.4)$$

where $A$ is the region of the interfacial plane $z = 0$ occupied by the crack and $S_1, S_2, S_3$ are the tractions that would be transmitted across the interface in the absence of the crack.

We shall seek a potential function solution of the problem in the half-space, so we supplement (24.1–24.4) by the continuity and equilibrium conditions

$$\sigma_{xz1}(x,y,0^+) = \sigma_{xz2}(x,y,0^-)\ ; \qquad (24.5)$$

$$\sigma_{yz1}(x,y,0^+) = \sigma_{yz2}(x,y,0^-)\ ; \qquad (24.6)$$

$$\sigma_{zz1}(x,y,0^+) = \sigma_{zz2}(x,y,0^-)\ ; \qquad (24.7)$$

$$u_{x1}(x,y,0+) = u_{x2}(x,y,0-)\ ; \qquad (24.8)$$

$$u_{y1}(x,y,0+) = u_{y2}(x,y,0-)\ ; \qquad (24.9)$$

$$u_{z1}(x,y,0+) = u_{z2}(x,y,0-)\ , \qquad (24.10)$$

in $\bar{A}$ — i.e. the uncracked part of the interface[3].

It is clear from (24.1–24.3) that (24.5–24.7) apply throughout the interface and are therefore global conditions. We can exploit this fact by expressing the fields in the two bonded half-spaces by separate sets of stress functions and then developing simple symmetric relations between the sets.

J.Appl.Mech., Vol. 42 (1975), 93–98.

[3]Notice that only those stress components that act on the interface (and hence have a $z$-suffix) are continuous across the interface, because of the equilibrium requirement. The remaining three components will generally be discontinuous because of (24.8–24.10) and the dissimilar elastic properties.

### 24.2.1 Global conditions

The global conditions involve the three stress components $\sigma_{zx}, \sigma_{zy}, \sigma_{zz}$ at the surface $z = 0$, so it is convenient to choose a formulation in which these components take simple forms at the surface and are as far as possible uncoupled. This can be achieved by superposing Solutions E,F,G of Chapter 16, for which

$$\sigma_{zxi}(x, y, 0) = \frac{\partial^2 \Psi_i}{\partial y \partial z} + \frac{\partial^2 \chi_i}{\partial x \partial z}\ ; \tag{24.11}$$

$$\sigma_{zyi}(x, y, 0) = -\frac{\partial^2 \Psi_i}{\partial x \partial z} + \frac{\partial^2 \chi_i}{\partial y \partial z}\ ; \tag{24.12}$$

$$\sigma_{zzi} = -\frac{\partial^2 \phi_i}{\partial z^2}\ , \tag{24.13}$$

from Tables 16.1, 16.2, where the suffix $i$ takes the value 1,2 for bodies 1,2 respectively.

The global condition (24.7) now reduces to

$$\frac{\partial^2 \phi_1}{\partial z^2}(x, y, 0^+) = \frac{\partial^2 \phi_2}{\partial z^2}(x, y, 0^-) \tag{24.14}$$

and, remembering that $\phi_1$ is only defined in $z > 0$ and $\phi_2$ in $z < 0$, we can satisfy it by imposing the symmetry relation

$$\phi_1(x, y, z) = \phi_2(x, y, -z) \equiv \phi(x, y, z) \tag{24.15}$$

throughout the half-spaces. In the same way[4], the two remaining global conditions can be satisfied by demanding

$$\Psi_1(x, y, z) = -\Psi_2(x, y, -z) \equiv \Psi(x, y, z)\ ; \tag{24.16}$$

$$\chi_1(x, y, z) = -\chi_2(x, y, -z) \equiv \chi(x, y, z)\ , \tag{24.17}$$

where the negative sign in (24.16, 24.17) arises from the fact that symmetry of the derivatives $\partial\Psi/\partial z, \partial\chi/\partial z$ implies *antisymmetry* of $\Psi, \chi$.

### 24.2.2 Mixed conditions

We can now use the remaining boundary conditions (24.1–24.3) and (24.8–24.10) to define a mixed boundary-value problem for the three potentials $\phi, \Psi, \chi$ in the half-space $z > 0$.

[4]This idea can also be extended to steady-state thermoelastic problems by adding in Solution P of Table 17.1, which leaves the expressions (24.11–24.13) for the interface stresses unchanged and requires a further symmetry relation between $\psi_1, \psi_2$ to satisfy the global condition of continuity of heat flux, $q_{z1}(x, y, 0^+) = q_{z2}(x, y, 0^-)$.

For example, using the expressions from Tables 16.1, 16.2 and the definitions (24.15–24.17), the boundary condition (24.8) reduces to

$$\frac{1}{\mu_1}\frac{\partial\psi}{\partial y}+\frac{(1-2\nu_1)}{2\mu_1}\frac{\partial\phi}{\partial x}+\frac{(1-\nu_1)}{\mu_1}\frac{\partial\chi}{\partial x}=-\frac{1}{\mu_2}\frac{\partial\psi}{\partial y}+\frac{(1-2\nu_2)}{2\mu_2}\frac{\partial\phi}{\partial x}-\frac{(1-\nu_2)}{\mu_2}\frac{\partial\chi}{\partial x} \quad (24.18)$$

in $\bar{A}$, $z=0$.

Rearranging the terms and dividing by the non-zero factor $[(1-\nu_1)/\mu_1+(1-\nu_2)/\mu_2]$ (which is $4A$ in the terminology of equation (12.65) for plane strain), we obtain

$$\gamma\frac{\partial\Psi}{\partial y}+\beta\frac{\partial\phi}{\partial x}+\frac{\partial\chi}{\partial x}=0 \ ; \ \text{in } \bar{A} \ , \quad (24.19)$$

where $\beta$ is Dundurs' constant (see §§4.4.5, 12.7) and

$$\gamma=\left[\frac{1}{\mu_1}+\frac{1}{\mu_2}\right]\Bigg/\left[\frac{(1-\nu_1)}{\mu_1}+\frac{(1-\nu_2)}{\mu_2}\right] \ . \quad (24.20)$$

A similar procedure applied to the boundary conditions (24.9, 24.10) yields

$$-\gamma\frac{\partial\Psi}{\partial x}+\beta\frac{\partial\phi}{\partial y}+\frac{\partial\chi}{\partial y} = 0 \ ; \quad (24.21)$$

$$\frac{\partial\phi}{\partial z}+\beta\frac{\partial\chi}{\partial z} = 0 \ , \quad (24.22)$$

in $\bar{A}$.

In addition to (24.19, 24.21, 24.22), three further conditions are obtained from (24.1–24.3), using (24.11–24.13) and (24.15–24.17). These are

$$\frac{\partial^2\Psi}{\partial y\partial z}+\frac{\partial^2\chi}{\partial x\partial z} = -S_1(x,y) \ ; \quad (24.23)$$

$$-\frac{\partial^2\Psi}{\partial x\partial z}+\frac{\partial^2\chi}{\partial y\partial z} = -S_2(x,y) \ ; \quad (24.24)$$

$$-\frac{\partial^2\phi}{\partial z^2} = -S_3(x,y) \ , \quad (24.25)$$

in $A$. The six conditions (24.19, 24.21–24.25) define a well-posed boundary-value problem for the three potentials $\phi, \Psi, \chi$, when supplemented by the requirements that (i) displacements should be continuous at the crack boundary and (ii) the stress and displacement fields should decay as $R\to\infty$ in accordance with (24.4).

The two boundary conditions (24.19, 24.21) involve only two independent functions $\Psi$ and $(\beta\phi+\chi)$ and hence we can eliminate either by differentiation with the result

$$\frac{\partial^2\Psi}{\partial x^2}+\frac{\partial^2\Psi}{\partial y^2}=-\frac{\partial^2\Psi}{\partial z^2} = 0 \ ; \quad (24.26)$$

$$\left(\frac{\partial^2}{\partial x^2}+\frac{\partial^2}{\partial y^2}\right)(\beta\phi+\chi)=-\beta\frac{\partial^2\phi}{\partial z^2}-\frac{\partial^2\chi}{\partial z^2} = 0 \quad (24.27)$$

in $\bar{A}$, where we have used the fact that the potential functions are all harmonic to express the boundary conditions in terms of a single derivative normal to the plane.

In the same way, (24.23, 24.24) yield

$$\frac{\partial^3 \Psi}{\partial z^3} = \frac{\partial S_1}{\partial y} - \frac{\partial S_2}{\partial x} ; \tag{24.28}$$

$$\frac{\partial^3 \chi}{\partial z^3} = \frac{\partial S_1}{\partial x} + \frac{\partial S_2}{\partial y} , \tag{24.29}$$

in $A$.

Equations (24.26, 24.28) now define a two-part boundary-value problem for $\Psi$, similar to those solved in Chapter 22 for the axisymmetric geometry. It should be noted however that it is not necessary, nor is it generally possible, to satisfy the requirement of continuity of in-plane displacement $u_x, u_y$ at the crack boundary in the *separate* contributions from the potential functions $\phi, \Psi, \chi$, as long as the *superposed* fields satisfy this requirement. Thus, the homogeneous problem for $\Psi$ obtained by setting the right hand side of (24.28) to zero has non-trivial solutions which are physically unacceptable in isolation, since they correspond to displacement fields which are discontinuous at the crack boundary, but they are an essential component of the solution of more general problems, where they serve to cancel similar discontinuities resulting from the fields due to $\phi, \chi$. This problem also arises for the crack in a homogeneous medium with shear loading[5] and in non-axisymmetric thermoelastic problems[6]

The remaining boundary conditions define a coupled two-part problem for $\phi, \chi$. As in the two-dimensional contact problems of Chapter 12, the coupling is proportional to the Dundurs' constant $\beta$. If the material properties are such that $\beta = 0$ (see §12.7), the problems of normal and shear loading are independent and (for example) the crack in a shear field has no tendency to open or close.

## 24.3 The penny-shaped crack in tension

We now consider the special case of the circular crack $0 \le r < a$, subjected to uniform tensile loading $S_3 = S,\ S_1 = S_2 = 0$. The boundary-value problem for $\phi, \chi$ is then

[5]See for example J.R.Barber, The penny-shaped crack in shear and related contact problems, Int.J.Engng.Sci., Vol. 13, (1975), 815–829.

[6]L.Rubenfeld, Non-axisymmetric thermoelastic stress distribution in a solid containing an external crack, Int.J.Engng.Sci., Vol. 8 (1970), 499–509, J.R.Barber, Steady-state thermal stresses in an elastic solid containing an insulated penny-shaped crack, J.Strain Analysis, Vol. 10 (1975), 19–24.

defined by the conditions[7]

$$\begin{aligned}
\frac{\partial^3\chi}{\partial z^3} &= 0 \;\; ; \;\; z=0,\; 0\le r<a\; ; && (24.30)\\
\frac{\partial^2\phi}{\partial z^2} &= S \;\; ; \;\; z=0,\; 0\le r<a\; ; && (24.31)\\
\frac{\partial\phi}{\partial z}+\beta\frac{\partial\chi}{\partial z} &= 0 \;\; ; \;\; z=0,\; r>a\; ; && (24.32)\\
\beta\frac{\partial^2\phi}{\partial z^2}+\frac{\partial^2\chi}{\partial z^2} &= 0 \;\; ; \;\; z=0,\; r>a\, . && (24.33)
\end{aligned}$$

From Table 22.1, we see that we can satisfy (24.32, 24.33) by writing[8]

$$\beta\phi+\chi=\phi_1 \;\; , \;\; \phi+\beta\chi=\phi_3 \tag{24.34}$$

— i.e.

$$\phi=\frac{\phi_3-\beta\phi_1}{(1-\beta^2)} \;\; , \;\; \chi=\frac{\phi_1-\beta\phi_3}{(1-\beta^2)}\; , \tag{24.35}$$

where $\phi_1, \phi_3$ are defined by equation (22.21).

The remaining boundary conditions then define two coupled integral equations of Abel-type for the unknown functions $g_1(t), g_3(t)$ in the range $0 \le t < a$. To develop these equations, we first note that the condition $\nabla^2\chi = 0$ and the fact that the solution is axisymmetric enable us to write (24.30) in the form

$$\frac{1}{r}\frac{d}{dr}r\frac{d}{dr}\frac{\partial\chi}{\partial z}(r,0)=0 \;\; ; \;\; z=0,\; 0\le r<a\; , \tag{24.36}$$

which can be integrated within the plane $z=0$ to give

$$\frac{\partial\chi}{\partial z}=C \;\; ; \;\; z=0,\; 0\le r<a\; , \tag{24.37}$$

where $C$ is a constant which will ultimately be chosen to ensure continuity of displacements at $r=a$ and we have eliminated a logarithmic term to preserve continuity of displacements at the origin.

Table 22.1 and (24.35, 24.37) now enable us to write

$$\int_0^r\frac{g_1(t)dt}{\sqrt{r^2-t^2}}+\beta\int_r^a\frac{g_3(t)dt}{\sqrt{t^2-r^2}}=(1-\beta^2)C \;\; ; \;\; 0\le r<a\; , \tag{24.38}$$

[7]This problem was first considered by W.I.Mossakowskii and M.T.Rybka, Generalization of the Griffith-Sneddon criterion for the case of a non-homogeneous body, J.Appl.Math., Vol. 28 (1964), 1061–1069, and by F.Erdogan, Stress distribution on bonded dissimilar materials containing circular or ring-shaped cavities, ASME J.Appl.Mech., Vol. 32 (1965), 829–836. A formulation for the penny-shaped crack with more general loading was given by J.R.Willis, The penny-shaped crack on an interface, Q.J.Mech.Appl.Math., Vol. 25 (1972), 367–382.

[8]This method would fail for the special case $\beta = \pm 1$, but materials with positive Poisson's ratio are restricted by energy considerations to values in the range $-\frac{1}{2} \le \beta \le \frac{1}{2}$.

whilst (24.31, 24.35) and Table 22.1 give

$$\frac{1}{r}\frac{d}{dr}\int_0^r \frac{tg_3(t)dt}{\sqrt{r^2-t^2}} - \frac{\beta}{r}\frac{d}{dr}\int_r^a \frac{tg_1(t)dt}{\sqrt{t^2-r^2}} = (1-\beta^2)S \ ; \ 0 \le r < a \ , \tag{24.39}$$

which can be integrated to give

$$\int_0^r \frac{tg_3(t)dt}{\sqrt{r^2-t^2}} - \beta\int_r^a \frac{tg_1(t)dt}{\sqrt{t^2-r^2}} = \frac{(1-\beta^2)Sr^2}{2} + B \ ; \ 0 \le r < a \ , \tag{24.40}$$

where $B$ is an arbitrary constant.

We now treat (24.38) as an Abel equation for $g_1(t)$, carrying the $g_3$ integral onto the right hand side and using the inversion rules (Table 22.2) to obtain

$$g_1(x) = \frac{2\beta}{\pi}\int_0^a \frac{tg_3(t)dt}{(x^2-t^2)} + \frac{2C(1-\beta^2)}{\pi} \ , \tag{24.41}$$

where we have simplified the double integral term in $g_3$ by changing the order of integration and performing the resulting inner integral.

In the same way, treating (24.40) as an equation for $tg_3(t)$, we obtain

$$xg_3(x) = -\frac{2\beta}{\pi}\int_0^a \frac{t^2 g_1(t)dt}{(x^2-t^2)} + \frac{2B}{\pi} + \frac{2S(1-\beta^2)x^2}{\pi} \ . \tag{24.42}$$

The function $g_3(x)$ must be bounded at $x=0$ and hence $B$ must be chosen so that

$$B = -\beta\int_0^a g_1(t)dt \ . \tag{24.43}$$

Thus, (24.42) reduces to

$$g_3(x) = -\frac{2\beta x}{\pi}\int_0^a \frac{g_1(t)dt}{(x^2-t^2)} + \frac{2S(1-\beta^2)x}{\pi} \ . \tag{24.44}$$

### 24.3.1 Reduction to a single equation

We observe from equations (24.41, 24.44) that $g_1$ is an even function of $x$, whereas $g_3$ is odd. Using this result and expanding the integrands as partial fractions, we arrive at the simpler expressions[9]

$$g_1(x) = \frac{\beta}{\pi}\int_{-a}^a \frac{g_3(t)dt}{(x-t)} + \frac{2C(1-\beta^2)}{\pi} \ ; \ -a < x < a \ ; \tag{24.45}$$

$$g_3(x) = -\frac{\beta}{\pi}\int_{-a}^a \frac{g_1(t)dt}{(x-t)} + \frac{2S(1-\beta^2)x}{\pi} \ ; \ -a < x < a \ . \tag{24.46}$$

[9] *cf.* equation (22.26) and the associated discussion.

Two methods are available for reducing (24.45, 24.46) to a single equation. The most straightforward approach is to use (24.46) (for example) to substitute for $g_3$ in (24.45), resulting after some manipulations in the integral equation

$$\begin{aligned} g_1(x) &+ \frac{\beta^2}{\pi^2(1-\beta^2)}\int_{-a}^{a}\left(\log\left|\frac{a-t}{a+t}\right| - \log\left|\frac{a-x}{a+x}\right|\right)\frac{g_1(t)dt}{(t-x)} \\ &= -\frac{2\beta S}{\pi^2}\left(2a + x\log\left|\frac{a-x}{a+x}\right|\right) + \frac{2C}{\pi}\ ; \ -a < x < a\ . \end{aligned} \tag{24.47}$$

The kernel of this equation is bounded at $t = x$, the limiting form being obtainable by L'Hôpital's rule as $2a/(x^2 - a^2)$. Equations of this kind with definite integration limits and bounded kernels are known as *Fredholm integral equations*[10].

However, the identity of kernel and range of integration in equations (24.45, 24.46) permits a more direct approach. Multiplying (24.46) by $i$ and adding the result to (24.45), we find that both equations are contained in the complex equation

$$g(x) + \frac{i\beta}{\pi}\int_{-a}^{a}\frac{g(t)dt}{(x-t)} = \frac{2(1-\beta^2)}{\pi}(C + iSx)\ , \tag{24.48}$$

where $g(x) \equiv g_1(x) + ig_3(x)$. Furthermore, this equation can be inverted explicitly[11] in the form of an integral of the right hand side, leading to a closed-form solution for $g(x)$. The tractions at the interface can then be recovered by separating $g(x)$ into its real and imaginary parts and substituting into the expressions

$$\sigma_{zr} = \frac{1}{(1-\beta^2)}\frac{d}{dr}\int_0^a \frac{g_1(t)dt}{\sqrt{r^2-t^2}}\ ; \ z = 0,\ r > a\ ; \tag{24.49}$$

$$\sigma_{zz} = -\frac{1}{(1-\beta^2)r}\frac{d}{dr}\int_0^a \frac{tg_3(t)dt}{\sqrt{r^2-t^2}}\ ; \ z = 0,\ r > a\ . \tag{24.50}$$

## 24.3.2 Oscillatory singularities

An asymptotic analysis of the integrals (24.48–24.50) at $r = a+s$, $s \ll a$, shows that they have the form

$$\sigma_{zz} + i\sigma_{zr} = SK\left(\frac{s}{2a}\right)^{-\frac{1}{2}+i\epsilon} + O(1) \tag{24.51}$$

$$= SK\left(\frac{s}{2a}\right)^{-\frac{1}{2}}\left[\cos\left(\epsilon\log\frac{s}{2a}\right) + i\sin\left(\epsilon\log\frac{s}{2a}\right)\right] + O(1)\ , \tag{24.52}$$

where

$$\epsilon = \frac{1}{2\pi}\log\left(\frac{1+\beta}{1-\beta}\right) \tag{24.53}$$

[10] F.G.Tricomi, ***Integral Equations***, Interscience, New York, (1957), Chapter 2.

[11] N.I.Muskhelishvili, ***Singular Integral Equations***, (English translation by J.R.M.Radok, Noordhoff, Groningen, (1953))

and Kassir and Bregman[12] have found the complex constant $K$ to be

$$K = \frac{\Gamma(2+i\epsilon)}{\pi^{\frac{1}{2}}\Gamma(\frac{1}{2}+i\epsilon)} . \tag{24.54}$$

Equation (24.51) shows that the stresses are square-root singular at the crack tip, but that they also oscillate with increasing frequency as $r$ approaches $a$ and $\log(s/2a) \to -\infty$. This behaviour is common to all interface crack problems for which $\beta \neq 0$ and can be predicted from an asymptotic analysis similar to that of §11.2.2, from which a complex leading singular eigenvalue $\lambda = \frac{1}{2} \pm i\epsilon$ is obtained.

## 24.4 The contact solution

A more disturbing feature of this behaviour is that the crack opening displacement, $(u_{z1}(r,0) - u_{z2}(r,0))$, also oscillates as $r \to a^-$ and hence there are infinitely many regions of interpenetration of material near the crack tip.

This difficulty was first resolved by Comninou[13] for the case of the plane crack, by relaxing the superficially plausible assumption that the crack will fully open and hence have traction-free faces in a tensile field, using instead the more rigorous requirements of frictionless unilateral contact — i.e. permitting contact to occur, the extent of the contact zones (if any) to be determined by the inequalities (21.11, 21.12). She found that very small contact zones are established adjacent to the crack tips, the extent of which is of the same order of magnitude as the zone in which interpenetration is predicted for the original 'open crack' solution. At the closed crack tip, only the shear stresses in the bonded region are singular, leading to a 'mode II' (shear) stress intensity factor, by analogy with equation (13.39). However, the contact stresses show a proportional compressive singularity in $r \to a^-$. Both singularities are of the usual square-root form, without the oscillatory character of equations (24.51).

Solutions have since been found for other interface crack problems with contact zones, including the plane crack in a combined tensile and shear field[14]. These original solutions of the interface crack problem used numerical methods to solve the resulting integral equation, but more recently an analytical solution has been found by Gautesen and Dundurs[15].

The problem of §24.3 — the penny-shaped crack in a tensile field — was solved

[12] M.K.Kassir and A.M.Bregman, The stress-intensity factor for a penny-shaped crack between two dissimilar materials, ASME J.Appl.Mech., Vol. 39 (1972), 308–310.

[13] M.Comninou, The interface crack, ASME J.Appl.Mech., Vol. 44 (1977), 631–636.

[14] M.Comninou and D.Schmueser, The interface crack in a combined tension-compression and shear field, ASME J.Appl.Mech., Vol. 46 (1979), 345–358.

[15] A.K.Gautesen and J.Dundurs, The interface crack in a tension field, ASME J.Appl.Mech., Vol. 54 (1987), 93–98; A.K.Gautesen and J.Dundurs, The interface crack under combined loading, *ibid.*, Vol. 55 (1988), 580–586.

in a unilateral contact formulation by Keer et al[16]. They found a small annulus of contact $b<r<a$ adjacent to the crack tip. In fact, the formulation of §24.3 is easily adapted to this case. We note that the crack opening displacement must be zero in the contact zone, so the range of equation (24.32) is extended inwards to $b<r$, whilst (24.31) — setting the total normal tractions at the interface to zero — now only applies in the open region of the crack $0\leq r<b$. This can be accommodated by replacing $a$ by $b$ in the definitions of $\phi_3$, so that (24.38, 24.39) are replaced by

$$\int_0^r \frac{g_1(t)dt}{\sqrt{r^2-t^2}} + \beta\int_r^b \frac{g_3(t)dt}{\sqrt{t^2-r^2}} = (1-\beta^2)C \;;\; 0\leq r<a\,, \quad (24.55)$$

$$\frac{1}{r}\frac{d}{dr}\int_0^r \frac{tg_3(t)dt}{\sqrt{r^2-t^2}} - \frac{\beta}{r}\frac{d}{dr}\int_r^a \frac{tg_1(t)dt}{\sqrt{t^2-r^2}} = (1-\beta^2)S \;;\; 0\leq r<b\,, \quad (24.56)$$

and (24.45, 24.46) by

$$g_1(x) = \frac{\beta}{\pi}\int_{-b}^{b}\frac{g_3(t)dt}{(x-t)} + \frac{2C(1-\beta^2)}{\pi} \;;\; -a<x<a\,; \quad (24.57)$$

$$g_3(x) = -\frac{\beta}{\pi}\int_{-a}^{a}\frac{g_1(t)dt}{(x-t)} + \frac{2S(1-\beta^2)x}{\pi} \;;\; -b<x<b\,. \quad (24.58)$$

The difference in range in the integrals in (24.56, 24.57) now prevents us from combining them in a single complex equation, but it is still possible to develop a Fredholm equation in either function[17] by substitution, as in equation (24.47).

## 24.5 Implications for Fracture Mechanics

When an interface crack is loaded in tension, the predicted contact zones are very much smaller than the crack dimensions and in many cases will be smaller than the fracture process zone defined in §13.3.1. In such cases, it must be possible to characterize the conditions for fracture in terms of features of the elastic field further from the crack tip, where the open and contact solutions are practically indistinguishable.

If the Griffith fracture criterion applies, fracture will occur when the strain energy released per unit crack extension — known as the *energy release rate* — exceeds a critical value for the material. For a homogeneous material under in-plane loading, the energy release rate is proportional to $K_I^2+K_{II}^2$, where $K_I$ is the stress intensity factor defined by equation (13.39) and $K_{II}$ is the corresponding factor associated with the shear stress on the crack plane. For the open solution of the interface crack, the oscillatory behaviour prevents the stress intensity factors being given their usual

[16] L.M.Keer, S.H.Chen and M.Comninou, The interface penny-shaped crack reconsidered, Int.J.Engng.Sci., Vol. 16 (1978), 765–772.

[17] The best choice here is to eliminate $g_1(x)$, since the opposite choice will lead to an integral equation on the range $-a<x<a$, which might have discontinuities in the kernel at the points $x=\pm b$. These could cause problems with convergence in the final numerical solution.

meaning, but the energy release rate depends on the asymptotic behaviour of $\sigma_{zr}^2+\sigma_{zz}^2$, which is non-oscillatory. The contact solution gives almost the same energy release rate, as indeed it must do, since the two solutions only differ in a very small region at the tip.

Experiments with homogeneous materials show that the energy release rate required for fracture varies significantly with the ratio $K_{II}/K_I$, which is referred to as the *mode mixity*[18]. Generally it is harder to make a crack propagate in a shear field than in a tension field. This is probably more indicative of our over-idealization of the crack geometry than of any limitation on Griffith's theory, since real cracks are seldom plane and the attempt to shear them without significant opening will probably lead to crack face contact and consequent frictional forces.

It is difficult to extend the concept of mode mixity to the open interface crack solution, since, for example, the ratio $\sigma_{zr}/\sigma_{zz}$ passes through all values, positive and negative, in each cycle of oscillation. However, the change in the ratio with $s$ is very slow except in the immediate vicinity of the crack tip, leading Rice[19] to suggest that some specific distance from the tip be agreed by convention at which the mode mixity be defined.

An interesting feature of the asymptotics of the open solution is that, in contrast to the case $\beta = 0$ (including the crack in a homogeneous material), all possible fields can be mapped into each other by a change in length scale and in one linear multiplier (e.g. the square root of the energy release rate). Thus, an alternative way of characterizing the conditions at the tip for fracture experiments would be in terms of the energy release rate and a characteristic length, which could be taken as the maximum distance from the tip at which the open solution predicts interpenetration. Conditions approach most closely to the classical mode I state when this distance is small.

It must be emphasised that these arguments are all conditional upon the characteristic length being small compared with all the other dimensions of the system, *including the anticipated process zone dimension.* If this condition is not satisfied, it is essential to use the unilateral contact formulation.

---

[18]See for example, H.A.Richard, Examination of brittle fracture criteria for overlapping mode I and mode II loading applied to cracks, in G.C.Sih, E.Sommer and W.Dahl, eds., *Application of Fracture Mechanics to Structures*, Martinus Nijhoff, The Hague, (1984), 309–316.

[19]J.R.Rice, Elastic fracture mechanics concepts for interfacial cracks, ASME J.Appl.Mech., Vol. 55 (1988), 98–103.

# Chapter 25

# THE RECIPROCAL THEOREM

Energy or variational methods have an important place in Solid Mechanics both as an alternative to the more direct method of solving the governing partial differential equations and as a means of developing convergent approximations to analytically intractable problems. They are particularly useful in situations where only a restricted set of results is required — for example, if we wish to determine the resultant force on a cross-section or the displacement of a particular point, but are not interested in the full stress and displacement fields. Indeed, such results can often be obtained in closed form for problems in which a solution for the complete fields would be intractable.

In this Chapter, we shall restrict attention to one such energy theorem — the Reciprocal Theorem associated with the names of Maxwell and Betti — which is a very fruitful source of useful results in linear elasticity[1].

## 25.1 Maxwell's Theorem

The simplest form of the theorem states that if a linear elastic body is kinematically supported and subjected to an external force $\boldsymbol{F}_1$ at the point $P$, which produces a displacement $\boldsymbol{u}_1(Q)$ at another point $Q$, then a force $\boldsymbol{F}_2$ at $Q$ would produce a displacement $\boldsymbol{u}_2(P)$ at $P$ where

$$\boldsymbol{F}_1.\boldsymbol{u}_2(P) = \boldsymbol{F}_2.\boldsymbol{u}_1(Q) \; . \tag{25.1}$$

To prove the theorem, we consider two scenarios — one in which the force $\boldsymbol{F}_1$ is applied gradually[2] and then held constant whilst $\boldsymbol{F}_2$ is applied gradually and the other

[1]For a more comprehensive review of the energy theorems in elasticity, the reader is referred to S.G.Mikhlin, *Variational Methods in Mathematical Physics*, Pergamon, New York, (1964) and S.P.Timoshenko and J.N.Goodier, *loc. cit.*, Chapter 8.

[2]When we state that a force is applied gradually, we mean that the rate of application of the force is sufficiently slow for dynamic effects to be negligible, so that the stress field is always quasi-static in the sense of Chapter 7. This in turn implies that all the work done by the force appears as elastic strain energy in the body, rather than partially as kinetic energy.

in which the order of application of the forces is reversed. Denoting the displacement fields due to $\boldsymbol{F}_1, \boldsymbol{F}_2$ acting alone by $\boldsymbol{u}_1, \boldsymbol{u}_2$, we find that the work done in the first scenario is

$$W_1 = \frac{1}{2}\boldsymbol{F}_1.\boldsymbol{u}_1(P) + \frac{1}{2}\boldsymbol{F}_2.\boldsymbol{u}_2(Q) + \boldsymbol{F}_1.\boldsymbol{u}_2(P) \ . \tag{25.2}$$

In this expression, the first term represents the work done by $\boldsymbol{F}_1$ whilst it was being applied, the second term is the work done by $\boldsymbol{F}_2$ whilst *it* was being applied and the final term is the additional work done by $\boldsymbol{F}_1$ as it moved 'involuntarily' through the additional displacement caused by $\boldsymbol{F}_2$.

In the second scenario, the work done is obtained by interchanging the suffices 1,2 and the locations $P, Q$ in equation (2) — i.e.

$$W_2 = \frac{1}{2}\boldsymbol{F}_2.\boldsymbol{u}_2(Q) + \frac{1}{2}\boldsymbol{F}_1.\boldsymbol{u}_1(P) + \boldsymbol{F}_2.\boldsymbol{u}_1(Q) \ . \tag{25.3}$$

In each case, since the forces were applied gradually, the work done will be stored as strain energy in the deformed body. But the final state in each case is the same, so we conclude that $W_1 = W_2$. Comparing the two expressions, we see that they differ only in the last term and hence

$$\boldsymbol{F}_1.\boldsymbol{u}_2(P) = \boldsymbol{F}_2.\boldsymbol{u}_1(Q) \ , \tag{25.4}$$

proving the theorem.

## 25.2 Betti's Theorem

Maxwell's theorem is useful in Strength of Materials, where point forces are used to represent force resultants (e.g. on the cross-section of a beam), but in Elasticity, the reciprocal theorem is more useful in the generalized form due to Betti, which relates the surface displacement fields $\boldsymbol{u}_1$,$\boldsymbol{u}_2$ due to two different surface traction distributions $\boldsymbol{T}_1$,$\boldsymbol{T}_2$ respectively.

The work done during the application of these tractions can be found by performing an integral around the surface $S$ of the body. Thus, if the tractions $\boldsymbol{T}_1$ are applied first, we find

$$W_1 = \frac{1}{2}\int\int_S \boldsymbol{T}_1.\boldsymbol{u}_1 dS + \frac{1}{2}\int\int_S \boldsymbol{T}_2.\boldsymbol{u}_2 dS + \int\int_S \boldsymbol{T}_1.\boldsymbol{u}_2 dS \tag{25.5}$$

where the double integral recognizes that the surface of a three-dimensional body is a two-dimensional curvilinear surface.

When the traction distributions are applied in the opposite order, the work done $W_2$ is obtained by interchanging suffices 1,2 in equation (25.5) and the condition $W_1 = W_2$ then leads to the result

$$\int\int_S \boldsymbol{T}_1.\boldsymbol{u}_2 dS = \int\int_S \boldsymbol{T}_2.\boldsymbol{u}_1 dS \tag{25.6}$$

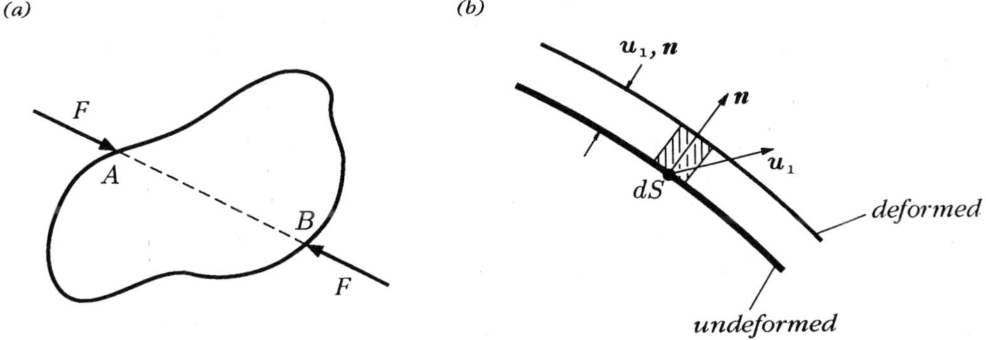


Figure 25.1: Increase in volume of a body due to two colinear forces

which is a statement of Betti's Theorem[3]

## 25.3 Use of the theorem

Betti's theorem defines a relationship between two different stress and displacement states for the same body. In most applications, one of these states corresponds to the problem under investigation, whilst the second is an *auxiliary solution*, for which the tractions and surface displacements are known. Often the auxiliary solution will be a simple state of stress for which the required results can be written down by inspection.

The art of using the reciprocal theorem lies in choosing the auxiliary solution so that equation (25.6) yields the required result. For example, suppose we wish to determine the change in volume of a body due to a pair of equal and opposite colinear concentrated forces, as shown in Figure 25.1($a$). Figure 25.1($b$) shows an enlarged view of a region of the boundary in the deformed and undeformed states, from which it is clear that the change in volume associated with some small region $dS$ of the boundary is $\boldsymbol{u}_1 . \boldsymbol{n} dS$, where $\boldsymbol{n}$ is the unit vector normal to the surface at $dS$. Thus, the change in volume of the entire body is

$$\delta V_1 = \int\int_S \boldsymbol{u}_1 . \boldsymbol{n} dS \; . \tag{25.7}$$

Now this expression can be made equal to the right hand side of (25.6) if we choose

[3]Notice that the theorem can be generalized to problems involving body forces by the inclusion of a volume integral through the body, but this is not done here in the interests of simplicity.

$\boldsymbol{T}_2 = \boldsymbol{n}$. In other words, the auxiliary solution corresponds to the case where the body is subjected to a purely normal (tensile) traction of unit magnitude throughout its surface.

The solution of this auxiliary problem is straightforward, since the tractions are compatible with an admissible state of uniform hydrostatic tension

$$\sigma_{xx} = \sigma_{yy} = \sigma_{zz} = 1 \;\; ; \;\; \sigma_{xy} = \sigma_{yz} = \sigma_{zx} = 0 \tag{25.8}$$

throughout the body. The stress-strain relations then give

$$e_{xx} = e_{yy} = e_{zz} = \frac{(1-2\nu)}{E} \;\; ; \;\; e_{xy} = e_{yz} = e_{zx} = 0 \tag{25.9}$$

and hence, in state 2, the body will simply deform into a larger, geometrically similar shape, any linear dimension $L$ increasing to $L(1+(1-2\nu)/E)$.

It remains to calculate the *left hand side* of (25.6), which we recall is the work done by the tractions $\boldsymbol{T}_1$ in moving through the surface displacements $\boldsymbol{u}_2$. In this case, $\boldsymbol{T}_1$ consists of the two concentrated forces, so the appropriate work is $F\delta_2$, where $\delta_2$ is the amount that the length $L$ of the line $AB$ shrinks *in state 2*, permitting the points of application of the forces to approach each other.

From the above argument,

$$\delta_2 = -\frac{(1-2\nu)L}{E} \tag{25.10}$$

and hence we can collect results to obtain

$$\begin{aligned} \delta V_1 &= \int\!\!\int_S \boldsymbol{u}_1.\boldsymbol{n}dS = \int\!\!\int_S \boldsymbol{u}_1.\boldsymbol{T}_2 dS \\ &= \int\!\!\int_S \boldsymbol{u}_2.\boldsymbol{T}_1 dS = F\delta_2 = -\frac{F(1-2\nu)L}{E}\,, \end{aligned} \tag{25.11}$$

which is the required expression for the change in volume.

Clearly the same method would enable us to determine the change in volume due to any other self-equilibrated traction distribution $\boldsymbol{T}_1$, either by direct substitution in the integral $\int\!\int_S \boldsymbol{u}_2.\boldsymbol{T}_1 dS$ or by using the force pair of Figure 25.1($a$) as a Green's function to define a more general distribution through a convolution integral. It is typical of the reciprocal theorem that it leads to the proof of results of some generality.

### 25.3.1 A tilted punch problem

For any given auxiliary solution, Betti's theorem will yield only a single result, which can be the value of the displacement at a specific point, or more often, an integral quantity such as a force resultant or an average displacement. Indeed, when interest in the problem is restricted to such integrals, the reciprocal theorem should always be considered first as a possible method of solution.

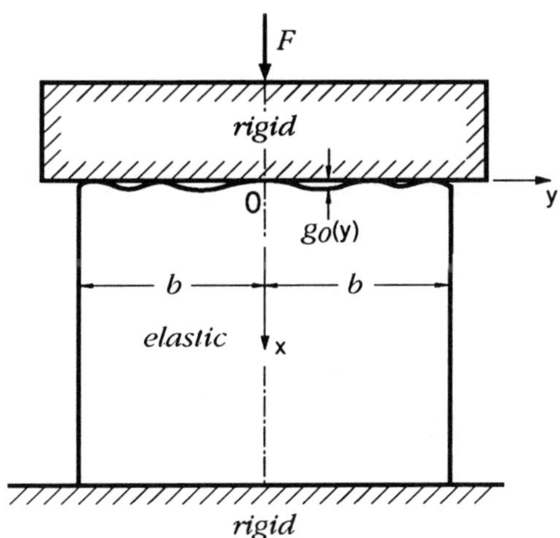


Figure 25.2: The tilted punch problem

As a second example, we shall consider the two-dimensional plane stress contact problem of Figure 25.2, in which a rectangular block with a slightly irregular surface is pressed down by a frictionless flat rigid punch against a frictionless rigid plane. We suppose that the force $F$ is large enough to ensure that contact is established throughout the end of the block and that the line of action of the applied force passes through the mid-point of the block. The punch will therefore generally tilt through some angle, $\alpha$, and the problem is to determine $\alpha$.

We suppose that the profile of the block is defined by an initial gap function $g_0(y)$ when the punch is horizontal and hence, if the punch moves downward a distance $u_0$ and rotates clockwise through $\alpha$, the contact condition can be stated as

$$g(y) = g_0(y) + u_1(y) - u_0 - \alpha y = 0 \; , \tag{25.12}$$

where $u_1(y)$ is the downward normal displacement of the block surface, as shown in Figure 25.3($a$).

Since the punch is frictionless, the traction $\boldsymbol{T}_1$ is also purely normal, as shown.

The key to the solution lies in the fact that the line of action of $\boldsymbol{F}$, which is also the resultant of $\boldsymbol{T}_1$, passes through the centre of the block. Taking moments about this point, we therefore have

$$\int_{-b}^{b} yT_1(y)dy = 0 \; . \tag{25.13}$$

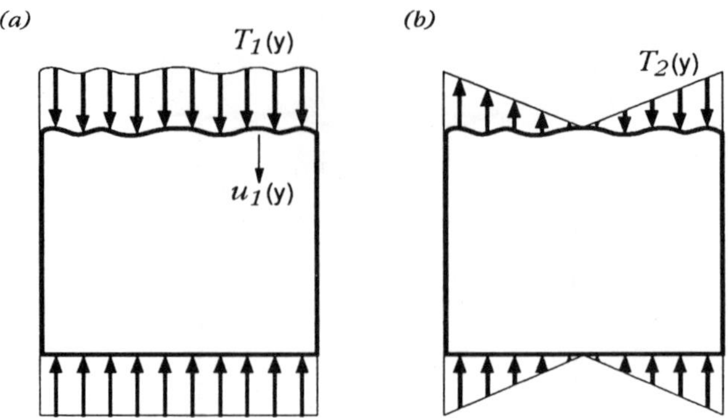


Figure 25.3: Tractions for the original and auxiliary problems for the tilted punch

We can make this expression look like the left hand side of (25.6) by choosing the auxiliary solution so that the normal displacement $u_2(y) = Cy$, which, of course, corresponds to a simple bending distribution with

$$T_2(y) = \frac{ECy}{a} , \tag{25.14}$$

as shown in Figure 25.3($b$).

Betti's theorem now gives

$$\int_{-b}^{b} \frac{ECyu_1(y)dy}{a} = \int_{-b}^{b} CyT_1(y)dy = 0 \tag{25.15}$$

and hence, treating (25.12) as an equation for $u_1(y)$ and substituting into (25.15), we get

$$\int_{-b}^{b} [u_0 + \alpha y - g_0(y)]ydy = 0 \tag{25.16}$$

— i.e.

$$\alpha = \frac{3}{b^3} \int_{-b}^{b} yg_0(y)dy , \tag{25.17}$$

which is the required result.

This result is relevant to the choice of end conditions in the problem of §9.1, since it shows that the end of the beam can be restored to a vertical plane by a self-equilibrated purely normal traction if and only if

$$\int_{-b}^{b} y g_0(y) dy = 0 , \tag{25.18}$$

which is equivalent to the third condition of (9.21).

However, unfortunately we cannot rigorously justify the complete set of conditions (9.21) by this argument, since the full built-in boundary conditions (9.16) also involve constraints on the local extensional strain $e_{yy} = \partial u_y / \partial y$. It is clear from energy considerations that if these constraints were relaxed, any resulting motion would involve the end load doing positive work and hence the end deflection obtained with conditions (9.21) must be larger than that obtained with the strong conditions[4] (9.16).

In the above solution, we have tacitly assumed that the only contribution to the two integrals in (25.6) comes from the tractions and displacements at the top surface of the block. This is justified, since although there are also normal tractions $\boldsymbol{T}_1, \boldsymbol{T}_2$ at the lower surface, they make no contribution to the work integrals because the corresponding normal *displacements* ($\boldsymbol{u}_2, \boldsymbol{u}_1$ respectively) are zero. However, it is important to consider all surfaces of the body in the application of the theorem, including a suitably distant boundary in problems involving infinite or semi-infinite regions.

## 25.3.2 Indentation of a half-space

We demonstrated in §21.2 that the contact area $A$ between a frictionless rigid punch and an elastic half-space is that value which maximizes the total load $P(A)$ on the punch. Shield[5] has shown that this can be determined without solving the complete contact problem, using Betti's theorem.

The total load on the punch is

$$P(A) = \int \int_A p_1(x, y) dA , \tag{25.19}$$

where $p_1(x, y)$ is the contact pressure. We can make the left hand side of (25.6) take this form by choosing the auxiliary solution such that the normal displacement $u_2(x, y)$ in $A$ is uniform and of unit magnitude.

The simplest result is obtained by using as auxiliary solution that corresponding to the indentation of the half-space by a frictionless flat rigid punch of plan-form $A$,

[4]J.D.Renton, Generalized beam theory applied to shear stiffness, Int. J. Solids Structures, Vol. 27 (1991), 1955–1967, argues that the correct shear stiffness for a beam of any cross-section should be that which equates the strain energy associated with the shear stresses $\sigma_{xz}, \sigma_{yz}$ and the work done by the shear force against the shear deflection. For the rectangular beam of §9.1, this is equivalent to replacing the multiplier on $b^2/a^2$ in (9.24$b, c$) by $2.4(1 + \nu)$. This estimate differs from (9.24$c$) by at most 3%.

[5]R.T.Shield, Load-displacement relations for elastic bodies, Z.angew.Math.Phyz. (ZAMP), Vol. 18 (1967), pp. 682–693.

defined by the boundary conditions

$$\begin{aligned} u_2 &\equiv u_z(x,y) = 1 \ ; \ \text{in } A \ ; \\ p_2 &\equiv -\sigma_{zz}(x,y) = 0 \ ; \ \text{in } \bar{A} \ ; \\ \sigma_{xz} &= \sigma_{yz} = 0 \ ; \ \text{in } A \text{ and } \bar{A} \end{aligned} \qquad (25.20)$$

Since the only non-zero tractions in both solutions are the normal pressures in the contact area, $A$, equation (25.6) reduces to

$$P(A) \equiv \int\int_A p_1(x,y)dA = \int\int_A p_2(x,y)u_1(x,y)dA \ , \qquad (25.21)$$

where $u_1(x,y)$ is the normal displacement under the punch, which is defined through the contact condition (21.3) in terms of the initial punch profile $u_0(x,y)$ and its rigid-body displacement.

In problems involving infinite domains, it is important to verify that no additional contribution to the work integrals in (25.6) is made on the 'infinite' boundaries. To do this, we consider a hemispherical region of radius $b$ much greater than the linear dimensions of the loaded region $A$. The stress and displacement fields in both solutions will become self-similar at large $R$, approximating those of the point force solution (§18.2.1). In particular, the stresses will tend asymptotically to the form $R_{-2}f(\theta,\alpha)$ and the displacements to $R^{-1}g(\theta,\alpha)$.

If we now construct the integral $\int\int \boldsymbol{T}_1 . \boldsymbol{u}_2 dS$ over the hemisphere surface, it will therefore have the form

$$\int_0^{\frac{\pi}{2}} \int_0^{2\pi} b_{-2}f_1(\theta,\alpha)b_{-1}g_2(\theta,\alpha)b^2 d\theta d\alpha \ , \qquad (25.22)$$

which approaches zero with $b^{-1}$ as $b \to \infty$, indicating that the distant surfaces make no contribution to the integrals in (25.6).

Of course, we can only use (25.21) to find $P(A)$ if we already know the solution $p_2(x,y)$ for the flat punch problem with plan-form $A$ and exact solutions are only known for a limited number of geometries, such as the circle, the ellipse and the strip[6]. However, (25.21) does for example enable us to write down the indenting force for a circular punch of arbitrary profile[7] and hence, using the above theorem, to determine the contact area for the general axisymmetric contact problem.

## PROBLEMS

1. Figure 25.4 shows an elastic cube of side $a$ which just fits between two parallel

[6] An *approximate* solution to this problem for a punch of fairly general plan-form is given by V.I.Fabrikant, Flat punch of arbitrary shape on an elastic half-space, Int.J.Engng.Sci., Vol. 24 (1986), 1731–1740. Fabrikant's solution and the above reciprocal theorem argument are used as the bases of an approximate solution of the general smooth punch problem by J.R.Barber and D.A.Billings, An approximate solution for the contact area and elastic compliance of a smooth punch of arbitrary shape, Int.J.Mech.Sci., Vol. 32 (1990), 991-997.

[7] R.T.Shield, *loc. cit.*.

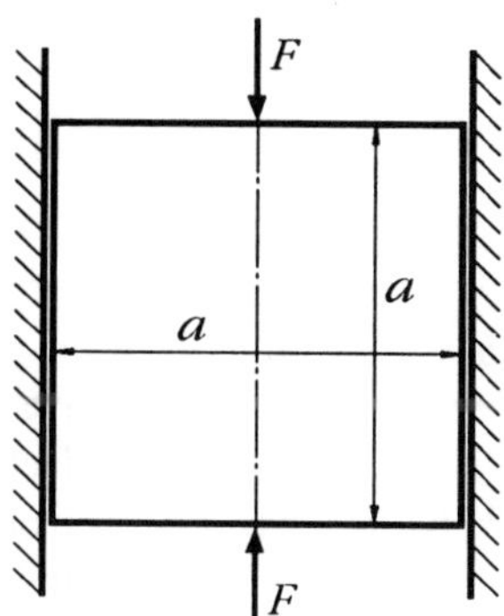


Figure 25.4: Block restrained from lateral expansion

frictionless rigid walls. The block is now loaded by equal and opposite concentrated forces, $F$, at the mid-point of two opposite faces, as shown. 'Poisson's ratio' strains will cause the block to exert normal forces $P$ on the walls. Find the magnitude of these forces.

Would your result be changed if (i) the forces $F$ were applied away from the mid-point or (ii) if their common line of action was inclined to the vertical?

2. Use the method of §§21.2, 25.3.2 to find the relations between the indentation force, $F$, the contact radius, $a$ and the maximum indentation depth, $\delta$, for a rigid frictionless punch with a fourth order profile $u_0 = -Cr^4$ indenting an elastic half-space.

# Index


Abel integral equations, 254–258, 262, 264, 266, 275–276
inversion, 257
Airy stress function, 40–42, 67–69, 84, 107, 179–180, 212
alternating tensor, 16, 23
auxiliary solution, 283
axisymmetric problems, 230–233, 235–239, 249
plane thermoelastic, 213–214
torsion, 225, 237
asymptotic methods, 125–134
bending, 47–48
Betti's theorem, 282–288
biharmonic equation, 42, 45, 61, 75, 84, 121, 180
body force, 21, 41, 67–81, 167, 181, 197
bonded interfaces, 132–133, 269–280
boundary conditions, 5
global, 222, 262, 271, 272
mixed, 242, 249, 262, 272
weak, 34, 49, 54, 59, 74, 235–236
Boussinesq potentials, 199–203
Boussinesq problem, 221–224
bubble model, 169–170
bulk modulus, 19
Burgers' vector, 169
Cartesian tensors, 8
Cauchy integral equations, 146, 176, 277
centre of compression, 218
centre of dilatation, 218
Cesaro integrals, 24
climb dislocation, 167, 175
change in volume, 281
circular cylinder, 213–214, 234, 235–237, 239
circular disk, 76, 83, 180, 214
circular hole, 83, 87–89, 92–94, 173
circular sector, 135–136
closure conditions, 171, 172, 176
coefficient of friction, 156
Collins' method, 250–258
compatibility equations, 22–27, 69, 189
completeness, 56, 63, 193–197
complex variable formulation, 206–209
axisymmetric complex potentials, 250–254
concentrated force, 139–143, 165–167
conservative vector fields, 68
constraint equations, 46
contact problems, 69, 134, 145–162, 185, 204, 241–246, 284–288
frictionless contact, 145–151, 154, 204, 241–246, 284–288
unilateral inequalities, 148, 243
continuity of displacement, 105, 107, 166, 259, 262, 271, 275
coördinate transformation, 7–10, 14–15
corrective solution, 33–34, 59–62, 88, 158, 160, 174–177, 238, 261–263, 264, 271–274
Coulomb friction, 156, 252
crack opening displacement, 177, 263, 278–279
cracks, 130–131, 173–177, 185, 261–280
creep velocity, 160, 162
curved beams, 107–119
D'Alembert's principle, 70
degenerate cases, 90–92, 230
special solutions, 90–91, 113–114,

117–118
dilatation, 18, 19
centre of, 218
surface dilatation, 223
dislocations, 168–177, 183
dislocation derivatives, 172
displacement,
calculation from stress components, 97–105
notation for, 10
rigid-body displacements, 10, 99–100, 102, 144, 145
single-valued displacements, 24–25, 105, 107, 165–167
displacement gradient, 145
Dundurs' constants, 43, 155, 273, 274
Dundurs' theorem, 183–185
eigenvalue problems, 62–65, 112, 125–135
eigenfunction expansion, 63, 65, 128–130, 136
elastodynamics, 71
end effects, 59–65
energy release rate, 279, 280
equilibrium equations, 21–22, 27, 77, 181, 189, 190, 211
Flamant solution, 140–143, 165
flat punch indentation, 147–148, 250–252
fracture mechanics, 173, 277–278
fracture toughness, 173
Fredholm integral equations, 277, 279
friction, 156
Fourier series methods, 54, 85–87, 146–147, 209, 227
Fourier transform methods, 54–56, 135
Galerkin vector, 190, 199
gap function, 151–153, 285
glide dislocation, 169
global boundary conditions, 224, 264, 273, 274
gravitational loading, 11–12, 67, 70
Green-Collins representation, 252–259
Green's functions, 128, 142, 144, 151, 165, 171, 218, 242, 284
Griffith fracture theory, 173, 279–280
gross slip, 159
half-plane, 125, 132, 141–145
surface displacements, 142–144
half-space, 202–205, 214, 221, 241, 271
coupling between normal and tangential effects, 154, 274
surface displacements, 223, 241
Hankel transform, 249–250
heat conduction, 181
Hertzian contact, 148–151, 156, 258
history-dependence, 156–162
Hooke's law, 3, 12, 17, 179
hydrostatic stress, 8, 284
inclusions, 105, 173, 239
incompressible material, 19, 154
index notation, 6
inertia forces, 70
integral equations, 146, 242, 250, 254, 277
interface crack, 269–280
contact solution, 278–280
irrotational vector field, 68, 190
Kelvin problem, 165–167, 219–221
Kronecker delta, 18
Lamé's constants, 18
Legendre's equation, 228
Legendre functions, 229–230
recurrence relations, 233
Legendre polynomials, 224, 229, 236, 238
Love's strain potential, 191, 201
Maxwell's theorem, 281–282
Mellin transform, 135–136
Michell's solution, 92, 93, 103–105
microslip, 159, 162
Mindlin's problem, 156–159
mixed boundary-value problem, 242–243, 249–259, 262, 264, 273–274

mode mixity, 280
Mohr's circle 8
multiply-connected bodies, 25–27, 105, 183, 212
notches, 125–126, 128, 132
oscillatory singularities, (*see* singular stress fields)
Papkovich-Neuber solution, 192, 199–201, 207–208
particular solution, 75, 76
path independent integrals, 24, 68
penny-shaped crack, 261–280
  at an interface, 269–280
  obstructing heat conduction, 264–267
  in tension, 261–263
phase transformation, 219
plane crack in tension, 174–177
plane strain, 31–34
  relation to plane stress, 36, 180
  solution in complex variables, 206–210
plane stress, 34–36
  generalized plane stress, 36
Poisson's ratio, 17
polynomial solutions, 45–47, 59, 79
process zone, 174, 280
reciprocal theorem, 281–288
rectangular beams, 48–56, 73–75, 78–80, 97–101
  end conditions, 59–64, 99–101, 287
  shear deflection, 101
rigid-body displacements, (*see* displacements)
rolling contact, 159–162
rotation of a line, 13–14
rotation vector, 16, 190
rotational acceleration, 76–80
Saint-Venant's principle, 34, 59
self-equilibrated tractions, 33–34, 59
self-similarity, 139, 220, 222
separated variable solutions, 56, 61, 64, 139, 220
shift, 155
simple radial distribution, 141
singular stress fields, 126–128, 148, 165, 170, 217–225, 269, 273–278
  oscillatory singularities, 277, 279
slip, 156–162
source solution, 217, 224–225, 252
specific heat, 182
sphere, 218, 227, 240
spherical harmonics, 227–234
spherical hole, 225, 234, 237–239
spherical polar angle, 229
stick, 156–161
strain,
  notation for, 11
  shear strain, definition, 16
  tensile strain, 11–12
strain-displacement relations, 12, 16
  in polar coördinates, 84–85
strain energy, 127, 130
strain potential, 189, 192, 200
strain suppression, 181
strain transformation relations, 17
stress,
  hydrostatic stress, 8, 284
  mean stress, 19
  notation for, 4
  principal stress, 8, 17
  shear stress, 5
stress concentrations, 87–89, 92–94, 109, 126, 134, 172, 237–239
stress concentration factors, 89, 94, 239
stress intensity factors, 173, 177, 263, 267, 278, 279
stress-strain relations, 17, 179
stress transformation relations, 8, 78
summation convention, 6
surface dilatation, 223
surface energy, 173
symmetry, 17, 51, 62, 76, 79, 124, 166, 204, 218, 235, 266, 272

thermal conductivity, 181
thermal capacity, 181
thermal diffusivity, 182
thermal distortivity, 183, 216
thermal expansion coefficient, 179
thermoelastic displacement potential, 211–212, 218
thermoelasticity, 179–185, 211–216
    axisymmetric stress in the cylinder, 213–214
    heat flow obstructed by a crack, 263–267
    plane problems, 179–185, 212–214
    steady-state temperature, 182–185, 214–216
    Williams' solution, 214
    thick plate, 216
thermoelastic plane stress, 214
tilted punch, 284–287
uniform rotation, 73–75
unilateral contact (*see* contact problems — unilateral inequalities)
uniqueness, 193–197
vector notation 7
wedge problems, 121–137
Williams' asymptotic method, 125–134
Young's modulus, 17

# Mechanics

## *SOLID* MECHANICS AND ITS APPLICATIONS

*Series Editor:* G.M.L. Gladwell

*Aims and Scope of the Series*

The fundamental questions arising in mechanics are: *Why?*, *How?*, and *How much?* The aim of this series is to provide lucid accounts written by authoritative researchers giving vision and insight in answering these questions on the subject of mechanics as it relates to solids. The scope of the series covers the entire spectrum of solid mechanics. Thus it includes the foundation of mechanics; variational formulations; computational mechanics; statics, kinematics and dynamics of rigid and elastic bodies; vibrations of solids and structures; dynamical systems and chaos; the theories of elasticity, plasticity and viscoelasticity; composite materials; rods, beams, shells and membranes; structural control and stability; soils, rocks and geomechanics; fracture; tribology; experimental mechanics; biomechanics and machine design.

1. R.T. Haftka, Z. Gürdal and M.P. Kamat: *Elements of Structural Optimization*. 2nd rev.ed., 1990 ISBN 0-7923-0608-2
2. J.J. Kalker: *Three-Dimensional Elastic Bodies in Rolling Contact*. 1990 ISBN 0-7923-0712-7
3. P. Karasudhi: *Foundations of Solid Mechanics*. 1991 ISBN 0-7923-0772-0
4. N. Kikuchi: *Computational Methods in Contact Mechanics*. (forthcoming) ISBN 0-7923-0773-9
5. Y.K. Cheung and A.Y.T. Leung: *Finite Element Methods in Dynamics*. (forthcoming) ISBN 0-7923-1313-5
6. J.F. Doyle: *Static and Dynamic Analysis of Structures*. With an Emphasis on Mechanics and Computer Matrix Methods. 1991 ISBN 0-7923-1124-8; Pb 0-7923-1208-2
7. O.O. Ochoa and J.N. Reddy: *Finite Element Modelling of Composite Structures*. (forthcoming) ISBN 0-7923-1125-6
8. M.H. Aliabadi and D.P. Rooke: *Numerical Fracture Mechanics*. ISBN 0-7923-1175-2
9. J. Angeles and C.S. López-Cajún: *Optimization of Cam Mechanisms*. 1991 ISBN 0-7923-1355-0
10. D.E. Grierson, A. Franchi and P. Riva: *Progress in Structural Engineering*. 1991 ISBN 0-7923-1396-8
11. R.T. Haftka and Z. Gürdal: *Elements of Structural Optimization*. 3rd rev. and exp. ed. 1992 ISBN 0-7923-1504-9; Pb 0-7923-1505-7
12. J.R. Barber: *Elasticity*. 1992 ISBN 0-7923-1609-6

Kluwer Academic Publishers – Dordrecht / Boston / London

# Mechanics

## *FLUID* MECHANICS AND ITS APPLICATIONS

*Series Editor:* R. Moreau

*Aims and Scope of the Series*

The purpose of this series is to focus on subjects in which fluid mechanics plays a fundamental role. As well as the more traditional applications of aeronautics, hydraulics, heat and mass transfer etc., books will be published dealing with topics which are currently in a state of rapid development, such as turbulence, suspensions and multiphase fluids, super and hypersonic flows and numerical modelling techniques. It is a widely held view that it is the interdisciplinary subjects that will receive intense scientific attention, bringing them to the forefront of technological advancement. Fluids have the ability to transport matter and its properties as well as transmit force, therefore fluid mechanics is a subject that is particularly open to cross fertilisation with other sciences and disciplines of engineering. The subject of fluid mechanics will be highly relevant in domains such as chemical, metallurgical, biological and ecological engineering. This series is particularly open to such new multidisciplinary domains.

1. M. Lesieur: *Turbulence in Fluids*. 2nd rev. ed., 1990 ISBN 0-7923-0645-7
2. O. Métais and M. Lesieur (eds.): *Turbulence and Coherent Structures*. 1991 ISBN 0-7923-0646-5
3. R. Moreau: *Magnetohydrodynamics*. 1990 ISBN 0-7923-0937-5
4. E. Coustols (ed.): *Turbulence Control by Passive Means*. 1990 ISBN 0-7923-1020-9
5. A. A. Borissov (ed.): *Dynamic Structure of Detonation in Gaseous and Dispersed Media*. 1991 ISBN 0-7923-1340-2
6. K.-S. Choi (ed.): *Recent Developments in Turbulence Management*. 1991 ISBN 0-7923-1477-8

Kluwer Academic Publishers – Dordrecht / Boston / London

# Mechanics

From 1990, books on the subject of *mechanics* will be published under two series:
***FLUID*** **MECHANICS AND ITS APPLICATIONS**
*Series Editor:* R.J. Moreau
***SOLID*** **MECHANICS AND ITS APPLICATIONS**
*Series Editor:* G.M.L. Gladwell

Prior to 1990, the books listed below were published in the respective series indicated below.

MECHANICS: DYNAMICAL SYSTEMS
Editors: L. Meirovitch and G.Æ. Oravas

1. E.H. Dowell: *Aeroelasticity of Plates and Shells*. 1975 ISBN 90-286-0404-9
2. D.G.B. Edelen: *Lagrangian Mechanics of Nonconservative Nonholonomic Systems*. 1977 ISBN 90-286-0077-9
3. J.L. Junkins: *An Introduction to Optimal Estimation of Dynamical Systems*. 1978 ISBN 90-286-0067-1
4. E.H. Dowell (ed.), H.C. Curtiss Jr., R.H. Scanlan and F. Sisto: *A Modern Course in Aeroelasticity*. *Revised and enlarged edition see under Volume 11*
5. L. Meirovitch: *Computational Methods in Structural Dynamics*. 1980 ISBN 90-286-0580-0
6. B. Skalmierski and A. Tylikowski: *Stochastic Processes in Dynamics*. Revised and enlarged translation. 1982 ISBN 90-247-2686-7
7. P.C. Müller and W.O. Schiehlen: *Linear Vibrations*. A Theoretical Treatment of Multi-degree-of-freedom Vibrating Systems. 1985 ISBN 90-247-2983-1
8. Gh. Buzdugan, E. Mihăilescu and M. Radeş: *Vibration Measurement*. 1986 ISBN 90-247-3111-9
9. G.M.L. Gladwell: *Inverse Problems in Vibration*. 1987 ISBN 90-247-3408-8
10. G.I. Schuëller and M. Shinozuka: *Stochastic Methods in Structural Dynamics*. 1987 ISBN 90-247-3611-0
11. E.H. Dowell (ed.), H.C. Curtiss Jr., R.H. Scanlan and F. Sisto: *A Modern Course in Aeroelasticity*. Second revised and enlarged edition (of Volume 4). 1989 ISBN Hb 0-7923-0062-9; Pb 0-7923-0185-4
12. W. Szemplińska-Stupnicka: *The Behavior of Nonlinear Vibrating Systems*. Volume I: Fundamental Concepts and Methods: Applications to Single-Degree-of-Freedom Systems. 1990 ISBN 0-7923-0368-7
13. W. Szemplińska-Stupnicka: *The Behavior of Nonlinear Vibrating Systems*. Volume II: Advanced Concepts and Applications to Multi-Degree-of-Freedom Systems. 1990 ISBN 0-7923-0369-5
Set ISBN (Vols. 12–13) 0-7923-0370-9

MECHANICS OF STRUCTURAL SYSTEMS
Editors: J.S. Przemieniecki and G.Æ. Oravas

1. L. Frýba: *Vibration of Solids and Structures under Moving Loads*. 1970 ISBN 90-01-32420-2
2. K. Marguerre and K. Wölfel: *Mechanics of Vibration*. 1979 ISBN 90-286-0086-8

# Mechanics

3. E.B. Magrab: *Vibrations of Elastic Structural Members.* 1979 ISBN 90-286-0207-0
4. R.T. Haftka and M.P. Kamat: *Elements of Structural Optimization.* 1985
   *Revised and enlarged edition see under* Solid Mechanics and Its Applications, Volume 1
5. J.R. Vinson and R.L. Sierakowski: *The Behavior of Structures Composed of Composite Materials.* 1986 ISBN Hb 90-247-3125-9; Pb 90-247-3578-5
6. B.E. Gatewood: *Virtual Principles in Aircraft Structures.* Volume 1: Analysis. 1989
   ISBN 90-247-3754-0
7. B.E. Gatewood: *Virtual Principles in Aircraft Structures.* Volume 2: Design, Plates, Finite Elements. 1989 ISBN 90-247-3755-9
   Set (Gatewood 1 + 2) ISBN 90-247-3753-2

---

MECHANICS OF ELASTIC AND INELASTIC SOLIDS

Editors: S. Nemat-Nasser and G.Æ. Oravas

---

1. G.M.L. Gladwell: *Contact Problems in the Classical Theory of Elasticity.* 1980
   ISBN Hb 90-286-0440-5; Pb 90-286-0760-9
2. G. Wempner: *Mechanics of Solids with Applications to Thin Bodies.* 1981
   ISBN 90-286-0880-X
3. T. Mura: *Micromechanics of Defects in Solids.* 2nd revised edition, 1987
   ISBN 90-247-3343-X
4. R.G. Payton: *Elastic Wave Propagation in Transversely Isotropic Media.* 1983
   ISBN 90-247-2843-6
5. S. Nemat-Nasser, H. Abé and S. Hirakawa (eds.): *Hydraulic Fracturing and Geothermal Energy.* 1983 ISBN 90-247-2855-X
6. S. Nemat-Nasser, R.J. Asaro and G.A. Hegemier (eds.): *Theoretical Foundation for Large-scale Computations of Nonlinear Material Behavior.* 1984 ISBN 90-247-3092-9
7. N. Cristescu: *Rock Rheology.* 1988 ISBN 90-247-3660-9
8. G.I.N. Rozvany: *Structural Design via Optimality Criteria.* The Prager Approach to Structural Optimization. 1989 ISBN 90-247-3613-7

---

MECHANICS OF SURFACE STRUCTURES

Editors: W.A. Nash and G.Æ. Oravas

---

1. P. Seide: *Small Elastic Deformations of Thin Shells.* 1975 ISBN 90-286-0064-7
2. V. Panc: *Theories of Elastic Plates.* 1975 ISBN 90-286-0104-X
3. J.L. Nowinski: *Theory of Thermoelasticity with Applications.* 1978
   ISBN 90-286-0457-X
4. S. Łukasiewicz: *Local Loads in Plates and Shells.* 1979 ISBN 90-286-0047-7
5. C. Fiřt: *Statics, Formfinding and Dynamics of Air-supported Membrane Structures.* 1983 ISBN 90-247-2672-7
6. Y. Kai-yuan (ed.): *Progress in Applied Mechanics.* The Chien Wei-zang Anniversary Volume. 1987 ISBN 90-247-3249-2
7. R. Negruţiu: *Elastic Analysis of Slab Structures.* 1987 ISBN 90-247-3367-7
8. J.R. Vinson: *The Behavior of Thin Walled Structures.* Beams, Plates, and Shells. 1988
   ISBN Hb 90-247-3663-3; Pb 90-247-3664-1

---

# Mechanics

MECHANICS OF FLUIDS AND TRANSPORT PROCESSES
Editors: R.J. Moreau and G.Æ. Oravas

1. J. Happel and H. Brenner: *Low Reynolds Number Hydrodynamics.* With Special Applications to Particular Media. 1983 ISBN Hb 90-01-37115-9; Pb 90-247-2877-0
2. S. Zahorski: *Mechanics of Viscoelastic Fluids.* 1982 ISBN 90-247-2687-5
3. J.A. Sparenberg: *Elements of Hydrodynamics Propulsion.* 1984 ISBN 90-247-2871-1
4. B.K. Shivamoggi: *Theoretical Fluid Dynamics.* 1984 ISBN 90-247-2999-8
5. R. Timman, A.J. Hermans and G.C. Hsiao: *Water Waves and Ship Hydrodynamics.* An Introduction. 1985 ISBN 90-247-3218-2
6. M. Lesieur: *Turbulence in Fluids.* Stochastic and Numerical Modelling. 1987 ISBN 90-247-3470-3
7. L.A. Lliboutry: *Very Slow Flows of Solids.* Basics of Modeling in Geodynamics and Glaciology. 1987 ISBN 90-247-3482-7
8. B.K. Shivamoggi: *Introduction to Nonlinear Fluid-Plasma Waves.* 1988 ISBN 90-247-3662-5
9. V. Bojarevičs, Ya. Freibergs, E.I. Shilova and E.V. Shcherbinin: *Electrically Induced Vortical Flows.* 1989 ISBN 90-247-3712-5
10. J. Lielpeteris and R. Moreau (eds.): *Liquid Metal Magnetohydrodynamics.* 1989 ISBN 0-7923-0344-X

MECHANICS OF ELASTIC STABILITY
Editors: H. Leipholz and G.Æ. Oravas

1. H. Leipholz: *Theory of Elasticity.* 1974 ISBN 90-286-0193-7
2. L. Librescu: *Elastostatics and Kinetics of Aniosotropic and Heterogeneous Shell-type Structures.* 1975 ISBN 90-286-0035-3
3. C.L. Dym: *Stability Theory and Its Applications to Structural Mechanics.* 1974 ISBN 90-286-0094-9
4. K. Huseyin: *Nonlinear Theory of Elastic Stability.* 1975 ISBN 90-286-0344-1
5. H. Leipholz: *Direct Variational Methods and Eigenvalue Problems in Engineering.* 1977 ISBN 90-286-0106-6
6. K. Huseyin: *Vibrations and Stability of Multiple Parameter Systems.* 1978 ISBN 90-286-0136-8
7. H. Leipholz: *Stability of Elastic Systems.* 1980 ISBN 90-286-0050-7
8. V.V. Bolotin: *Random Vibrations of Elastic Systems.* 1984 ISBN 90-247-2981-5
9. D. Bushnell: *Computerized Buckling Analysis of Shells.* 1985 ISBN 90-247-3099-6
10. L.M. Kachanov: *Introduction to Continuum Damage Mechanics.* 1986 ISBN 90-247-3319-7
11. H.H.E. Leipholz and M. Abdel-Rohman: *Control of Structures.* 1986 ISBN 90-247-3321-9
12. H.E. Lindberg and A.L. Florence: *Dynamic Pulse Buckling.* Theory and Experiment. 1987 ISBN 90-247-3566-1
13. A. Gajewski and M. Zyczkowski: *Optimal Structural Design under Stability Constraints.* 1988 ISBN 90-247-3612-9

# Mechanics

MECHANICS: ANALYSIS

Editors: V.J. Mizel and G.Æ. Oravas

1. M.A. Krasnoselskii, P.P. Zabreiko, E.I. Pustylnik and P.E. Sbolevskii: *Integral Operators in Spaces of Summable Functions*. 1976 ISBN 90-286-0294-1
2. V.V. Ivanov: *The Theory of Approximate Methods and Their Application to the Numerical Solution of Singular Integral Equations*. 1976 ISBN 90-286-0036-1
3. A. Kufner, O. John and S. Pučík: *Function Spaces*. 1977 ISBN 90-286-0015-9
4. S.G. Mikhlin: *Approximation on a Rectangular Grid*. With Application to Finite Element Methods and Other Problems. 1979 ISBN 90-286-0008-6
5. D.G.B. Edelen: *Isovector Methods for Equations of Balance*. With Programs for Computer Assistance in Operator Calculations and an Exposition of Practical Topics of the Exterior Calculus. 1980 ISBN 90-286-0420-0
6. R.S. Anderssen, F.R. de Hoog and M.A. Lukas (eds.): *The Application and Numerical Solution of Integral Equations*. 1980 ISBN 90-286-0450-2
7. R.Z. Has'minskiĭ: *Stochastic Stability of Differential Equations*. 1980 ISBN 90-286-0100-7
8. A.I. Vol'pert and S.I. Hudjaev: *Analysis in Classes of Discontinuous Functions and Equations of Mathematical Physics*. 1985 ISBN 90-247-3109-7
9. A. Georgescu: *Hydrodynamic Stability Theory*. 1985 ISBN 90-247-3120-8
10. W. Noll: *Finite-dimensional Spaces*. Algebra, Geometry and Analysis. Volume I. 1987 ISBN Hb 90-247-3581-5; Pb 90-247-3582-3

MECHANICS: COMPUTATIONAL MECHANICS

Editors: M. Stern and G.Æ. Oravas

1. T.A. Cruse: *Boundary Element Analysis in Computational Fracture Mechanics*. 1988 ISBN 90-247-3614-5

MECHANICS: GENESIS AND METHOD

Editor: G.Æ. Oravas

1. P.-M.-M. Duhem: *The Evolution of Mechanics*. 1980 ISBN 90-286-0688-2

MECHANICS OF CONTINUA

Editors: W.O. Williams and G.Æ. Oravas

1. C.-C. Wang and C. Truesdell: *Introduction to Rational Elasticity*. 1973 ISBN 90-01-93710-1
2. P.J. Chen: *Selected Topics in Wave Propagation*. 1976 ISBN 90-286-0515-0
3. P. Villaggio: *Qualitative Methods in Elasticity*. 1977 ISBN 90-286-0007-8

# Mechanics

MECHANICS OF FRACTURE

Editors: G.C. Sih

1. G.C. Sih (ed.): *Methods of Analysis and Solutions of Crack Problems.* 1973 ISBN 90-01-79860-8
2. M.K. Kassir and G.C. Sih (eds.): *Three-dimensional Crack Problems.* A New Solution of Crack Solutions in Three-dimensional Elasticity. 1975 ISBN 90-286-0414-6
3. G.C. Sih (ed.): *Plates and Shells with Cracks.* 1977 ISBN 90-286-0146-5
4. G.C. Sih (ed.): *Elastodynamic Crack Problems.* 1977 ISBN 90-286-0156-2
5. G.C. Sih (ed.): *Stress Analysis of Notch Problems.* Stress Solutions to a Variety of Notch Geometries used in Engineering Design. 1978 ISBN 90-286-0166-X
6. G.C. Sih and E.P. Chen (eds.): *Cracks in Composite Materials.* A Compilation of Stress Solutions for Composite System with Cracks. 1981 ISBN 90-247-2559-3
7. G.C. Sih (ed.): *Experimental Evaluation of Stress Concentration and Intensity Factors.* Useful Methods and Solutions to Experimentalists in Fracture Mechanics. 1981 ISBN 90-247-2558-5

MECHANICS OF PLASTIC SOLIDS

Editors: J. Schroeder and G.Æ. Oravas

1. A. Sawczuk (ed.): *Foundations of Plasticity.* 1973 ISBN 90-01-77570-5
2. A. Sawczuk (ed.): *Problems of Plasticity.* 1974 ISBN 90-286-0233-X
3. W. Szczepiński: *Introduction to the Mechanics of Plastic Forming of Metals.* 1979 ISBN 90-286-0126-0
4. D.A. Gokhfeld and O.F. Cherniavsky: *Limit Analysis of Structures at Thermal Cycling.* 1980 ISBN 90-286-0455-3
5. N. Cristescu and I. Suliciu: *Viscoplasticity.* 1982 ISBN 90-247-2777-4

Kluwer Academic Publishers - Dordrecht / Boston / London